To Ruru
with

Linde

November 2018

Basic Biology for Born Engineers

Basic Biology for Born Engineers:

Living Mosaics

By

Guenter Albrecht-Buehler

Cambridge
Scholars
Publishing

Basic Biology for Born Engineers: Living Mosaics

By Guenter Albrecht-Buehler

This book first published 2018

Cambridge Scholars Publishing

Lady Stephenson Library, Newcastle upon Tyne, NE6 2PA, UK

British Library Cataloguing in Publication Data
A catalogue record for this book is available from the British Library

Copyright © 2018 by Guenter Albrecht-Buehler

All rights for this book reserved. No part of this book may be reproduced, stored in a retrieval system, or transmitted, in any form or by any means, electronic, mechanical, photocopying, recording or otherwise, without the prior permission of the copyright owner.

ISBN (10): 1-5275-1673-3
ISBN (13): 978-1-5275-1673-1

To my wife Veena,
whose encouragement and never less than loving criticism was
invaluable.
To our children, Ananya and Ishan,
and with no less love
to my older children, Christine, Conrad, and Carl.

The opposite of a profound truth may very well be another profound truth.

(Niels Bohr, 1885 – 1962)

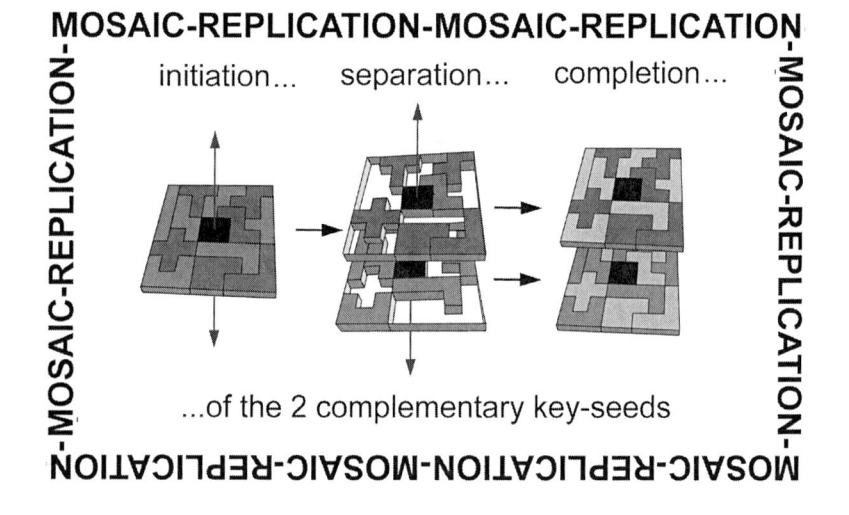

MOSAIC-REPLICATION-MOSAIC-REPLICATION

initiation...   separation...   completion...

...of the 2 complementary key-seeds

TABLE OF CONTENTS

Summary x

Preface xi
How Evolution Must Create Superb Engineering

Chapter One 1
Introduction

A Survey of the World of Mosaics

Chapter Two 24
A Cursory View of Living Mosaics

Chapter Three 37
A Simple Model Mosaic to Be Used Throughout This Book

Chapter Four 45
Some Common Properties of Mosaics and Biological Objects

Chapter Five 58
The Diversity of 'Fitting'

Chapter Six 76
Movements and Growth of Living Mosaics

Fractal Mosaics

Chapter Seven 88
A Most Important Functional Mosaic: The Iteration

Chapter Eight 91
Nested Mosaics

A Role for Teleology in Living Mosaics?

Chapter Nine 116
The Tasks and Interactions of Living Mosaics

Chapter Ten 139
A Task-Based Taxonomy of Living Mosaics

The 'Seeds' of Living Mosaics

Chapter Eleven 146
The 'Seeds' of Mosaics

The Variation of Living Mosaics

Chapter Twelve 174
Re-Direction of the Assembly

Chapter Thirteen 177
'Mutation' of the Seeds

The Replication of Living Mosaics

Chapter Fourteen 182
A Role of Seed- and Scaffold-Mosaics in Replication

Chapter Fifteen 194
The Strategies of Mosaic Replication

The Expression of Living Mosaics

Chapter Sixteen 202
Partial Replication of Fitted Mosaics ('Coding' and 'Expression')

The Cycles of Living Mosaics

Chapter Seventeen 210
The Imperfect Cycles of Living Mosaics

Chapter Eighteen 213
Iteration-Driven Mosaic Cycles

Chapter Nineteen 229
The 'Almost'-Periodic Behaviour of Iterated Cycles

Numerical Descriptions of the Functions of Mosaics and Tiles

Chapter Twenty 242
Measuring Distances, Contacts, and Functions of Tiles and Mosaics

Speculations about Living Mosaics

Chapter Twenty-One 264
Notes About the Origin of Living Mosaics

Chapter Twenty-Two 286
A Critical View of Mosaics

References/Search Terms 321

Glossary 324

Appendix A 332
Numerical Methods for the Assembly of Mosaics

Appendix B 338
Fitting as a Matrix Operation

Appendix C 354
Algorithms to Find All Standardized Solutions of Pentomino Mosaics

Appendix D 366
Computation of the Fractal Dimension of Fractal Mosaics Created
by Recursive Replacements

Appendix E 375
Outline of a Task-Based Taxonomy of Living Mosaics

Summary

Like all technologies, terrestrial life must comply with fundamental engineering principles, lest it would fail. In particular, all life forms must consist of **modules**, i.e. components that are

(a) **discrete** (there is no continuous transition between them), and

(b) **semi-autonomous** (most of their functions are completely independent, whereas a few need the cooperation of other components).

Only then, an organism can assemble, service, repair, or replace parts and functions predictably without disrupting or disabling the rest.

The discreteness of the modules implies that all life forms are **'mosaics'** of their specific, distinct modules (**'tiles'**) that operate under certain constraints (**'frame'**), which limits them, but also hold them together.

The semi-autonomy of the modules implies that they cannot fit together automatically. On the contrary, their largely self-reliant actions tend to create conflicts between them. Hence, the modules of each living mosaic must find and maintain ways to **fit together** with respect to their space, timing, forces and meaning.

Mosaics may contain **tangible** (material) modules or **symbolic** (mental) ones. The modules may interact and fit **physically** or **logically**.

This book interprets living things as **'living mosaics'**. Its perspective intends to provide a **novel, intuitive approach** for students of biology, medicine, engineering, and other disciplines, if they have a passion for the practical application of scientific insights. It intends to help **unify** the bewildering variety of biological phenomena, to **simplify** their classification, and to further their **understanding**. Most importantly, it offers a **unifying description of biological phenomena that is independent of** their sizes ranging from molecules to ecologies.

The key concepts towards the understanding of living mosaics are **'tasks'**, **'key-seeds'**, assembly-, replication-, and expression-mechanisms, cyclic behaviour, and their intriguing **'almost'-repeats**. Alas, like all real phenomena, they also give rise to a number of **paradoxes**.

Preface

How Evolution Must Create Superb Engineering

Today's biologists understate all too frequently the logic, elegance, and boldness of the engineering of living things. I suppose, their reason is the fear that any emphasis on engineering would supply ammunition to the defenders of 'Intelligent Design' and 'Creationism'.

This is unfortunate, because teaching how biological structures explain their functions helps the students' retention and understanding. More importantly, recognizing intermediary stages and different levels of perfection of biological engineering provides convincing arguments *for* evolution, but not against it.

While proposing a unifying view of biological systems, the book tries to dispel the fear that permitting engineering principles into biology would invite religion into science. It does not. The existence of a supernatural engineer is neither provable nor disprovable by the finding that living systems comply with these engineering principles. They result from an evolution of their own, because the compliance with engineering principles creates quite large selective advantages for living systems.

Engineering principles also present important didactic advantages for science teachers by their remarkable power to unify our understanding of living things, regardless of their size and diversity. The present book will try to show this by presenting and analyzing in detail one specific engineering principle with which all living things must and do comply.

Based on my training as a physicist and many decades of experimental work in cell biology I decided to write about the engineering principle of 'modularity'. It states that all life forms must be composed of **modules**, i.e. components that are

(a) **discrete** (there is no continuous transition between them), and

(b) **semi-autonomous** (most of their functions are completely independent, whereas a few need the cooperation of other components). Only then, can the organism or machine assemble, service, repair, or replace parts and functions predictably without disrupting or disabling the rest.

Since all living things obey this principle, it offers novel ways of classifying them, but also allows us to unify the explanations of their forms and actions.

I wrote this book for students of biology, medicine, and engineering. Besides logical thinking and an open mind, I expect the students to have at least high school knowledge of biology, histology, physiology, and arithmetic.

In return, biologists can rightfully expect from me certain documentations of scholarship, such as a complete reference list, and a thorough collection of accompanying notes. With my apologies to the experts, I risked the criticism of naiveté and of stating the obvious by forgoing those.

Instead, for the benefit of simplicity and clarity, I spent the lion's share of my efforts on designing didactic models and writing the computer programs needed for the 120 illustrations to support my lines of reasoning. I hope the results will be able to persuade experts and students alike to give it a chance. After all, today's easy access of students to powerful search engines renders the completeness of scholarly documentations rather obsolete. More importantly, in my experience of some forty years of teaching students of biology and medicine, the impact of eruditeness could intimidate them while they are still trying to slash their personal – and initially crude - trail through the seemingly impenetrable jungle of biology.

Still, after following the arguments of this book, the student should turn to the excellent, authoritative report of The National Academies of Sciences, Engineering, and Medicine [13] about the engineering principles of biology in all their forms, including the principle of modularity.

Chapter One

Introduction

Everything is a Mosaic; However, the Tiles of Living Mosaics are Modules

The world consists of elementary particles, atoms, chemical compounds, sediments, rocks, tectonic plates, planets, stars, galaxies, and galaxy clusters, which all appear, far from being single homogenous things, as mosaics composed of discrete, largely independent parts.

Describing these parts or mosaic 'tiles' as discrete objects does not imply that two different tiles of a mosaic cannot share material, symbolic, or functional components. They may very well share components, and yet be quite different objects.

The same applies to all living things. They, too, are mosaics, which consist of a wide-reaching hierarchy of sizes. Their discrete 'tiles' may have sizes as different as macro-molecules, polymers, organelles, membranes, cells, tissues, organs, organisms, and populations. However, there is a huge difference between inanimate and living mosaics. The 'living tiles' have 'tasks' and they are semi-autonomous in carrying them out.

Many consequences of the mosaic-character of living things need still more exploration. Here is a crude rationale for studying one of them.

The conceptual pillars of biology are the theory of evolution, molecular genetics, biochemistry, and the electrical activity of neurons. However, there is a significant divide between them. Their dominant mechanisms are very different from each other and seem to split biology into conceptual domains that are discrete and even may seem incompatible with each other.

The mechanisms of molecular genetics and biochemistry are governed by thermodynamics and quantum mechanics, which only apply to the microscopic world of biological molecules.

On the other hand, the mechanisms behind the electrical pulse-storms of neurons that reverberate through cellular networks and brains, originate from membranes, synapses, cells, and brain domains that are many orders of magnitude larger than molecules. Their dominant concepts and mechanisms relate to membrane biology, cell biology and histology.

Finally, the mechanisms of natural selection only make sense in the even larger world of organisms and ecologies. In exceptional cases, such as in the case of immune cells, there may be a natural selection among rapidly proliferating cells. However, there is no struggle for survival among molecules. They do not need to survive in order to proliferate, as the actions of messenger RNAs and ribosomes guarantee their unchallenged reproduction.

In other words, biology itself is a mosaic of at least three very different conceptual domains, namely the molecular, the cellular and the organismal size levels. Of course, in spite of their differences these 'conceptual tiles of biology' have countless mutual contacts and interactions.

Considering that this book is about mosaics, should we not be delighted to find biology itself to be a mosaic that is divided into at least three discrete fields that are semi-autonomous?

Actually, no. At their most fundamental level, all sciences strive to find common ground and unified formulations for all their phenomena. The described divide seems to present an obstacle for the goal of fundamental unity.

Sometimes turning a problem on its head may point to the solution. The concepts and rationales of mosaics may be able to provide the very commonalities that the mosaic character seems to exclude. After all, mosaics are not restricted to certain sizes. Whatever vocabulary describes mosaics, size plays no role in it.

To be sure, another set of size-bridging concepts exist already, namely the concepts and mechanisms of communication and information processing. The advances in neurobiology have demonstrated convincingly how these principles apply equally well to all size levels, be they molecules, neurons, or brains.

At closer inspection, however, we notice that all communication and data processing systems are in particular also mosaics. For example, microprocessors, computers, synapses, neural networks, or entire brains are composite objects. Hence, mosaics may belong to an even deeper level of biological foundation than the principles of communication and data processing.

Their ranking close to the foundations of biology justifies the study of mosaics, their 'tiles', and 'frames' in considerable detail. They will be called 'living mosaics', because they carry out meaningful 'tasks' such as metabolism, communication, searches, hunts and other targeted movements.

Obviously, inanimate mosaics such as the rings of Saturn are not capable of any such tasks.

The tiles of living mosaics are parts of these tasks. To be sure, they will always be discrete objects, but as parts of a common task they can never be entirely independent. Therefore, I will consider them as 'semi-autonomous', and call them no longer 'tiles', but use the name of 'modules'. They play pivotal roles in all mosaics that carry out tasks. They are at the centre of a fundamental principle of biology, namely the 'principle of modularity'.

This principle is size-independent. Therefore, our goal of exploring the size-independent vocabulary of mosaics will focus primarily on modules, their properties, actions, and interactions.

The Rarely Mentioned Principle of Modularity

Modern biology has discovered a large number of fundamental principles. Some of the best known among them are the principles of evolution, the laws of genetics, the principle of base-pairing, the so-called 'central dogma of molecular biology', the genetic code, nerve excitation, synaptic transmission, ATP-hydrolysis, the ligand-receptor binding, post-translational modifications of proteins, and several others.

Naturally, their claim of universality is - so far at least - restricted to planet Earth. In this limited sense of 'terrestrial universality', we need them as the pillars of our scientific explanations. The more such universals we find, the more phenomena of terrestrial life we can explain.

This book focuses on one of the rarely mentioned ones, namely the principle that *all known life forms are composites of distinct and largely autonomous, yet cooperating modules*. In spite of its universality, I could not find any textbooks dedicated to the principle of modularity. Among other goals, the present book is an attempt to close this gap. I hasten to add, though, that the principle of modularity is presented and discussed in the larger context of modern theoretical approaches to biology in the above mentioned excellent report of The National Academies of Sciences, Engineering, and Medicine [13]

The Principle of Modularity as a Requirement of Engineering

The principle of modularity is in plain sight. One can easily observe that all living things have discrete, clearly identifiable, semi-autonomous, yet cooperating parts, such as macromolecules, cells, organs, organisms, or populations.

Why are living things built like that? Building the opposite, namely homogeneous, seamlessly integrated units should serve much better the need of living things to implement body-wide cooperation, integration, communication and the creation of emergent macroscopic properties, should it not?

Actually, it would not. The reason is simple. Eventually, all constructed objects must malfunction or prove inadequate some time. However, once such a failure has happened, any attempt to update or repair a homogeneous object would cripple all its inseparable parts at once. In short, they would violate the vital engineering principle of modularity.

We can observe how dominant this principle is even in our human world. Every successful machine that humans ever built consists of independent modules. For example, every car is a collection of modules such as the engine, battery, transmission, wheels, headlights, etc. All of them are discrete objects, whose removal for servicing will not damage the rest of the car. In addition, they are semi-autonomous, and can be operated and tested outside the car. It is not only true for mechanical objects. Every computer has discrete and semi-autonomous, yet cooperating modules such as a screen display, memory banks, a power supply, circuit boards, microprocessors, integrated circuits, cooling fans, etc. even the computer software contains classes, subroutines, functions, data bases, interrupt routines, patches, etc. Removing them for repairs, upgrades, or diagnosis will not damage the rest.

The definition of modules is ambiguous. Sometimes they contain sub-modules. For example, a car engine, already a module, may contain as its own modules a generator and a starter that are firmly connected and precisely fitted to it. Yet, removing either of them from the engine and operating them separately will not damage the engine.

Alternatively, a complete machine may become itself a mere module of a larger machine. For example, a computer can operate as the single module of a server bank, or a car can function as one 'detail' of an entire police fleet. Regardless, modules operate always as identifiable, discrete, and semi-autonomous parts of a larger system.

Living things are no exception. They cannot function for long, let alone survive for millions of years, unless they comply with the principle of modularity. And they all do comply, of course.

Examples of Other Engineering Principles

Modularity is not the only engineering principle. Another important one is the provision of **back-up systems**. For example, the glycogen supply in muscle and liver is a vital back-up for emergency needs of metabolic energy. The fat deposits around the heart have a similar function. During starvation, they are understandably the last fat deposits, which the body depletes.

Already the bilateral symmetry of the body-plan of many organisms provides one back-up system for many body parts. Examples are the second kidney and the second lung we all have, even though a functional body only needs one. The much higher symmetries of the body-plans of most plants and of the parts of plants provide multiple back-ups for their vital organs.

There are exceptions, though, where a back-up system is not feasible. For example, a second independent heart could not pump blood through the vascular system without creating necrotic domains where the pumping of one heart would counteract that of the other.

Yet another important engineering principle is the **storage of spare parts** that may be needed at a moment's notice, whereas their *de novo* construction would be very time-consuming (e.g. the storage of platelets in the spleen; the rotating teeth of sharks). Yet others are the **standardization of parts and processes** that facilitate construction and repair (e.g. the universality of ATP hydrolysis, glycosylation, and phosphorylation), **the shielding from noise** (e.g. the anchoring of enzymes to huge 'cytoskeletal' polymers that are little affected by Brownian movements), the **preparedness for failure** (e.g. the clotting machinery of blood, regeneration of limbs), and so on.

These principles of engineering are universal, and they play prominent roles in biological systems and human-made machines alike. However, this book will focus exclusively on the principle of modularity, because its implementation seems to be a fundamental pre-condition for all the others.

The Size Independence of the Principle of Modularity

One of its earliest formulations of the principle of modularity was the cell theory of Schleiden and Schwann. It states that all tissues, organs, and ultimately all living organisms consist of either a single cell, or else are the composite of multiple single cells as their distinct and semi-autonomous modules.

But cells are not the smallest modules of organisms. Cells contain smaller, distinct, and semi-autonomous sub-components and sub-compartments, such as sarcomeres, mitochondria, chloroplasts, cell membranes, nuclear pore complexes, chromosomes, cytoskeletal domains, etc. Even their macromolecules, oligomers and monomers alike, consist of molecular sub-units or domains that are able to exist and function in isolation, and can be modified or even recycled without incapacitating the cell.

Similarly, the principle of modularity applies to the various coding and non-coding segments of genomes that sub-divide into long series of fractal sequences [02]. The student of biology will undoubtedly be able to think of many more such examples.

Cells are not the largest modules, either. Entire organisms may function as discrete and semi-autonomous modules of a population of interacting organisms. Every coral reef, herd of caribou, wolf pack, or colony of algae represent a semi-autonomous module within their well-fitted ecology. Ultimately, entire populations of organisms are but modules within Earth's gigantic biosphere.

Modularity as the Source of Emergent Properties of Living Things

Present day biology considers macromolecules as the ultimate elements of explanation. To be sure, they are important modules of cells, and their broad success is undeniable. Yet, the exclusive focus on molecular levels must fail to explain some of their emergent properties.

In biology, we are facing some of the most mysterious emergent phenomena that exist anywhere, such as aging, language, intelligence, meaning, and even consciousness, and creativity. Macromolecular mechanisms are light-years removed from the levels of compounded complexity that may someday explain these ultimate expressions of life.

Perhaps the principle of modularity can point the way. Emergence describes how a collection of units creates a larger object that expresses new properties, which none of the component units had. The phenomenon creates hierarchies of increasing sizes that span many orders of magnitude. In short, *where there is modularity, there is likely emergence as well.*

Modules as Universal Instruments of Scientific Reduction and Explanation

Biological macromolecules can and do combine into larger groups that acquire new functions, individuality, and even autonomy, which elevates them to the level of modules. Examples for such next-higher-level modules are ribosomes, chromosomes, chloroplasts, centrioles, nuclei and so forth.

These, in turn, can group into even larger modules such as cells, which in turn combine into even larger ones such as tissues and organs, and so forth.

Since nature herself offers us a hierarchy of natural modules of living mosaics, they may turn out to be the most appropriate elements for our scientific analyses, even when they are much larger than biological macromolecules [103]. In other words, *macromolecules are powerful instruments of biological explanations, but they are not the only ones. All higher level modules may qualify, as well.*

Of course, many biologists have put this concept to use. Especially, neurobiologists have pioneered elements of analysis that are much larger than neurotransmitter molecules, receptor molecules, or ion channels. The search for the explanation of emergent phenomena such as pain, perception, and cognition employs successfully not only whole neurons, but also groups and networks of neurons and even whole brain domains.

Similarly, the students of DNA have long moved beyond the study of nucleotides, base pairs, polymerases, or transcription factors. They place equal importance on the interactions between the entire double helix and nucleosomes. They have formulated an entire hierarchy of coiled pairings of DNA and nucleosomes, as well as centromeres, telomeres, chromosomes, all the way up to the properties and actions of the entire chromatin and its interactions with the nuclear envelope.

Also, cell biologists have turned their attention to the study of the shapes, dynamics, and functions of a family of very large, linear protein polymers that carry the – misleading - name of the 'cytoskeleton'.

To be sure, it may not take more than single macromolecules to trigger the *failure* of any of these higher-level modules. On the other hand, single macromolecules, can never explain the complex *functions* of these modules.

There is no danger that such higher-level approaches will neglect the molecular levels. On the contrary, such analyses will necessarily climb down a ladder of emergent properties all the way back to the properties of single molecules.

The Problem of Identifying Modules

The omnipresence of modules does not mean that they are easy to identify. In spite of their discrete nature, determining the actual borders of modules poses occasionally very difficult problems.

As an example, think of the enormously complex outlines of a Purkinje cell. Another example is our vascular system with all its arteries, veins, and capillaries. It is actually a single contiguous object, and if seen in its entirety, it occupies a space of almost incomprehensible complexity. Or consider an ant nest as one of the modules that comprise the living mosaic of a forest. If one includes all the foraging ants, where does the nest begin and end?

The search for new staining techniques may help. Since Camillo Golgi's most revealing silver stain of neurons, there has been considerable progress of our staining techniques, such as fluorescent in situ hybridization (FISH), immuno-fluroescence in combination with confocal laser microscopy, computerized tomography, imaging of proteins in the live state by green fluorescent protein (GFP), and others. Methods like these are able to outline biological components, in spite of their enormously complex shapes and their intimate intertwining with others. Further improvements of our probing and staining techniques should help outline biological modules better and better.

The principle of modularity as the most universal feature of all life forms

Considering the enormous diversity of modules, it seems that modular architecture not only is one among other basic principles of biology. Arguably, it is the most universally shared property of all life forms on Earth. There is no living thing whose architecture does not comply with it.

In addition, all life forms depend on each other in some way or another, which makes them all tiles of an even larger living 'mega'-mosaic.

The 'Tasks' of Modules

What earns the tiles and whole mosaics of living things the attribute of 'living' is the ability to carry out tasks. i.e. an action with an intent of their own. Therefore, if we encounter *a mosaic that carries out a task, we can decide to describe it as a living mosaic*. It may seem reasonable, because mosaics that could not possibly have tasks, such as moon craters, or sun flares cannot possibly have intentions, because they are not living.

On the face of it, this definition seems to exclude mosaics that carried out tasks in the past, but are no longer functional (e.g. a skeleton), or that will carry out tasks in the future (e.g. pro-insulin), but are not yet active. Therefore, we will call *all mosaics as living mosaics, which are used, have been used, or will be used to carry out tasks*. This includes former tiles of a living mosaic, as well as objects that were collected, fabricated, or modified by a living mosaic in order to be used as 'tools', provided their task is to participate as one of its tiles in the performance of the living mosaic's task.

Philosophically speaking, we are moving here on very thin ice. The definitions of words like 'living', 'task', and 'intent' are ambiguous and even circular. Still, their common sense meaning is clear. In chapters 9, 10, and 28 we will discuss the concept of tasks in more detail. For the time being, we notice that tasks are using material objects, but they themselves are not material objects. Instead, *tasks are composed of instructions, data, and 'markers' for processes and functions that are placed in specific order.* All living mosaics are associated with one or several such tasks. The non-material components of tasks may be called 'symbols' as they stand for certain material objects.

Hence, tasks are composed of parts, which are not material and, therefore, they may be categorizes as 'symbolic' mosaics, themselves. Examples are patterns, texts, languages, and programs, whose discrete tiles are symbols such as signals, body markings, pulses, letters, numbers, functions, etc. In his recent book, Harari [108] describes them as organic 'algorithms'.

Symbolic mosaics do not stand alone, but interact with members of yet other categories of mosaics, which use logical functions as their nodules. They process information by discrete sets of rules that constitute yet other symbolic mosaics. And so forth.

The Problematic Concept of 'Tasks' and Other Forms of Teleology

Maybe, engineering is in our blood. From our earliest childhood on we love building things. Conversely, when we chance upon an unknown object, we cannot help but trying to decide whether it was built intentionally or whether it is an accidental product of nature.

All this applies to mosaics. Regardless whether the mosaic is a painting, a symphony, or an unfamiliar machine, if all the parts fit perfectly well, one is immediately convinced that there was great effort behind it, and therefore the mosaic must have a purpose.

These are, of course, teleological thoughts, which science banned long ago. However, to be fair, only the sciences about inanimate nature rejected teleology. Should biology ban it, too?

It will not be easy to do. As long as we only look at inanimate nature, we can easily agree that (say) moon craters have no purpose, and inventing one will not help us explaining their properties, let alone the moon.

The situation is quite different in biology. We cannot effectively investigate biological objects without considering their interlinked functions, because biology is all about engineering; and engineering is all about purpose, albeit about human purposes. Even Rube Goldberg machines 'have the purpose' of demonstrating none is needed.

For instance, teaching the histology of the human kidney without reference to the purpose of the medulla, the basement membrane of the glomeruli, the shape of the podocytes, the proximal convoluted tubules, the loop of Henle, etc. would be an endless stream of unrelated, meaningless statements. Certainly, no medical student reading it would be inspired to devote his/her life to nephrology and kidney pathology. Teaching the kidney organization without linking it to its biological purpose would be as effective as teaching students the entries in a telephone book.

However, the concept of 'purpose' cannot exist in isolation. Every purpose is supposed to be in the service of a superior level purpose. Conversely, whenever we claim to know the 'purpose' of a machine, we look at the component parts and look for the way, in which they contribute to the purpose.

In other words, one cannot ascribe a purpose to a 'part' without justifying it by the purpose of the higher-ranking 'whole'. For example, each nephron, whose purpose it is to serve the kidney function, would ultimately be useless, if the whole kidney had no purpose. Surely, the steering wheel of a car is useless, if the car is a dysfunctional wreck.

Hence, every time we want to justify the purpose of an object, we must go up the scale of sizes, and find out whether the next higher level still has a purpose. Unfortunately, along the way even biology reaches quickly a limit where teleology becomes meaningless.

Take again the example of the kidney. Moving up one level of organization, we can identify its purpose to filter blood plasma and to adjust the blood pressure. We will also be able to find a 'purpose' for the maintenance of 'healthy' blood conditions of a whole mammal like a horse. However, when it comes to naming a purpose for the whole horse, most people will struggle to name one in objective terms. However, even if a cavalry enthusiast could name an objective purpose for horses, we would have to continue further up the scale of sizes. Eventually, we would have to demonstrate the purpose of mammals in general, or of animals, or of life on Earth per se. At the latest at this level in the hierarchy of life, we have to admit that we cannot name a purpose for life on Earth that science can prove. There are too many counter-examples of planets without life, even in our own solar system, and they seem to be doing quite well without it.

Although biology must eventually reject teleology as a global concept, it remains indispensable for our understanding, teaching, and heuristics of biology on countless levels of size and evolution. Adopting a compromise, we may resort to 'local' versions of teleology, i.e. versions, which work up to certain levels of biological organization, but stop making sense at the next higher levels. The compromise allows us to interpret biological configurations and networks of interactions by their 'local' purposes, and consider them correspondingly as tools, scaffolds, developmental precursors, food sources, signals, etc., while we do not try to identify a 'global' purpose for their higher levels in the hierarchy.

Jigsaw Puzzles as the Most Trivial Examples of Fitted Mosaics

The simplest examples of fitting and fitted objects are mosaics of rigid, flat tiles with variable shapes. Before any misunderstandings arise, this book is not comparing life forms with jigsaw puzzles. *Fitting a mosaic is a much more challenging task than restoring a set of fragments to its original order!*

In the first place, unlike the pieces of a jigsaw puzzle, the tiles of a fitted mosaic are not fragments, but autonomous or at least semi-autonomous individuals. Most of all, they do not carry any attached clues of 'the final picture' on their faces that may help decide whether a piece was put in its right place.

Worse, *there are no single 'right' places for any of them.* Many different mosaics may contain each of the tiles in many different locations and functions, as the same set of components can yield very different functional and perfectly fitted mosaics. Only the conditions of tessellation must remain fulfilled:

1. Every tile must fit perfectly its neighbours.
2. All tiles combined must fill the frame exactly.

Differential Calculus, the Antithesis of Modularity?

The discontinuity between their precisely fitted parts distinguishes mosaics quite dramatically from the favourite conceptual tools of physics. Hamiltonians, entropy, electro-chemical potentials, wave functions, electromagnetic and gravitational fields etc. are infinitely smooth and even differentiable mathematical functions and fields. Even *particles* are formulated as quantities that are inseparably linked to differentiable fields and wave functions. Their very requirement of differentiability guarantees that most points of such infinitely smooth objects expand all the way to infinity.

Mosaics are not only dramatically different objects, but differentiable functions and fields are not important for biology. At best they play peripheral and auxiliary roles. For example, they may occur as time courses, temperature dependencies, concentration gradients, or current-voltage relationships. Although important for specific disciplines, they are not central objects of biology.

The central objects of biology are almost the opposite. They are heterogeneous, complex, discrete, individual objects such as evolutionary trees, skeletons, protein structures, nucleotide sequences, nucleosomes, chromosomes, nuclear pores, microtubules, cells, organs such as lungs, networks of blood vessels, of neurons, brains, individual organisms, ecologies, and so forth. Invariably, other, similarly discontinuous objects and environments surround them, *not* forming graded, continuous transitions between them.

These core objects of biology may interface and interact with others like them. The interactions may rely directly on contact or indirectly on surrounding, often circulating media, or a chain of other individuals. Each may have their own specific evolution, development, replication, and molecular machinery. Yet, no matter how complex their compositions, shapes, and functions, they are always limited in size and reach, and can be recognized as distinct objects that are exquisitely well fitted together.

It is no surprise that the above line of reasoning justifies us again to treat the objects of biology as living mosaics.

Chemical Compounds, the Particularly Important Implementations of Mosaics

In contrast to physics and its world of mathematical functions, chemistry had always played a much closer role in biology. Already the alchemists of many centuries ago entertained strange ideas about matter linked to other strange ideas about life. Eventually, biochemistry and molecular biology dominated modern biology, both of which are, of course, sub-specialties of chemistry.

Seen from the vantage point of this book, the close relationships between chemistry and biology are not too surprising. After all, chemical compounds are a subset of mosaics, namely all the mosaics that are composed of atoms as their fitted tiles.

C, H

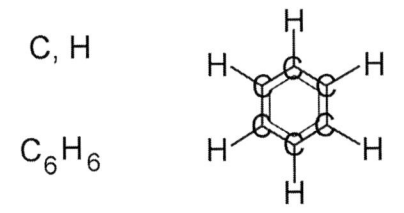

C₆H₆

Fig. 1-01. The method used by chemistry to describe a chemical compound (= spatial mosaic of atoms) such as benzene as mosaics. The example defines as tiles the common elements C and H, describes their content by a summary formula C₆H₆ of the numbers of linked atoms, and depicts their spatial fitting by a stick-graph that images the presence or absence of their interactions. As to its classification, benzene is assigned to the class of aromatic hydrocarbons.

As far back as the eighteenth centuries, chemistry developed a language to describe and classify chemical compounds as composite objects, i.e. as mosaics.
1. Chemistry assigns each compound (mosaic) to a class of compounds such as salts, acids, metals, polymers, hydrocarbons, etc.
2. It describes each such mosaic by a summary formula that lists the atoms and counts their numbers.
3. It describes the fitting of the tiles using a stick-graph that depicts which atom forms bonds with which other by what valence (figure 1-01).

Chemistry expresses the fitting of the tiles through the concept of chemical valence: The stick-graphs in the figure depict the valence requirements for each atom in well-known ways by the number of edges (including 'partial' edges), which emanate from each node (figure 1-01).

As much as it would be tempting to take advantage of the century-old legacy of chemical mosaics, the chemical approach is too specialized to apply to all mosaics: Chemical valence is the consequence of the laws of quantum mechanics, whereas the fitting mechanisms of mosaics in general share no such common natural laws. Their fitting employs mechanisms as diverse as physical laws, geometry, formal logic, celestial periodicities, accidents of evolution, and numerous others.

The Principle of Modularity as Justification for Biological Research

Whether biologist are aware of it or not, they frequently apply the idea that organisms consist of discrete, semi-autonomous units whenever they take the organisms apart in the name of research. In these situations, researchers assume that - at least in principle - the parts of living things work in their natural context essentially the same way, as they do in isolation after an experimenter has removed and purified them. What other justification is there, but to assume that living systems consist of discrete, semi-autonomous units?

Researchers are often rather reluctant to mention the assumption explicitly, because it is controversial and quite vulnerable. Nevertheless, its best justification is the principle of modularity. Indeed, if the investigated objects are modules of a mosaic and thus be at least semi-autonomous, many of their functions should work in isolation the same way as in their natural context. Isolating them or at least observing them in isolation may not alter fundamentally the results.

Nevertheless, it is a questionable assumption. In the first place, the experimenter cannot know *a priori* which of the functions of a living module are truly autonomous and will work unaltered in isolation, and which are not.

In addition, there is a logical danger. By splitting the living mosaics conceptually into semi-autonomous parts, their modularity becomes a self-fulfilling prophecy. What, if there are living mosaics and expressions that defy the logical or practical isolation of modules?

In principle, there are. The most obvious exceptions are some of the fractal mosaics, which we will discuss in chapter 8. These are all the mosaics that are made *entirely* of tiles that are mosaic themselves, which in turn only contain tiles that are also mosaics, and so forth. In the limit, such mosaics clearly contain no more distinct or semi-autonomous tiles. Instead, their 'inside' is completely continuous. Experience tells us that no other mosaics violate the modularity principle. Besides, they only violate it at their infinite limit, whereas all real living mosaics only have a finite number of components.

Holistic approaches to biology such as traditional Chinese medicine or the Indian Ayurvedic medicine present a number of different kinds of exception to modularity. There are also accepted examples for holistic approaches in Western medicine, such as hypnosis and the so-called 'placebo effect'.

In a sense, even these approaches define their own kind of modules, such as a life force chi, a meridian, and selected points, as in the example of acupuncture. However, as long as there are no reproducible assays and experiments to isolate these postulated modules, we consider these systems as having no modules, at least for the time being.

The most radical exceptions to the principle of modularity in biology are presumably emotions like love, faith, fear, wrath, or grief, but also expressions of creativity in art, music, poetry, or mathematics, as they have no isolatable parts. On the contrary, they emerge as indivisible, new whole entities from their underlying sources and carriers.

To be sure, their sources are brain domains, and thus mosaics of neurons. Yet, one may argue that they have no discrete parts, but express variable intensities and are subject to undercurrents, similar to bodies of water or air. Nevertheless, psychologists and poets alike have offered analyses of these very elusive feelings, although their nature as perceptions makes the objective and reproducible identification of parts very difficult.

Obviously, it does not help trying to exclude these exceptions as unbiological, because, naturally, we only find these phenomena with certainty in the living mosaics of biology, and only there. Yet, for millennia, they have resisted successfully every attempt to break them into distinct, tangible elements that would allow us to study them in isolation.

For the time being, these most challenging properties of living mosaics must remain unexplained, at least in experimentally verifiable terms. The present treatise will settle for much more practical questions, such as questions about some of the logical consequences, if a system *is verifiably,* a mosaic composed of modules.

Natural Selection as a Driving Force behind the Principle of Modularity

Naturally, the definition of modules also raises the question of their origin. At first, it may seem that we can settle this question rather quickly. The origin of discrete, autonomous objects needs no explanation, unless life began in a featureless continuum. There is no evidence for such beginnings. On the contrary, a billion years ago, as today, all matter on Earth consisted of discrete objects that had existed and acted in isolation for some time before groups of them aggregated to initiate life. Hence, like all other initial matter, the seeds of life had also to be discrete, and associated with each other from the start.

However, this argument does not explain why the manifestations of life remained modular for millions of years, and even evolved their modules to ever-higher levels of complexity and efficiency. Therefore, we should ask whether there is a selective advantage of a living mosaic for having a modular architecture. If so, evolution would quite likely have improved and enhanced modularity in all its manifestations.

As was mentioned multiple times before, effective engineering always uses modules, because systems with a modular architecture are more robust, effective and reliable than non-modular ones. The latter are prone to continuous cross-talking and cross-reacting. They cannot confine the damage caused by perturbations, and therefore tend to freeze, disintegrate, or flip into detrimental actions upon suffering trauma.

In addition, the evolution of modular systems can proceed faster than non-modular ones, because the modular system can evolve its modules one at a time. In contrast, non-modular systems need to revamp their entire body at the risk of rendering it dysfunctional, even after a relatively insignificant change.

Therefore, generally speaking, systems with a modular architecture operate more predictable, less error-prone, and adapt faster to changing conditions. Hence, modularity is most certainly a major selective advantage.

There are several other definitions of modules. For example, one may define them as the morphological characters and homologues of organisms [25]. As these quantities are the markers and road signs of evolution, they offer much greater objectivity. Supported by the fossil record, and as natural evolutionary concepts, they suggest right from the start quite detailed scenarios of their evolution. On the other hand, they are less obvious choices to describe the inner architecture of these morphological characters and homologues, especially the molecular architecture of single cells, organelles, viruses, and prions, for which there are few morphological characters and no fossil records.

Yet other definitions of modules can be derived from biological networks [04]. Every living mosaic is in particular a network of interacting components. By counting the local density of nodes, one can derive a definition of modules that seems entirely objective. On the other hand, its linkage to known evolutionary pathways appears more difficult to establish. Nevertheless, a linkage to evolution can be argued quite credibly, especially by a model study [25], which demonstrated the emergence of modularity in such networks depending on the 'cost' of its connections.

Living Mosaics: A New Vantage Point and a Wealth of New Questions

Considering organisms and their components with the metaphor of living mosaics offers a number of new perspectives. Here are some of them in no particular order.

The 'Islands' of the Fitted Solutions in a Gigantic 'Ocean' of Failed Ones

Anybody who tries to assemble manually a mosaic will soon realize that the task is a surprisingly difficult and counter-intuitive. And it is not an entirely dispassionate one.

Actually, the fitting of mosaics can evoke quite strong emotions. Initially, one is easily frustrated. One may not always be aware of it, but 'fitting' requires the adjustment of huge numbers of linked, multi-dimensional variables. Hence, the number of 'failures', i.e. the number of non-fitting tile configurations is usually enormous. In other words, failure *is the rule*.

Conversely, the completion of a perfectly fitted mosaic solution, and even the mere encounter with one may evoke profound delight. The completely fitted solution projects a strong sense of inner logic. One may even feel that only this solution is a possible one.

Yet, in most case, this impression of uniqueness would be entirely wrong. There may be hundreds or even thousands of different 'sibling'-solutions for the same fitting task, in spite of seemingly impossible odds for finding even a single one.

Still, regardless how many perfectly fitted solutions exist, the non-solutions always outnumber them. They are like tiny islands, distant from each other, and surrounded by an 'ocean' of non-solutions.

The 'Seeds' of Mosaics and a New Kind of Determinism

The surprising number of different sibling solutions of most mosaics poses an intriguing question. If the assembly is successful, what decides which sibling-solution it yields? The decisive factors cannot be the frame or the tiles, because they are the same for all sibling-solutions. Do the location and orientation of the first tile decide whether the assembly leads to a solution, and which one it is? Or is it the second tile? Do all tiles matter equally much for the solution?

The answer is surprising, too. Each fitted mosaic gives rise to several small groups of initial tiles that have a very special property: If assembly starts with any of them, the fitting mechanism leads unerringly to one and only one mosaic. Chapter 11 will describe these determinant groups as '*key-seeds*' of their mosaics in greater detail. It will turn out that there are many more key-seeds than there are sibling solutions.

If it starts with a key-seed, the assembly of a mosaic is entirely a matter of necessity, no matter how large and complex the mosaic may be. Therefore, it presents a special kind of determinism that is neither intuitive, nor is it a matter of some universal law. Instead, it reflects intriguing and highly individual implications of the fitting process.

The driving forces of fitting are *logical* rather than *physical*. For example, the fitting may apply trial-and-error methods, which offer no force-intensity or force-distance laws, or any other explicit formula. Instead, they use information gleaned from look-up tables and other ways of relating individual actions to individual consequences.

Nevertheless, during assembly, the tiles experience and create some peculiar kind of 'action at a distance' that applies to the inside of the mosaics. For example, the key-seeds contain tiles in locations all over the frame, far away from the place where the next tile is to be added. Yet, they influence the location and orientation of all others. Actually, this 'action at a distance' of tiles is not exactly an 'action'. In reality, the distant tiles merely restrict the possible locations and orientations of tiles elsewhere. In general, no single tile from the earlier stages of the assembly is able to *dictate* the placement of the successor tiles. After all, the very existence of multiple sibling-solutions proves that fitting can follow many alternate paths. Only all pre-placed tiles together are able to restrict the possible choices so much that they amount to enforcement.

By definition, the logical relationships between mosaics and their key-seeds leave no room for alternative paths, but lead from key-seed to mosaic in a progression of necessity. Thus, it represents a peculiar kind of mechanism that drives the assembly: It follows a *path of necessity without any physical mechanism that involves the interactions between the tiles*. At this stage, the statements may sound rather abstract, but chapter 11 will offer several simple examples.

The progressions by necessity involve the exact fitting requirements, the boundary conditions imposed by the frames, and the exact consequences of misplacing any one of the tiles. Obtaining this knowledge may be no problem in cases of human-designed mosaics. Nevertheless, it is the very essence of biological research.

Iterations of Mosaics and Fractal Mosaics

Naturally, the tiles of a mosaic may consist of smaller tiles, and thus be mosaics, too. Composing the smaller tiles from even smaller ones can go on ad infinitum. If one drives this kind of iteration towards infinity, it may lead to mosaics whose tiles are so small and dense they have no more discrete tiles. Are they still mosaics? What kinds of objects do mosaics become if we submit them to an infinite series of iterations? (In chapter 8 we discuss this possibility, but also the argument that we need not answer such questions, because nothing in biology is ever infinite).

The 'Almost' Repetition Cycles of Mosaic Patterns

It seems that there are countless identical copies of the anatomy and actions of organisms, organs, tissues, cells, organelles, etc. However, if we look closer, then we find without exception, that they are not precise copies or repetitions. On the contrary, no two organism or their actions are ever, have ever been, or will ever be precisely the same. In some cases the reason for this universal individuality is the effect of random noise or some other exogenous perturbations present in the systems; in other cases the causes are the amazing creativity of nature: *No living creature is ever exclusively the result of an automatic, template-indifferent copying process, but it involves rebuilding at least major parts of it, 'from scratch'.*

This effective re-creation of living things from much simpler seeds and building blocks yields individuals that are basically, but never precisely the same, because the number of conceivable variations of the re-building process is hugely larger than the number of actual individuals, which ever existed or ever will exist. Unless the re-building mechanism is so unrealistically detailed as to exclude all possible deviations and errors, it is free to turn out every time a novel variation of forms or actions. Using our 'mosaic-terminology', in view of the astronomically large number of 'tiles' in every living mosaic, the number of conceivable 'sibling-solutions' exceeds by far the number of mosaics that exist now and existed ever before.

Take the example of the human genome with its approximately $3 \cdot 10^9$ nucleotides. In principle, each of the nucleotides can be mutated, which would generate $3 \cdot 10^9$ conceivable single point mutations, $9 \cdot 10^{18}$ conceivable dual point mutations, $3 \cdot 10^{28}$ conceivable triple point mutations, $9 \cdot 10^{37}$ conceivable quadruple point mutations, and so forth. Of course, in reality much, much larger differences exist between the genomes of different people than a mere 4 nucleotides.

It is true; genome replication relies on the seemingly invariant copying of templates and thus guarantees an extremely high level of precision. Yet, even this mechanism involves a certain amount of 're-creation from building blocks', because it needs to link up billions of Okazaki-fragments as it synthesizes the so-called lagging DNA strand.

But assume that the mechanism of genome replication were absolutely precise, nature does not allow them to remain this way. Exposure to sunlight causes thousands of point mutations every second in our genomes. Even if there were never more than 4 mutations, the number $9 \cdot 10^{37}$ of conceivable quadruple mutants of the human genome would exceeds by astronomical proportions the number 10^{23} of all genome copies of all living people, and is larger than the number who ever lived and ever will. And even if a replicated genome would be completely identical to its template, radiation and chemical damages would not allow the identity to last long. That applies especially to cells inside the body, where the ultraviolet photons of the sunlight cannot reach. Of course, there are control- and repair-mechanisms, but they are not error-free, either, as malignant mutations prove tragically. Therefore, there can never be two exactly equal genomes, never will, and never have been.

Mutatis mutandis, the same impossibility of identical repetitions applies to the occurrence of identical fingerprints, ridge patterns on the surface of oral mucosa cells, the patterns of hair follicles on skin, the pulse patterns in brains, the vibrational states of every protein molecule in every cell, and countless other details of living systems.

As we will see in chapters 17 to 19, the same applies to mosaics. Whenever mosaics cycle they tend to proceed without actually repeating themselves precisely. Yet, like living mosaics, they produce configurations and periods that are very similar to each other. We may call them as 'almost-identical'.

The 'Pseudo-Specificity' of Response and Repair Mechanisms

Assume that a 'mutated' tile no longer fits anywhere in its 'parent' mosaic. In response, the surrounding modules may scramble to restore the perfect mutual fitting. If this 'repair' is successful, it may appear as if it was a programmed action, focused on the damaged region.

Yet, this apparent targeting may be misleading. Many modules remain passive or neutral during such rescue efforts, while only a minority is active. The modular character and semi-autonomy allows many tiles to ignore the damage and the repair, as long as they are not directly affected. The resulting concentration of the action on selected tiles may sometimes mimic 'intelligent' actions during development and even during evolution where none existed.

A Supplemental Taxonomy

The classical Linnaean taxonomy characterizes organisms by their anatomical features. Since the anatomy of an organism is nothing but a mosaic, the Linnaean taxonomy is the taxonomy of mosaic *structures*.

However, as discussed earlier, living mosaics have not only structures; they also fulfil certain *tasks*. As we shall discuss in chapter 9, tasks can be described as non-material mosaics consisting of their own functions. I believe it makes it necessary to add functional taxonomy to the classical structural one (see chapters 10, 28).

The Set of Sibling-Solutions

As mentioned before, the often-large number of sibling-solutions will surprise everyone who had trouble of assembling even a single one. However, once one has gotten over it, another surprise is in waiting. If a particular living mosaic exists in nature, one should expect that its sibling solutions should exist, too. However, as it turns out, this is rarely the case.

Consider the classes of different skeletons! One can think of many more variations of functional skeletons than seem to exist in nature. For example, one could imagine sibling-skeleton that are not mirror symmetrical, or do not feature a separate head, or do not place the food intake machinery at the head, and so forth.

Why do so few truly different sibling classes exist in reality? The difficulty of assembling alternative solutions can hardly explain it. Looking at the boldness, diversity and elegance of nature's creations, technical difficulties do not seem to be serious obstacles.

One can think of a great many factors that exclude the realizations of drastically different sibling-classes. For example, evolution must begin with one or few common ancestors and thus entail the accidental dominance of particular solutions over their siblings.

Next, there is nature's choice of development as her universal assembly technique. If some robots along an assembly line would assemble the mosaics from pre-fabricated parts, one could vary the order of the assembly steps to arrive with ease at different sibling-solutions. However, as mentioned before, nature predominantly builds the copies of mosaics from scratch. It means that they have the same beginnings, and thus disfavour alternate solutions.

Furthermore, nature enforces interaction between her living mosaics by squeezing them into the same living space, by feeding them off the same food chains, by requiring mates for reproduction, etc. Successful integration takes a long time for the living mosaic, but also for all interacting members of the environment. Once a mosaic is integrated, its sibling-mosaics are not likely to fit equally well, but find themselves in a challenging battle against natural selection. Hence, from the point of view of mosaics, development and natural selection are powerful methods of restricting alternative life forms.

Of course, scientists have reached this conclusion many times before, when they puzzled about the universality of DNA, RNA, amino acids and their uniform chirality in terrestrial life forms. Here we suggest that the above line of reasoning not only applies to these molecular mosaics, but also to cellular, organismal, ecological, and any other macroscopic form of living mosaics. Once we understand living systems as formal mosaics and derive how many classes of sibling-solutions are conceivable in each case, we may be able to formulate and explain the full extent of these constraints.

A Survey of the World of Mosaics

Chapter Two

A Cursory View of Living Mosaics

Living Mosaic Can Be Tangible or Symbolic; Like All Mosaics They Have Composition, Architecture, Texture, and Dynamics; But Unlike Inanimate Mosaics, They Also Have Tasks

In view of the huge diversity and ubiquity of mosaics, it is a daunting task to sort them into different categories. We begin with a quick survey of their most fundamental properties, namely

1. **composition,**
2. **architecture,**
3. **texture**,
4. **dynamics**.

Later in the book (chapter 9) we will discuss a fifth basic category, which only living mosaics have, namely their

5. **task**,

In chapters 10 and 28 we will propose a supplemental taxonomy, based on the tasks of living mosaics.

The following descriptions of mosaics are **idealizations**. Real mosaics express in whole or in part, some or all of these idealized properties.

The Composition of Mosaics

We begin by sorting the mosaics into two major groups. If their tiles are objects of the material world, we will consider the mosaics as 'tangible'. Otherwise, if their tiles only exist in a mental world of ideas, concepts, meanings, information, words, numbers, etc., we will consider them 'symbolic'.

Of course, in reality mosaics are rarely one or the other. The abstract tiles of symbolic mosaics rely on material carriers, while most tangible mosaics communicate information and signals, which are composed of symbols. Symbolic elements link specifically to tangible mosaics, at least as the 'names' and 'road signs' of their tiles to facilitate the navigation of their hugely diverse physical manifestations. Conversely, the reading, transmission, and storage of the modules of symbolic mosaics require material carriers and energy supplies.

All this applies especially to living mosaics, which link tangible and symbolic mosaics inseparably for the performance of their tasks. Take the example of a functional brain. It is clearly tangible, whereas the meanings of its pulse trains are symbolic. The tiles of a brain (e.g. synapses, neurons, glia, blood vessels) must fit *physically*, while its pulse trains create and obey immensely complex and demanding *logical* rules.

Similarly, the human-made living mosaics need both kinds of characterization. Consider integrated circuits such as modern microprocessors! They are clearly tangible, and their tiles (transistors, capacitors, resistors) must fit spatially and functionally into dense patterns, in order to function and to dissipate heat efficiently. On the other hand, symbolic criteria, such as the rules of coding and their arithmetic and logical relationships, dictate the patterns of their connections and pulses.

It is hardly worth mentioning that the distinction between 'tangible' and 'symbolic' and its many other versions is at least as ancient as Buddhism (600 BC). In modern Western philosophy, the perhaps most important formulation is that of René Descartes (1596-1650) who distinguished between 'res extensa' and the 'res cogitans'. With the most recent advent of computers, it resurfaced in the prosaic disguise of (tangible) 'hardware' versus the (symbolic) 'software'.

A. Tangible mosaics

1. Spatial mosaics

There is no need to explain spatial mosaics in any detail. After all, most people upon hearing the word 'mosaic' think of spatial ones such as the floor mosaics in Pompeii or the toy jigsaw puzzles. Furthermore, chapter 3 will introduce in detail a spatial mosaic whose tiles are called 'pentominoes', and will use it throughout the text for illustrations.

2. Temporal mosaics

Although 'time' is not often considered 'tangible', it is at least a measurable part of the real world. More specifically, 'temporal mosaics' are mosaics whose components measure time, react to time points, set time, or derive their order and structure from the flow of time.

Of course, the most familiar examples of temporal mosaics are clocks. Their three characteristic components are oscillators, counting devices for the oscillations, and displays of the counts. In their most elegant forms, the counting devices also apportion the energy needed to run the clock, and to supply it to the clock in a synchronous fashion. However, biology offers much more complex and much less regular examples of temporal mosaics, which range from beating hearts to reproductive cycles, or to evolutionary histories.

Other examples are calendars, circadian rhythms, sleep-wake rhythms, and other biological cycles, or systems whose main variables rise, stabilize, or vanish, or head asymptotically towards catastrophes.

Typical variables of temporal mosaics are time intervals, states or phases, temporal frequencies, delay times, onset and termination times, oscillations, resonances, frequency spectra, bandwidth, noise, degrees of irreversibility, strings of historical events, but also time derivatives such as speeds or growth rates.

3. Functional mosaics

These mosaics are predominantly composed of linked, coordinated functions and actions. Their components do not require spatial fitting, but their tiles must fit functionally.

As a simple example, think of the vacuum tubes, induction coils, capacitors, resistors, loudspeakers, indicator lamps etc. of an old-fashioned radio. They may look like a jumble because their spatial fitting is not very demanding. In contrast, the functional fitting of each part is mandatory and critical for the radio to work.

Turning to examples from biology, the various organs of every functional organism fit functionally exquisitely well. Nevertheless, they usually also fit spatially together. There is no empty space inside organisms, unless its function requires it. Naturally, there is empty space inside hearts, lungs, bladders, glands, and the intestines, which is obviously their function needs. Yet, otherwise, the shapes of the tissues and organs inside a human body match perfectly.

In some cases, functional mosaics dispense with the conditions of spatial fitting to such a degree, that they become essentially fluid. The immune system may provide an example. In this case, the individual immune cells and protein complexes can physically flow past each other inside the blood stream and the lymphatic, but also entire 'armies' of white blood cells, like foraging ants, may penetrate solid tissues while cooperating in the maintenance and defence of the organism.

In other cases, the functional, yet fluid mosaics are un-mixable liquids or colloids in the sense that they are actually solid objects, which are able to flow past each other.

(At this point I am tempted to add that my emphasis on mosaics does not contradict the famous principle 'παντα ρει' ('Everything flows') of Heraclitus of Ephesus (535-475 BC). As the existence of fluid mosaics shows, a mosaic can very well consist of solid tiles that are yet free to flow.)

B. Symbolic mosaics

Although the many members of symbolic mosaics may have material carriers, each of their tiles represent mental objects, which are ultimately symbols. In the majority of cases, they relate directly to information. Their components are elements of a language or any other communication system. Here, the term 'information' not only is used in the sense of Shannon's definition as the 'deviation from the expected', but also in the more elusive sense of context-dependent, semantic information. We will call these mosaics 'information-rich'.

There are other symbolic mosaics, which are devoid of all information. We will call them 'information-free'. Their components are abstract objects that may evoke emotional responses but do not relate to any empirical reality. Information-free mosaics may have a powerful appeal to various observers, but they do not aim directly at any specific recipient and they carry no identifiable information. Let us look at some examples.

1. Information-rich mosaics

The component symbols of information-rich mosaics aim for a particular recipient. Indeed, most symbolic mosaics we encounter in real life are information-rich. They are the workhorses for the handling of information. Their contents do not have to be entirely novel, although they must be worth the attention of the recipient.

Every information-rich mosaic it is directed at someone or something, be it a spoken word, a written sentence, a song, a painting, a computer code, an algorithm, a legal case, a judgement, a bank account, or a theory. Its tiles have clear borders and often universally recognized modules such as phonemes, letters, musical notes, numbers, and other graphic or mathematical symbols.

Whenever information-rich living mosaics carry, detect, transmit, store, encode, and manipulate vital information they serve animal survival, recognition, camouflage, and territorial claims. Human civilization without the information-rich mosaics of language, laws, and economical rules is unthinkable.

The requirement of fitting tiles takes on a special meaning for information-rich mosaics. It is not merely a technical pre-condition, but it is their very essence. After all, the information encoded in an information-rich mosaic is a direct product of the correct fitting of its symbols.

Take the example of the string of letters *'rm Mateylil h lata.bad'*, which is meaningless at least in English as long as it ignores the rules of fitting of the English language such as syntax and semantics. The same letters in a fitted order will read *'Mary had a little lamb.'*, and thus transmit information.

Although mechanical carriers and procedures may play important roles and impose certain constraints on information-rich mosaics, the fitting of their symbols is predominantly mental. Usually, there is some kind of balance between the material and the mental aspects of information-rich mosaics. As Bertrand Russell's formulated it, 'matter is not so material and mind not so mental as is generally supposed' [20].

In the course of history, humans have grown the rules of fitting of information-rich mosaics to extremely high levels. Examples are the many complex rules of grammar, meaning, musical harmony, logic, style, codes, social conventions, and encryption/decryption algorithms.

2. Information-free mosaics

If there are no discernable recipients for the symbols, if the symbols have no definite meaning, or if they have multiple, even contradictory meanings, then the patterns of symbols of a symbolic mosaic carry no information, and we may rightly call them information-free.

Most of such mosaics belong to the highest manifestations of faith and art. Examples may be religious texts, meditation rituals, prayers, the abstract patterns of Islamic art, Bach's 'Art of Fuge', and others like them that either transmit no information about the material reality or have no provable recipient for the information.

Still, we worship some of these mosaics as the highest expressions of a universal truths. They may have a powerful appeal to the observer, but often require a long learning process before s/he acquires a deep, emotional certainty that they carry a message after all. Yet, whatever that message may be, it is always a matter of individual perception, and never a provable reality.

In many cases, we cannot be sure whether a symbolic mosaic is actually information-free. It may only appear unreadable because encryption methods, unknown languages, forgotten symbols, or the wrong recipients distorted its information carriers. After we found the right Rosetta stone, the mosaic may very well turn out to be information-rich. To this day, people are trying to decipher the Voinich manuscript, or try to interpret cryptic texts such as the Revelation of Saint John the Divine, or the I Ching.

The Architecture of Mosaics

A. Mosaics with complex frames: Mazes, networks, and art forms

So far we have treated mosaic frames as if they only were featureless boundaries, which prevent the tiles from spilling into the surroundings. In living mosaics, the frames are far from featureless. Examples are the skin, the peritoneum, and the many capsules of organs. In some cases, a peripheral feature is the content-rich, even central property of the object.

In addition to their main outer boundary, mosaics can have one or many inner boundaries, as well. Figure 2-01a shows the example of a 2-dimensional frame that contains numerous, fragmented and randomly placed inner walls. The highly convoluted spaces between them (white space) accommodate various kinds of tiles. Common language describes such objects as 'mazes'. Their main feature is their large inner surface, which increases the chances for interactions between the filling tiles.

Yet another closed frame may surround the entire mosaic as some sort of an exoskeleton. Many living mosaics, e.g. insects have such mosaics as their only skeletons. Sometimes the tiles of vertebrates have frames of this kind. Examples may be lungs and gills.

These frames lead to two complementary structures. In one case, the inner walls are all connected forming an internal frame like the endoskeleton of vertebrates (figure 2-01b).

In the alternate, i.e. complementary version of such mazes, tiles and frames reverse roles. In this case, the formerly convoluted space turns into a convoluted wall whereas the former wall fragments remain open (figure 2-01c). If they were living mosaics, we would call them sponges or foams. Another example is the so-called basement membrane in mammalian organisms, which surrounds the organs and even tissue domains within them.

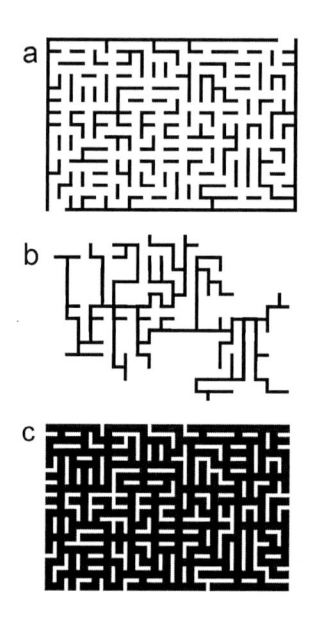

Fig. 2-01. Alternative interpretations of a 2-dimensional random maze. (a) It is a conventional maze: Fragmented portions of the outer frame fill the inside volume of the mosaic, leaving room between them for the tiles. The maze is held together by an outer frame like in an exoskeleton. (b) The inner walls are all connected forming an internal support structure like the endoskeleton of vertebrates. (c) It is a sponge: A highly convoluted, continuous frame encloses numerous empty cells that can be filled with tiles (in this case from above).

In the case of functional mosaics, the various inner wall segments may carry out different functions. Some may only interact with selected others, thus creating a specific network. Figure 2-02a illustrates how the various wall segments appear as vertices of the graph of a network, while the various interactions may function as directed edges (figure 2-02b).

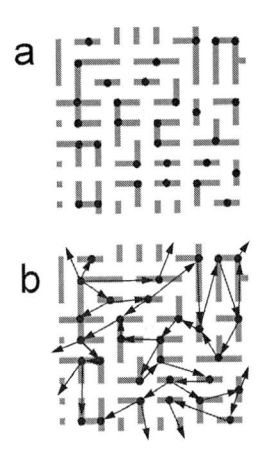

Fig. 2-02. Interpretation of a part of figure 2-01 as a network. (a) Interpretation of frame fragments as vertices of the graph of a network. (b) Depiction of the graph of the network by directed edges.

In cases of symbolic mosaics, the syntax of the letters or syllables, the outline of a text, the performance of a musical composition, etc. may play the roles of the restricting frames. They may take the form of the underlying abstract architecture. In the case of musical forms, these may be the architectural forms of sonatas, rondos, or fugues. In other cases, they may represent the variations and re-occurrences of a particular motif throughout the pictorial or musical composition. Figure 2-03a tries to illustrate such a situation by highlighting a motif as it migrates through a graphical composition.

Such a mosaic may also encode information. Figure 2-03b illustrates how the places of coincidence between a graphic composition and certain parallel lines may serve as keys for the encryption of a text.

Adding more frames to an already very complex universe of tiles opens up further dimensions of diverse and creative applications.

But the expansion is not without problems. We mentioned one of them already. As illustrated in Figure 2-01a and b, the distinction between tiles and frames is often ambiguous and hinders the scientific exploration of mosaics. On the other hand, it may add deeper meaning to the mosaics.

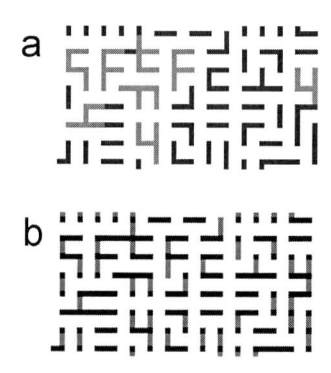

Fig. 2-03. Interpretation of a part of figure 2-01 as a symbolic mosaic representing artistic and information-technical 'texts'. (a) Artistic interpretation of frame fragments as variations on a pictorial or musical motif that is interspersed with related, fragmented motif alterations as in a fugue or complex painting. (b) Information-technical interpretation of frame fragments as encryption carriers: The decoded information (marked gray lines) is given by the lengths of the horizontal intersects with the pattern.

Letting complex frames become even more complex makes it more difficult to reach interior tiles, and to assemble or repair mosaics. Most living mosaics handle such problems by including separate assembly and repair centres inside their frames. They guarantee that tissues can grow and repair from within. Example of freestanding centres of growth and repair are the lens epithelium, the growth zones of bones, the resident stem cells in muscle tissues, ovaries, testicles, in the germinal layers of skin (*stratum basale*), in the imaginal disks of insects, and many others.

B. Meta-Mosaics

If mosaics become the tiles of larger mosaics, we may call the result 'meta-mosaics'. Like the more basic mosaics, they may be spatial, temporal, functional, information-rich, or information-free.

Composed of quite complex individuals, the composite mosaics of a meta-mosaic can reach formidable levels of complexity. Consider the meta-mosaic of the heart. It consists of the heart muscles, valves, blood vessels, and pacemakers. Similarly, the meta-mosaic of the tongue consists of a bewildering array of muscles, taste buds and salivary glands. The miraculous meta-mosaic of the eye contains retinal neurons, blood vessels, lens-forming epithelia, light adapting ciliary muscles, vitreous body, strongly shielding sclera, cleaning and anti-bacterial tear glands, and many more.

Particularly fascinating are seemingly paradoxical meta-mosaics that are capable of contradictory actions, such as counter flow multiplication. Take the example of the loops of Henle in the kidney. They pump water from the ultra-filtrate back into the blood stream, even though they consist to 85% of water. How can water extract water from a very watery solution?

Or consider the *rete mirabilis* that allows fish to pump gas into the swim bladder, regardless how high the outside water pressure is. Depending on the depth where the fish swims can amount to hundreds of atmospheres of pressure!

Another startling example is the vascular architecture of the skin of arctic animals. Here the blood flowing towards the skin would be in danger of losing all its precious heat as it contacts the very cold outside air. However, before this can happen, counter flow multiplication brings it into intimate contact with the cooled blood as it returns from the skin back into the body. Hence, the precious heat is not lost, but warms the returning cold blood, and returns with it to the body.

Meta-mosaics also play important roles in quite diverse matters of information. For example, the information-rich mosaics of linguistic elements may form meta-mosaics such as sentences, paragraphs, chapters, volumes, and libraries. They communicate information contents as readers and transmitters; they rearrange information contents as interpreters, encrypters and decoders; they create information contents as explanations, detailed arguments, case presentations, mathematical proofs, and many more.

Other meta-mosaics may apply logical algorithms to their information contents and thus express planning abilities, information processing, and other intelligent functions.

Meta-mosaics may interact with other meta-mosaics and form even higher levels of complexity as they create meta-...-meta-mosaics. One kind is especially frequent and important in biology, namely meta-mosaics that act as scaffolds or templates for other mosaics. Examples are egg shells, chrysalises, placentas, the hyaline cartilage templates of long bones, the hives of bees, the temporary memory needed in information storage and processing, and so forth.

Ultimately, meta-mosaics that contain structural, functional and information-rich mosaics may express emergent phenomena. Besides creating organizations and organisms, they may express intelligence.

Even a cursory look at the flora and fauna of our planet will suffice to realize that nature has taken this very path from mosaics to meta-mosaics, to meta-meta-mosaics, in overwhelming numbers, and that there is no limit to the stunning power of the emergent results.

The Texture of Mosaics

The Roman mosaics in Pompeii, the early Christian mosaics of San Vitale in Ravenna, or typical jigsaw puzzles leave no spaces between their tiles. In contrast, there are many mosaics, which also contain empty spaces that serve essential functions. For example, how could lungs function without their air sacs? If the leaves would fill the crowns of trees completely, how could the leaves catch any sunlight? How useful would cars and houses be without spaces for people?

Of course, the word 'empty' does not mean devoid of everything. It merely means the absence of mosaic-specific tiles.

The spaces between tiles need not be spatial. Many symbolic mosaics have to leave 'spaces' open for storage (memory) and other temporary processing functions. The pauses in musical compositions may offer another example. They can be as effective as the notes.

There are other kinds of empty spaces, such as the complex inner frames, which line the external spaces in certain mosaics. We mentioned the example of the lungs that are 'penetrated' by the external air space in a most convoluted, fractal pattern. We mentioned earlier the entire vascular system as the example of a module of immense complexity. Seen from the point of view of the muscular–skeletal part of the body, the blood plays the role of an exogenous medium, and the vascular walls become the inner frames that separate the muscle interior space from the 'external' blood.

There are also cases, where empty spaces are not permanent, but only open up temporarily by some kind of pulsating, breathing action. Examples are the ventricles of the heart, the spaces inside the cylinders of combustion engines. And so forth.

In order to categorize the relative amount of empty spaces of a mosaic, we will use the term 'texture', and distinguish between 'tight' and 'loose' mosaics.

A. Tight Mosaics

Typical examples of tight mosaics are jigsaw puzzles, walls, honey combs, microprocessors, or brains. In chapter 3 we will introduce a specific tight mosaic consisting of a 'model-frame' and twelve tiles, called 'pentominoes', which will provide most of the illustration throughout this book.

The most important characteristic of tight mosaics is their rigidity and lack of internal movement. Hence, we find them, where tiles or other objects need to be fixed or at least restricted in their mobility.

Assembling any one of them can pose extreme difficulties that border on seeming impossibilities. This difficulty of assembly makes them hard to imitate or duplicate. Hence, they are essential to guarantee uniqueness and security. In the course of the book, we will encounter many examples of tight mosaics.

B. Loose Mosaics

Loose mosaics are composite objects whose spaces occupy as much or more of their volume or function as do their tiles. The resulting mosaics may remain rigid, as in the case of the xylem of trees, which consists of countless, empty cellulose-lined cells. In other cases, the many spaces of loose mosaics may render them flexible. Depending on how much movement the spaces allow the tiles, the mosaics may become plastic or even fluid. Examples of fluid mosaics range from colloids to ant nests.

If the tiles are free to move relative to each other, the assembly of the mosaics is no longer a difficult task. As a result, the shape of such mosaics may be quite adaptable, and enable them to penetrate other mosaics. In cases of loose, symbolic mosaics, they are redundant or even wasteful and extravagant.

The Dynamics of Mosaics

Since loose mosaics allow their tiles to move relative to each other, it is possible to speak of the dynamic states of mosaics. Swarms of insects, large herds of cattle or horses, masses of pedestrians or cars may express quite distinct and predictable dynamic states and patterns. They may appear to be viscous, compressible, or even elastic in their response to external 'forces' that change their sizes and shapes.

A. Static (rigid, solid)

If mosaics have rigid tiles with fixed relationships between adjacent ones, they will appear solid and rigid. We will describe them as static mosaics.

B. Dynamic (fluid, flexible)

If the tiles of loose mosaics are free to move relative to each other, the mosaics may appear plastic and even fluid. It is possible that the tiles are rigid and have fixed locations, but are able to move in pre-scribed ways, nevertheless. Examples are railroad networks, but also the wheels and cogs of gears that are free to rotate, while remaining fixed in their bearings. These and similar mosaics will be called 'dynamic'.

CHAPTER THREE

A SIMPLE MODEL MOSAIC TO BE USED THROUGHOUT THIS BOOK

LIVING MOSAICS ARE NOT JIGSAW PUZZLES; BUT USING A MODEL SYSTEM SUCH AS PENTOMINOES CAN HELP GREATLY TO ILLUSTRATE THEIR HIDDEN COMPLEXITIES, AND HOW TO HANDLE THEM

Exploring a model mosaic best demonstrates the unfamiliar and often counter-intuitive properties of mosaics. Therefore, I need to ask the student to join me on a brief detour, where I introduce one that I consider a suitable candidate. The model will be an example of a 'tight' mosaic, because it illustrates many conceptual difficulties more easily and more accurately than a 'loose' mosaic, which would permit too much latitude of the interpretation.

Deliberately, I do not use a biological mosaic as model system, because they all have some unknown properties. Hence, the progress of biology may reveal new properties of a real living mosaic that disqualify it for one or the other illustration. An 'artificial' mosaic, as the one presented here, is unlikely to yield such unpleasant surprises.

Of course, an artificial mosaic cannot really represent the living mosaics of actual organisms. Yet, a great deal can be learned from it. I hope the student will grant me some leeway to introduce the model mosaic for this book right away, and to try a more elaborate justification/apology later on.

A Model Mosaic: Pentominoes

In order to be useful for our heuristics, illustrations, and explanations, but also for research into the properties of mosaics, our model system should not be too trivial, but also not too complex. The one I have been studying for years is the set of pentominoes.

Like many of my contemporaries, I was introduced to these brain children of Solomon Golomb's [106] by Martin Gardener of *Scientific American*. Pentominoes are the shapes one can form from 5 (pente (*Greek*) = five) squares that touch each other alongside (not merely at corners). Not counting 90°-rotations and mirror-reflections, there are exactly 12 of them. Traditionally their names are the capital letters, which they resemble the most (figure 3-01).

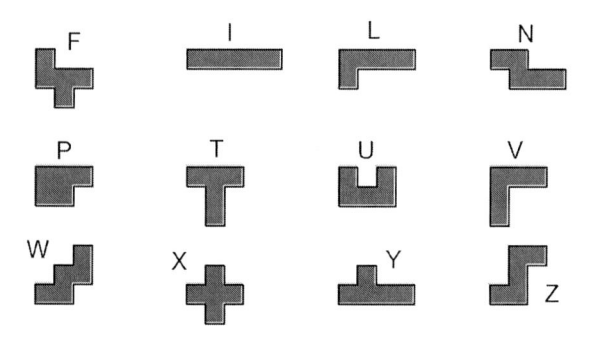

Fig. 3-01. The 12 pentominoes that are used as model tiles throughout the text.

Naturally, they are not merely different from each other, but they represent the set of *all* variations on their common theme. They can arise cyclically from one another by shifting a single square.

Over the years, the many pentomino enthusiasts have shown that pentominoes create perfect fittings for a great many different frames, as long as the areas inside the frames measure exactly 12x5 = 60 squares.

A Simple Model Mosaic to Be Used Throughout This Book   39

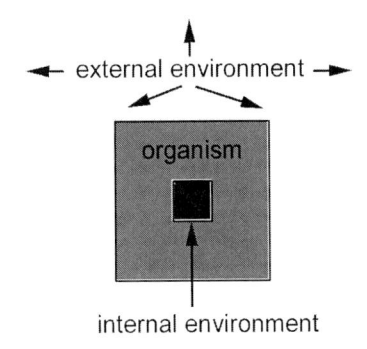

Fig. 3-02. The model-frame that is used for the illustrations and examples throughout the text and the possible interpretation of the various domains inside and around it.

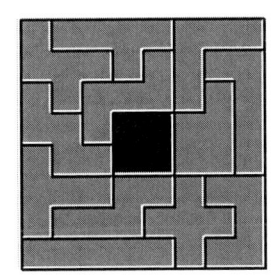

Fig. 3-03. One example of the 64 different solutions to fit the 12 pentominoes into the model-frame. (All solutions shown in figure 26-04).

In view of our biological perspective, we will use as the 'model-frame' an 8 x 8 square with a 2 x 2 small hole in the middle (figure 3-02). We chose this frame because it contains three biologically relevant compartments:

1. an external world in which object is embedded (external environment)
2. a space for the components (organism)
3. a space that is neither part of the object nor does it belong functionally to the surrounding world. We may consider it an 'internal environment'.

Figure 3-03 shows one of the solutions of fitting the 12 pentominoes into the model-frame. Appendix C describes more details of the algorithm I used to find all solutions of the model pentomino mosaic.

An Important Caveat for the Use of the Pentomino Model of Living Mosaics

I return now to my attempt to justify the use of the pentomino mosaic to model properties of living mosaics. Let me repeat, a handful of pentominoes fitted into a simple 2-dimensional frame are light-years away from the complex properties of living mosaics such as zebras, livers, cells, and mitochondria. Why then do I ask the student of biology or medicine to stay with me when I use them throughout the book as illustrations?

Understanding is a matter of logic, but mostly of hands-on experience. Richard Feynman famously said 'what I cannot create, I cannot understand'.

Obviously, the pentomino model offers hands-on experience of unfamiliar properties of living mosaics. Some of its properties are even counter-intuitive. Living mosaics share many of them with the pentomino model, no matter how much more complex they are. Therefore, I will employ the pentomino model system to illustrate and explain some of these shared properties. It helped me personally to accept a number of lessons and to appreciate the engineering requirements of actual living mosaics; I hope it will do the same for the student.

Naturally, I will not repeat this caveat later in the book, whenever I illustrate a mosaic property by the example of the pentomino mosaic. I will assume that *we both, student and author, remain fully aware of the limitations of the model.*

As a first lesson of the pentomino model, the student should experience the enormous difficulty of assembling fitted mosaics. It is most educational, trying to fit manually the 12 pentominoes into a rectangular or any other shaped frame that encloses an area of 12 x 5 = 60 squares. The encounter first hand of the technical and logical problems is most surprising. However, the powerful feeling of 'victory' after a successful assembly is also surprising. The understanding of the difficulties comes from a particular type of failure. Repeatedly one finds a pattern of 11 fitting pentominoes, while the 12th does not fill the remaining gap. Then the student's frustration can be a measure how far away the actual solution of the problem still is, although it seemed only to be one tile away. The experience will add a new dimension to the student's awe of the evolution of all exquisitely fitted living mosaics.

A Less Prominent Role of Averages in the Quantitative Description of Mosaics

Averaging individual data is a most successful practice in all fields of science where fluctuating data force the experimenter to decide which results to believe. Provided a sample of enough individuals is available, averaging data reduces dramatically the number of variables, which one needs to measure, and possibly explain.

More importantly, some of the averages are quite natural quantities, because nature herself is doing the averaging. Take the familiar example of the *naturally averaged* kinetic energy per molecule, i.e. the temperature! It causes the expansion of a column of mercury and countless other physical effects, but it is *not* the result of some human-created statistics program.

Often in biology, the averaging of data is the only way to demonstrate events and mechanisms, because the biological objects and processes concerned are intrinsically variable and to a large degree irreproducible.

In addition, similar to the example of temperature, many biological phenomena are also natural averages. For example, heartbeats, kidney functions, liver outputs, growth rates of embryos, viral infectivity and the likes are also the naturally averages of the actions of countless individual cells or viruses. Even the seemingly 'smooth' actions and properties of individual cells are actually the result of average interactions of some 10^{11} individual macromolecules that incessantly do very different things. If we were to look closer, the macromolecules are all discrete objects with different charges, post-translational modifications, linkages to oligomers and polymers, and so forth.

On the other hand, unique events, such as individual mutations, infections, seedlings, specific genomic configurations, or developmental alterations stimulated significant observations in biology. In these cases, averaging the data would obfuscate the underlying causes. Applying averaging procedures may jeopardize all the intricacies of the fitting of their tiles.

Let us look at the example of figure 3-04 that uses the 12 pentominoes as tiles of mosaics. Figure 3-04a depicts 8 such mosaics, and highlights two specific tiles, namely the 'F' and the 'X' pentominoes in order to make the differences between the mosaics more noticeable. In contrast, figure 3-04b shows an average of all 8 mosaics and their mirror images obtained by superimposing them on each other. Obviously, it resembles none of its contributors. Worse, it may obscure their most significant feature.

Assume a small detail of the mosaic structure is decisive for its function. For example, let the mosaics in figure 3-04a be unable to replicate, unless the 'F' and the 'X' tile touch each other. If the average mosaic figure 3-04b was our only source of information, we would never know why some mosaics replicate while others do not, as the averaging process leaves no clue whether the 2 tiles touch or not.

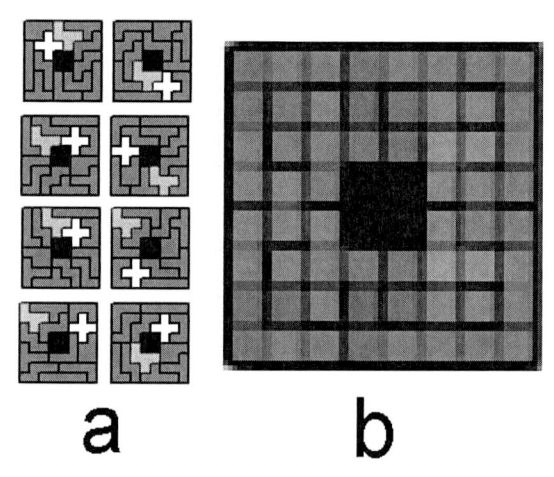

Fig. 3-04. The confounding of critical details by averaging mosaics.
(a) Eight sibling-solutions of the model frame mosaic. The 2 highlighted tiles contact each other in some and not in other cases.
(b) Average of the 8 mosaics and their mirror images. No contacts can be identified.

The Dependence of Mosaics on the Availability of Tiles ('Substitutions')

The availability of certain individual 'tiles' is often vital for the functionality and survival of living mosaics, but is far from guaranteed. Examples are food resources, living space, mates, opposable thumbs, oxygen, capillary beds, ATP, etc. In fact, the scarcity or the lack of specific components is one of the most important forces of natural selection.

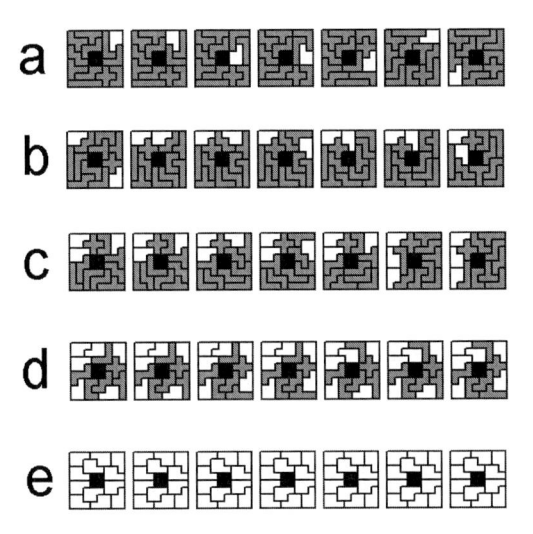

Fig. 3-05. Sibling-solutions with various numbers of 'P-pentominoes' replacing other tiles.

Let us consider the effect of the availability of tiles in more detail using the simple pentomino model. Figure 3-05 shows some examples of pentomino solutions of the model-frame where surplus 'P's substituted in increasing numbers the unavailable tiles.

For comparison, figure 3-05a shows a selection of 7 normal solutions where all pentominoes are different. As a first 'experiment', let us look at solutions that are missing an 'F', but substituted it with an additional 'P'.

Obviously, *the absence of the 'F' forced all other tiles to rearrange completely*, in order to create a 'hole' to accommodate the replacement 'P' (figure 3-05b). The same is true, if both 'F' and 'I' are unavailable (figure 3-05c). Most noticeable, the extra 'P'-tiles are no longer as freely distributed throughout the frame as they were in the above cases of fewer 'P's.

If even 3 'P's have to substitute for 'F', 'I', and 'L' (figure 3-05d), only one 'P' can be observed to jump to various places. The other 'P's remain frequently locked in fixed locations and orientations.

Eventually, we lose all variability: In the cases where the solutions contain 6 and 12 'P's (panels [3-05e and f), all components are effectively frozen in very few configurations. Only mirror symmetrical groups can flip one way or another.

If these mosaics were biological objects, we might interpret the progressive replacements as loss of diversity. The surviving variants differ dramatically from the original population, and therefore may be considered as mutants.

In the above example, we used the 'P', because it is able to fill the frame completely. Other pentomino shapes like the 'F', 'X', or 'Y' are not able to do this. Replacing unavailable tiles with any of them must fail beyond a certain number of substitutions, where a complete filling of the frame is impossible. In other words, once the number of unavailable tiles has exceeded a threshold, the original population of mosaics would not be able to adapt. Instead, it would become extinct.

The pentomino model will teach a great many more lessons, which we will present one at a time as the text leads to them.

CHAPTER FOUR

SOME COMMON PROPERTIES OF MOSAICS AND BIOLOGICAL OBJECTS

SOME OF THE STRANGE, COUNTER-INTUITIVE PROPERTIES OF MOSAICS ARE NATURALLY PROPERTIES OF LIVING MOSAICS

Fractality of Mosaics and Biological Objects

Fractals are sets of points with infinitely many layers of details. For example, the details shown in Figure 4-01b appear, if one enlarges an area of Figure 4-01a (see arrow),. Similarly, if one enlarges a portion of Figure 4-01b (see arrow) new details will appear as in Figure 4-01c. The details of this example differ from layer to layer, in contrast to other cases where the details of each layers are identical, endowing these fractals with the intriguing property of infinite *self-similarity*.

Mosaics become fractals quite easily, if one substitutes each tile with a sub-mosaic composed of even smaller tiles. In this way, the details of the mosaic, namely its tiles, have details of their own, namely *their* tiles, and so forth *ad infinitum*.

In the case of the pentomino tiles, one can construct each as a mosaic of 9 different pentomino shapes where each has 3x smaller sides. Let us call them '1/9th-pentominoes'. Each may even contain a smaller version of themselves in the same orientation (figure 4-02). The procedure can be carried on *ad infinitum*: By replacing each pentomino with one of these solutions, and each of the 1/9th-pentominoes with 9 pentominoes that are 81 times smaller, and so forth, each pentomino mosaic can be turned into a fractal.

46   Chapter Four

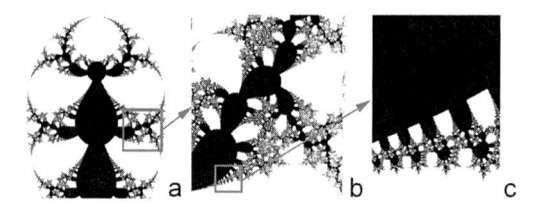

Fig. 4-01. The successive magnification of a fractal set that illustrates 3 of the infinitely many layers of details.

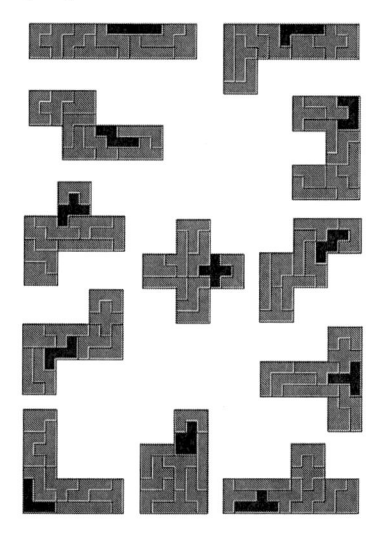

Fig. 4-02. The assembly each pentomino as a mosaic of nine 3-times smaller pentominoes. Such solutions may form the basis of an infinite regress which leads to fractal pentomino mosaics. There are many different solutions, including some where the mosaic contains its own shape in the same orientation (marked tiles).

There are many different solutions for the assembly of all 12 pentomino shapes from the $1/9^{th}$ pentominoes, although the number of different sibling-solutions depends on which piece one tries to assemble (table 4-01). The 'P'-shaped pentomino has by far the largest number of sibling-solutions, while the 'X' has the fewest. (My assembly program yielded these numbers of sibling-solutions.)

F	443
I	101
L	938
N	610
P	9144
T	382
U	444
V	482
W	202
X	40
Y	809
Z	395

Table 4-01. Number of different ways to assemble each pentomino shape (left column) as a mosaic of nine pentominoes with 3 times shorter sides (such as shown in figure 4-02).

Curiously, no other miniature pentomino shapes are able to tessellate all 12 pentomino shapes. For example, although it is possible to construct 10 pentomino shapes from four '½- pentominoes', but the 'X-' and the 'V-'pentominoes cannot be built in this way.

Of course, special cases like these are important only for the construction of mosaics that have the same shape as their tiles.

Components versus Environment in the Case of Mosaics and Biological Objects

In section C of chapter 2 we discussed the complexity and possible ambiguity between the tiles and frames of mosaics. Similar ambiguities may arise when we try to distinguish between the inner space of a mosaic and its external environment.

For example, draw an arbitrary line through the inside of a mosaic! As long as this line follows the interfaces between the tiles, one can always claim that the tiles to the left and right of the line are actually two different mosaics that share this line as their border. Since the two mosaics have no tiles in common, the line generates two mosaics with different tiles and different frames.

Similarly, we can excise any group of adjacent tiles and consider it as a new mosaic, while leaving the original mosaic with a hole.

Not only the definition of frames and surrounding spaces, but also the definition of tiles may be ambiguous. Actually, one can unite any arbitrary group of tiles into a new single tile (which still is a mosaic itself) for which the remaining tiles become part of its environment.

Biological objects share the same ambiguity. At one border line or another, biological objects must end and their environment must begin. However, how does one define the border? Is the ink-cloud surrounding an octopus still part of the animal or part of the environment?

Let us consider the nervous system of a fish to illustrate how complicated the situation can be. The nervous system of a fish is a biological object in itself, as it is embedded in its distinctly different environment of the rest of the fish. On the other hand, if we take the whole fish as our biological object, then all its body parts, including its nervous system appear as closely interacting tiles, whose environment is the distinctly different ocean.

The problem does not end there, because the ocean as an eco-system that teems with life is a biological object in its own right, which shares an important quality with the 'inner' compartments of every fish: The fish interacts and exchanges signals with it. Consequently, the biological object 'fish' does not really seem to end at its body surface. Instead, we have to expand it to a gigantic biological fish-ocean-object, whose environment is the rest of the Earth.

On the other hand, seen from the point of view of neurons the nervous system of the fish is actually their environment. Going further, even every neuron can be viewed as being a whole environment of something else, namely of its cytoskeleton, its mitochondria, etc.

In spite of the ambiguities, the distinction between a biological object and its environment is indispensable. How else can we ever formulate it that a living mosaic maintains itself, reproduces itself, or that it feeds on another? Whenever we claim that a living mosaic reproduced, we ascertain at the same time that there is an environment, which it did NOT reproduce. Whenever we claim that a living mosaic fed on another, we ascertain at the same time that we know who the prey was and who was predator.

This kind of inevitable ambiguity forces the student of a particular mosaic in nature to abandon any mechanical, formalistic approaches right from the start and, instead, define the appropriate tiles and frames in a creative, adaptive, even artistic way: Like the tessellated parts of nature, the language describing them is most demanding because it is capable of being alive, too.

The Huge Imbalance between the Numbers of Possible and Actual Individuals

In spite of its frequent use of pentominoes this book does not suggest seriously to use them as models for real life organisms. As was emphasized before, they are merely illustrations and tools for heuristics. Twelve pentominoes and one model-frame could not possibly model the seemingly endless variety of natural objects.

Yet, we should not underestimate how many variants of finite sized mosaics can result from relatively small numbers of finite sized tiles. In fact, the possibility to assemble fractal mosaics would have no problem to generate astronomically large numbers of variants, even after only a small number of rounds of fractal iterations.

Although biological objects are finite in size and numbers, they also have typically many more conceivable variants than there will ever be actual individuals. Hence, no two individual biological objects need to be the same and probably never have been and never will be identical.

One can get an idea of the number of possible variants of a system by counting the number of its variables and the number of different values each variable can assume. Take the example of a single protein like actin.

Variable 1: Its primary sequence contains 374 amino acids. Each of the approximately 20 molecular bonds in every of the amino acids is vibrating at the frequencies of the infrared light somewhere between 300 and 300,000 GHz. Consequently, the pattern of the states and the spatial configurations of the molecule change every 10^{-12} to 10^{-15} seconds.

Variable 2: It has a diameter of approximately 3 nm, and its 3-dimensional structure is arranged in various segments of beta-sheets and alpha helices that define at least 4 domains that are in constant relative thermal motion.

Variable 3: In addition, ATP, Ca^{++} or Mg^{++}-ions may modify each molecule by binding to it or dissociating from it.

Variable 4: Like all soluble proteins, it is encased in an approximately 1 nm thick hydration shell of water molecules that are in a permanent thermal exchange with the molecules of the bulk water. This shell corresponds to a swarm of some 12,000 water molecules that accompany each actin molecule in constantly changing configurations, which are part of the configuration of actin itself.

Variable 5: Inside a cell, each actin molecule may bounce around in complex Brownian pathways, or incorporate into one or the other of thousands of double-helical actin filaments in each cell.

Variable 6: Finally, like all other proteins, each actin molecule has a life span beginning with its 'birth' during the translation of its messenger RNA, and ending with its destruction by the inevitable turnover processes in a cell. The turnover processes, including ubiquitation and hydrolysis add yet more states of the actin molecules.

The actual number of possible states of real biological objects such as humans, of course, is very much larger, because we must consider the states of the some 100,000 different protein species in everyone of the 10^{12} cells of every one of the $5 \cdot 10^9$ different people on Earth. Assuming that during the past 6 million years the human population was always $5 \cdot 10^9$ – it certainly it was much lower for most of the time – and that each person lived about 60 years, no more than $5 \cdot 10^9 \cdot 6 \cdot 10^6/60 = 5 \cdot 10^{14}$ different humans have ever lived.

Even the number of possible states of each single protein in each person's body is much larger than that. Surely, the number of possible different fingerprints is even larger. After all, a fingerprint contains many, many more than one protein molecule. In short, the odds are extremely small that two people have ever been the same, even for a millisecond. In fact, each person is not the same one millisecond later, and will never be the same again. The same is true for any other organism on Earth.

Emergence of New Properties: Mechanical Stability of Mosaics

Once its last tile is in place, a finished mosaic often acquires additional properties that an unordered heap of the same tiles would not have. More specifically, a finished mosaic necessarily creates
1. a network of intimate contacts between all tiles,
2. novel constraints on each component's ability to move, and
3. a closed border, which consists of an uninterrupted chain of edge-tiles.

In the case of physically real mosaics, any or all of these properties may translate into emergent properties of the mosaics upon its completion. In the course of this book, we will come across a number of them. At this point we only discuss one example, namely the acquisition of a higher level of mechanical stability.

If the tiles are rigid, and if there is no room left between them in a finished mosaic, the tiles cannot yield to forces trying to move them. All other tiles and the frame hold every tile in place by. If any force acts upon a tile, it passes it on to the other linked tiles. Thus, they mutually stabilize their configuration.

The best examples are found in architecture, where walls may not become mechanically stable until the last stone is added. Take the function of the so-called 'keystone' of an arch. Without it, the arch would collapse, unless an appropriate scaffold holds the other stones in place during the building of mthe arch. However, after the addition of the last 'tile', namely the keystone ('*' in Figure 4-03), the arch becomes a very stable construct and remains so, even after the scaffold is removed.

Fig. 4-03. Self-stabilization of mosaics. Example: Stabilization of an arch after the addition of the key stone (marked '*')

It is worth noting that the key stone plays its exceptional role only at the termination of the assembly process. Immediately afterwards, the supporting function of the keystone does not differ from that of any other stone of the arch. Removing any of them would lead to the same collapse.

However, there is another significant change. At the instant of incorporation, when the keystone loses it exceptional 'status', inevitably the tiles lose some of their defining semi-autonomy: They can no longer be removed without damaging the mosaic. Hence, the mosaic gained a new quality at the expense of the autonomy of its tiles. This kind of trade-off between gains of the mosaic, and losses of its tiles, can be observed rather frequently in living mosaics (For a more biological example see chapter 22, 'How autonomous is autonomy?').

Other examples of the self-stabilization of finished mosaics are the famous fortress walls of the Sacsayhuaman ruins in Cuzco, Peru (figure 4-04). These megalithic structures are composed of irregularly shaped huge blocks that have no mortar between them. Yet, in spite of the frequent and strong earthquakes in the mountains of Peru, the walls have remained intact for at least 1000 years.

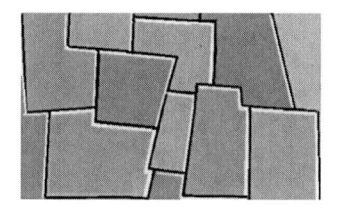

Fig. 4-04. Self-stabilization of mosaics. Example: The type of masonry of the fortress walls of the Sacsayhuaman ruins in Cuzco, Peru withstanding numerous earthquakes for at least 1000 years.

The reasons for their long-term stability are the irregular outlines of the blocks, because they only allow one fitting configuration. Thus, they guide each other back to their original configuration after an earthquake has momentarily opened up gaps between them. Naturally, there is a limit to the design principle. It no longer works if the earthquake, being too strong, extracted entire blocks from the wall.

Interlocking parts add stability and strength to living mosaics, as well. Well-known examples are the network of fine, mineralized trabeculae of bone, which give it its strength. Other examples are the layers of interlocking keratinocytes, which create the extraordinary strength and durability of human skin. As long as it is not torn, skin is amazingly resilient, even though the intact epidermis is barely 0.1 mm thick.

Yet another example may be an eggshell, which consists of the deposition of calcium carbonate crystals into a fibrous protein matrix that, in isolation, would offer very little mechanical protection. Yet, once the countless crystals have completely enclosed the surface of the egg, the thin shell acquires amazing strength.

Error Propagation during the Assembly of Mosaics and Biological Objects

As mentioned before, usually an 'ocean' of non-solutions surrounds the few and small 'islands' of fitting sibling-solutions. Therefore, it may take very little to disturb an ongoing fitting process and steer it into the 'ocean' towards failure.

Using the example of the pentomino mosaic, a single misplaced or disoriented piece can thwart all future attempts to find a solution. For example, if the misplaced pentomino creates inside the frame a space, which the remaining pentominoes cannot fill, the entire assembly is doomed to fail. Examples are depicted in figure 4-05 .

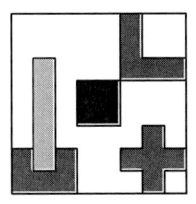

Fig. 4-05. Several placements of pentominoes, which would prevent the possibility of completing their configuration into a solution mosaic.

In the illustrations, the 'X' and 'V' pentominoes create spaces where no other pentomino can fit. The 'U'-'I' pair prevents a solution, because the space to the left of the 'I' (coloured pentomino) requires a second 'I' to generate a tessellation, which is not available.

Compared to the total mosaic, some of these changes may appear small. Yet, the continuation of the fitting process can never correct the consequences for the rest of the mosaic and, therefore, they remain 'lethal', regardless.

The reason for this inability of the pentomino mosaics to adapt or correct errors is the rigidity of the tiles. Figure 4-06 shows an example, where the flawless continuation of the assembly process propagates an initial error caused by a single wrongly oriented 'Z', all the way to the end: Eventually, a remaining 'P' has no place to go.

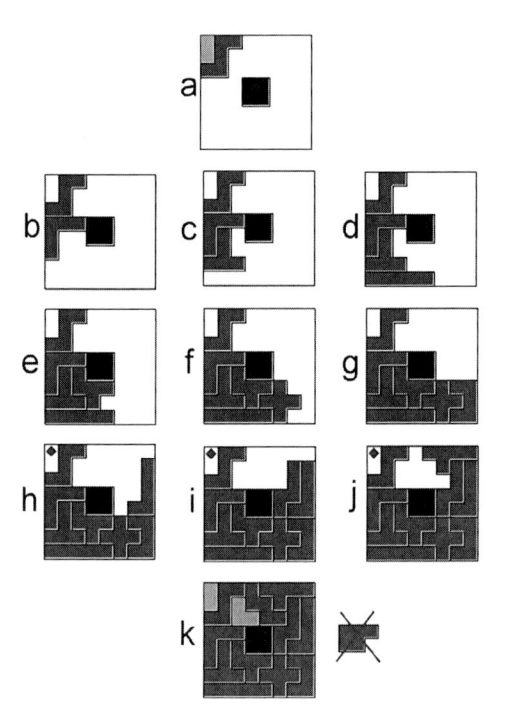

Fig. 4-06. Error propagation during mosaic assembly. Due to the rigidity of the tiles, an initial placement error (panel a) can never be corrected in spite of a flawless continuation of the assembly process (panels b-k)

Obviously, there only is a problem if we insist on using pentomino-shaped tiles to fill the frame. Otherwise, we could easily close the gaps with differently shaped tiles. Indeed, biological objects often resort to such 'unprincipled' methods to absorb an error of the assembly. They may not always present a solution. For example, seemingly 'small' organ defects or injuries may condemn embryonic developments to fatal outcomes.

In the face of assembly errors, living mosaics may express a third possibility besides adaptation and failure. There are situations where a misplaced or wrong tile in itself does no harm. Nevertheless, the organism's defence mechanisms tries to repair it, and it is this very repair that damages or even kills. The most obvious examples are food allergies, anaphylaxis, or sepsis.

The 'Logic' of Colloids. Simultaneous Presence of Conflicting Qualities

Assume you want to confine a strong and aggressive animal to a narrow space, but you do not want to cut off its air and food supply. The solution to the problem is obviously a cage, which actually is a mosaic. This particular mosaic alternates one kind of tiles, namely the solid and impenetrable bars, with a second kind of tiles, namely the openings, where air and food can come in. In short, this mosaic merges two conflicting qualities together into a bi-functional object without either quality suppressing the other.

It is not necessary that its tiles only have a single function such as blocking something. Like in this case, they may be the weight bearing elements, too, and provide the architecture for the conduits of the air.

The mechanism of the gas exchange chambers of our lungs follows this principle. The alveolar cells hold the blood capillaries in place near the airspaces. The capillary endothelial cells are strong enough to prevent the blood liquid from leaking into the air sacs, while being 'thin' enough to allow the free gas exchange across their bodies.

Or take the mosaic of an animal's fur. The structure of fur is extremely simple. The slender cylinders of the hairs consist of insulating materials and anchor with one end in the skin at small but regular intervals. However, let us look closer!

Each hair provides a very strong mechanical shield, yet by bending into the hair-free spaces around, it yields easily to external forces without giving up its protection. Its ability to bend combined with its connection to nerve endings turns each hair into a most sensitive touch sensor. In addition, each hair-free space traps a small pocket of air, which provides heat insulation. Yet, the hair-free spaces are large enough to permit temperature sensors in the skin access to the outside. There are tiny muscles at the base of each hair that are able to erect it, to increase the air spaces and with them the heat insulation. In addition, special glands near their roots coat each hair with water repellent oil. As the hairs are spaced densely, they keep water and other liquids from reaching the skin. All the while, their follicles replace the hairs with new ones, although the loss of the old hair could weaken the fur's multiple defence functions.

That is not the end of the multiplicity of mutually incompatible functions of fur. For example, the pigmentation of the hair creates camouflage patterns, mediates mate recognition, and ultimately serves the animal's behaviour on a much larger scale than single hair can.

This is an example of a very important principle of mosaics. *The larger the variety of tiles of a mosaic, the larger the number of conflicting functions that it may integrate into a functional whole without suppressing any of them.* Considering that every life form has to express countless different and often contradictory functions, is it surprising all life forms are mosaics, contain mosaics, combine into meta-mosaics,...?

Inevitably, there is a price to pay for this kind of conflicting functionality. There are always limitations and thresholds that determine where the technical compromise of the mosaic is no longer functional. However, we can expect such limitations with every kind of engineering, including the technology of terrestrial life.

We can go further by also endowing the interfaces between tiles with special properties. So far, we have treated the interfaces tacitly as if they were more a conceptual than a material thing, but they do not have to be.

Imagine moving each tile a bit away from its neighbours, and filling the resulting gaps with certain media or a special kind of very small tiles! We mentioned already the water-repellent oil that coats each hair. Still, the lining between tiles can be much more complex. The lining materials may modify the properties of the tiles, and thus the entire mosaic.

In addition, the filler material can serve as a scaffold to attach signalling and messaging tiles that can transmit information between the neighbouring tiles. For example, the so-called basement membrane is built from the very small molecular 'tiles' of collagen type IV. They are decorated with highly specific other kinds of molecules that help guide migrating cells during development and repair.

Let us imagine next that the tiles become smaller and smaller while their numbers and mutual distances grow larger and larger. As a result, we effectively turn the mosaic into a suspension of tiny solid particles in a medium such as water. If we give each particle the same electrical charge, their repulsion will tend to keep them apart. If the distances between the particles are much larger than the particles themselves, they can freely move past each other. We call the result a *colloid*. If the distances between the particles are comparable to the particle sizes, we call it a *gel*. If the particles are tiny portions of a liquid and the surrounding medium is a gas, it is called a *foam, and so forth.*

Unlike regular solutions, these kinds of materials have no boiling points. Similar to solutions, they may form precipitates or crystals, but unlike them they may form complex, highly ordered polymers. If they are flowing fast, they may become birefringent like crystals, even if none of the particles sits in a fixed crystal lattice.

However, like mosaics, they combine contradictory properties. They may flow like a liquid while being elastic like rubber. They may conduct electricity but not heat. They may adsorb salts and other solutes similar to the active surface of a solid, while flowing freely like a liquid, nevertheless.

The so-called Tyndall-effect provides a simple test for the colloidal nature of a material. If one sends a sharply bundled beam of light through it, the strong (and polarized) light scattering of the colloidal particles outlines the light path quite strongly.

In living mosaics, the use of colloids and gels outnumber the use of pure solutions by a large margin. The cytoplasm of every cell is colloidal in places and gel-like in others; blood is a colloid; all protein 'solutions' are colloidal; egg white and yolk are colloids; starch is a colloid; even urine is a dilute colloid. All glandular secretions are colloids. Mineralization employs colloidal mineral particles for the building of bones or shells. Many parts of animal organisms such as tendons are gels, and so forth.

Considering colloids, gels and foams as extreme cases of mosaics, it is not surprising to find them in pivotal roles of living things. Admittedly, they are no longer mosaics like the examples with which we started. Still, they are mosaics. We will expand the definition in chapter 6.

CHAPTER FIVE

THE DIVERSITY OF 'FITTING'

THE FITTING OF THE TILES OF A MOSAIC MEANS CONSIDERABLY MORE THAN FILLING A FRAME WITH COMPLEX SHAPES, WHILE LEAVING NO OPEN SPACES; IT CREATES INDIVIDUALITY, COMMUNICATION, COORDINATION, AND SPECIFICITY

The theory of evolution does not name a specific mechanism of natural selection, because there are far too many. Similarly, there is no single mechanism for the fitting of tiles in mosaics, because there are far too many as well. The various fitting processes create a correspondingly broad range of mosaic properties.

In order to forestall an easy misunderstanding, let me say that fitting of the tiles does not mean that solid objects fill wall-to-wall the spaces inside the frame. **Empty spaces may be included in order to guarantee the functionality of the tiles.** For example, a car must have empty space for the driver; a door must include space to open it; the space above a stove is included as part of the tile called 'stove', in order to make cooking possible.

When we talk about **'holes'** in a mosaic we will always mean **a missing tile**, including its own empty operational space.

'Fitting' Versus 'Docking'

There are three fundamentally different kinds of fitting. They are
1. the local fitting between adjacent tiles,
2. the local fitting between tiles and frames, and
3. the successful global fitting of all tiles as a frame-filling pattern.

So far, we had mostly discussed case (3) which describes the *global* result of all tile-to-tile and tile-to-frame contacts and interactions. We will continue to describe the totality of the matching contacts as 'fitting'. In contrast, the cases (1) and (2), which only apply to the local contacts, will be called 'docking'.

The Cumulative Effects of 'Fitting' and 'Docking'

Of course, the overall pattern of the fitted tiles is ultimately the result of a great many instances of local dockings of the tiles. Since the tiles are discrete objects, fitting them together locally as well as globally becomes necessarily a discontinuous process itself.

Compared to the time it takes during assembly to find the next fitting tile, the actual 'closing' of the next contacts between tiles is presumably quite short. It involves a large number of points of the contacting tiles (or frame) that match up with each other effectively at the same time, like a door that is snapping shut.

One may define a 'contact matrix' to express the process (see chapter 24). Upon the completion of docking, entire columns and rows of the contact matrix flip into a specific configuration. The configuration is symmetrical, because the docking of a tile 'X' to a tile 'Y' implies that the tile 'Y' also docked to the tile 'X'.

Although docking indicates a local process, the matching up of numerous contacts may spread their influence throughout the forming mosaic. One obvious consequence of successful docking is that all future tiles will have to fit into a correspondingly reduced space. At least in this indirect way, the chains of dockings during assembly create a non-local, cumulative effect on the mosaic.

If the tiles have directional properties, fitting creates further kinds of global effects. As in the cases of inanimate mosaics that consist of polarized elements, such as magnets or electrets, the surface of each tile can be oriented. Mathematicians and physicists formulate the situation by attaching a vector (arrow) to each point of its surface. In reality, the arrows may represent the orientation of electric or magnetic dipoles, or they may indicate streaming patterns that follow the tile surfaces.

Surfaces can be oriented in two fundamentally different ways. The arrows are attached *tangentially* to the surface as in Figures 5-01a and b, or *vertically* as in Figures 5-01c and d.

It is not always possible to orient the entire surface of an object in a self-consistent way. For example, if the object is 3- or higher dimensional, the attachment of tangential arrows leads to contradictions (Mathematicians call it the 'theorem of the well-combed hedgehog'). There will always be at least one point where one must assign two opposing arrows to the same surface point. Take the example of Figure 5-01e. The tangential vectors are all parallel to each other at the equator of the shown sphere. However, as they must clash with each other at the north pole, if they move in the 'North' direction while remaining tangential to the surface of the sphere and parallel to the longitudinal.

It sounds more abstract than it is. Assume that the arrows represent the direction of streaming water. There is no problem for water to stream in 2 dimensions parallel the perimeter of a circular shape. It would simply flow in a circle around it.

However, the same is not possible in 3 dimensions, e.g. in the case of a sphere: Water streaming from one pole down the lines of longitude along the surface of a sphere must produce turbulent swirls at least at the opposite poles.

In contrast, if the arrows are vertical to the surface, there is no such problem. A realistic example is the vertical orientation of electrical field lines emerging from a charged conductive sphere. They can point in a radial direction away from the surface without contradicting each other.

Upon docking, the arrows of the contacting 'surfaces' must point in opposite directions. Hence, they will compensate each other to some degree, or even neutralize each other, if their amplitudes are the same.

In this way, the accumulation of the fitted contacts can yield an internal space of the mosaic with neutral or at least reduced vectorial tile-interactions. However, as shown in Figure 5-01, the outside of the mosaic retains the direction and amplitude of the original tile polarization.

The 'Fitting' of 'Tasks'

For purely geometric mosaics, fitting meant the absence, or at least the minimization of any gaps between adjacent tiles. However, the need of ascribing 'purposes' and 'tasks' to living mosaics, forces us to not only consider the geometry of tiles, but their functions, too. Successful fitting means much more, than eliminating spatial gaps. It means that the *in- and output functions of the contacting tiles can be related to each other and co-operate optimally in the service of the whole mosaic.*

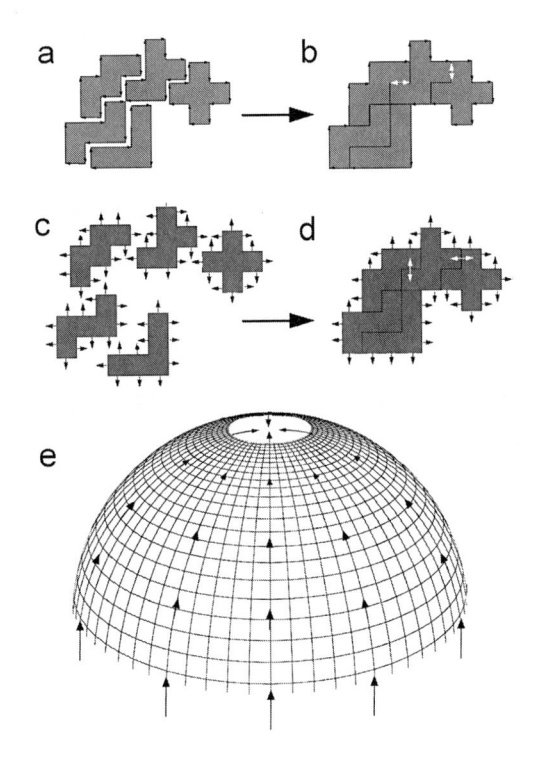

Fig. 5-01 Effect of fitting 2-dimensional shapes, whose perimeters are oriented either tangential or vertical as indicated by the arrows. (a) Example of a consistently clockwise *tangential* orientation of the perimeter of several pentomino shapes. (b) After docking the shapes together, the opposing arrows at the contacts neutralize each other. Two examples of opposing tangential arrows are indicated by the white arrows. (c) Example of a consistently *vertical* orientation of the perimeters of several pentomino shapes. (d) After fitting, the opposing arrows at the contacts neutralize each other. Two examples of opposing vertical arrows are indicated by the white arrows. (e) In three- or higher dimensional surfaces even tangential vectors may lead to contradicting directions. Example: Vectors that are tangential to the surface of a sphere and parallel to each other at the 'equator', must lead to colliding directions at the 'north pole', although they remain tangential to the surface of the sphere and parallel to their longitudinal.

Formally speaking, the tiles of mosaics with 'tasks' are functions t_1, t_2, t_3,.. with inputs j_1, j_2, j_3,.., and outputs o_1, o_2, o_3,…. Such tiles turn their mosaic from a spatial pattern into a *network* of functions $o_k = t_k(j_1, j_2, j_3,...)$ of its tiles ($k = 1,2,3,...$).

Chapter 9 (section IV) will discuss in more detail an important caveat. The inputs j_k and the outputs o_k are not necessarily numbers, and the tile functions $o_k = t_k(j_m)$ are not necessarily continuous, although some of them may be. The inputs may even be logical symbols, or perhaps tables of integers. The student familiar with programming languages may think of the variables as programming variables, and of the functions as lines of code.

In light of the functional properties of the tiles, the idea of 'fitting' leads to a new requirements for tile interactions.

In order to link the tile functions and implement their cooperation in the performance of the mosaic's task, *the output o_k of every tile function $o_k = t_k(j_m)$ must work as a valid input for some or all other tile function*s. The result must be a constructive interaction between maximal numbers of tiles. Formally speaking,

equ. 5-01

Let a mosaic perform task(s) T_M using as its tiles the functions $o_{Mk} = t_{Mk}(j_{M1}, j_{M2}, j_{M3},...)$, with their inputs j_{Mk}, and outputs o_{Mk}, for $Mk = 1,2,3,...$

Then, the mosaic is considered as fitted, if $o_{Mk} = t_{Mk}(o_{M1}, o_{M2}, o_{M3},...)$ describes an interactive network of functions with T_M as its emergent property.

Accelerated Natural Selection through the Fitting Requirements

The functional aspect of 'fitting' also plays an important role in the mechanisms of natural selection. Most of the time, we take it for granted that the confrontations between a living mosaic and its *external* environment decide its survival. However, it is also crucial for survival, how well the interactive functions and locations of its own tiles create an *internal* environment that is beneficial for the mosaic's task.

Hence, the arena for natural selection not only is the external environment, but the internal environment, as well: The survival of the fittest becomes also a function of the perfection of its internal fitting.

Obviously, the speed of evolution grows with the frequency of mutations. However, it also increases with the frequency of 'field-testing' them. It turns out that the latter frequency is much higher than the former.

Under normal circumstances, the critical confrontations between organisms and their external environment are a matter of chance and, therefore, they have to face them rather rarely. For example, if certain confrontations depend on the time of day, they only may occur once a day. If they depend on the seasons, they only may occur during (say) the winter. In addition, not all encounters lead immediately to a selective event. Therefore, some mutations may remain untested for considerable lengths of time.

In contrast, the 'field-testing' of the internal interactions between the tiles of a living mosaic (cells, tissues, organs) is very frequent. Confined to the inner space of the mosaic, the tiles cannot escape their mutual interactions. Hence, the quality of their fitting may be put to the test at every heartbeat or even at every nerve pulse. In view of such large testing frequencies, flaws or advances of the internal fitting will hardly remain undetected or inconsequential for long.

Therefore, the requirements of internal fitting constitute a powerful filter that pre-selects among variants. The requirements of internal fitting (equ. 5-01) are able to speed up evolution by offering advantages or weeding out detrimental mutants already in early developmental stages

The 'Globalization' of Tile Interactions through Fitting

The internal network of cooperating tile functions in living mosaics also changes dramatically by the distances that internal interactions are able to propagate in space and time.

Whenever fitting is a purely geometrical concept, the interaction between tiles is primarily local. In contrast, the existence of internal functional networks makes it possible that the output of every tile function 'spills over' into the entire inner volume of the mosaic. In this way, it can influence every other tile and even effect objects on the outside.

True, the result may not always be desirable. In fact, the potentially global effects of functional interactions often force organisms to confine them. They may build special channelling organs such as the vascular system to prevent that their materials penetrate the wrong spaces. Often they contain special collecting organs such as the urinary bladder, or insulating membranes such as the peritoneum in order to confine the spread or contacts of their products. Electrical insulators may block unwanted currents. For example, the myelin sheaths along the nerve axons confine the ion fluxes of action potentials to the nodes of Ranvier.

Restricting interaction to selected tiles creates specific interacting networks. The 'nodes' of these networks may only interact with specific other 'nodes', but not necessarily at all times. For example, the tiny volume of the synaptic cleft confines the synaptic transmission between neurons to a very small space and only occurs after an action potential arrived at the synapse... Thus, it decides locally which particular neuron is interacting with which other neuron. In contrast, the 'global' secretion of neuropeptides such as dopamine, serotonin, or epinephrine has dramatic global effects on the actions of many neurons located all over the brain.

Dynamic Fitting

As long as the tiles are rigid and fitted perfectly into the frame, they inhibit each other's movements. Their tight contacts all but eliminate their individual degrees of freedom. The mosaic only permits internal movements if its tiles lock into each other incompletely or only intermittently. The result may be described as 'dynamic fitting'.

Dynamic fitting of the tiles results from certain gaps between the tiles, from deformable tiles, or from rigid tiles that are able to escape their constraints for a brief moment. However, the gaps, deformations, and escape movements are supposed to be much smaller than the tiles.

Anyone who has ever watched densely packed ice floats on an almost frozen river, or the stop-and-go movements of cars during rush hour in a big city, has witnessed the strange and complex movements of this kind of 'almost fitted' tiles.

In these cases certain forces such as gravity, streaming water, or motors cause the tiles to move into the openings between them. In the process, they steal occasionally the wiggle spaces from their neighbours, thus tightening them up. The resulting increased displacements may propagate through the pattern, where they accumulate and amplify in some places - at least for a short period - while in other places the gaps disappear altogether and block locally all movements. At times, the waves of opportunistic displacements crisscrossing the pattern increase in amplitude. Eventually, they may create macroscopic rearrangements for a moment that change the entire pattern. As noted before, anybody stuck daily in rush-hour traffic has been 'victimized' by these waves many times.

However, thinking that 'sloppy' fitting would be the only method of movements in fitted mosaics, would underestimate the problem solving power of nature and of humans, too. It is quite possible that perfectly fitted tiles block each other's movement with great accuracy. Yet, they are able to break the condition at a precise time for a brief moment in perfect synchrony. During the brief release, the mosaic rearranges its tiles in precisely defined patterns.

The prototypes of such mosaics are mechanical clocks, especially their core, namely their escapement. It is a perfectly fitted set of cogs and wheels. They lock into each other and block any further movement of the wheels with great precision, while the clock's oscillator completes one of its periods. However, for a fraction of a second the wheels 'escape' their mutual inhibition, add one count to the number of oscillations, and supply the energy needed for the next identical action. The discovery of this escapement principle completely revolutionized the accuracy of timekeepers and, along with it, human navigation and science.

In order to illustrate the principle of synchronous release from a block of the fitted conditions, let us look closer at the example of gears (figure 5-02). The wheels of a gear have very little or no room to maneuver. In fact, after blocking one of the wheels, no other connected wheel would be able to move. Yet, if each cog exercises its residual degree of rotational freedom, the action synchronously move the cogs out of each other's way. This action not only permits their rotations, it is the very means by which clocks and other gears, such as car transmissions, work with great precision.

Living mosaics provide many examples of such precisely inhibited, yet periodically suspended constraints of their fitting. In these cases, the suppression of their degrees of freedom is vital for living mosaics, because they shield them from countless perturbations and interferences. Their sophisticated use of a few residual degrees of freedom allows them to control their actions with great accuracy.

Examples are the periodic opening and closing of the heart valves, the mechanics of insect thorax, wings, and flight muscle that enable them to beat their wings hundreds of times per second, and even the periodic way in which erythrocytes slip, one at a time, through a capillary that is barely as wide as the cells.

In many cases the oscillators of living mosaics that control the time when to 'stop' and when to 'go' do not need to have fixed periods. On the contrary, many are quite flexible, which enables them to adapt to the varying conditions of their life.

For example, the pulsation rate of the heart is quite variable, depending on conditions of stress and physical demands. In certain other cases, one cannot even recognize anymore the presence of an oscillator, because the release from the fitting block is indirect and entirely controlled by the changing outside world. For example, the ciliary muscle accommodates the focal length of the eye lens to the slow changes of distance of the objects in view.

The contact matrix (see chapter 24) can formally express the precise and patterned contact changes of the 'escapement' mechanics. Figure 5-02 and

Figure 5-03a show an example of the contacts between the cogs of a gear (marked dots). All possible degrees of freedom of the wheels are inhibited, except the rotation around their axes. The interlocking of the cogs renders the rotation of the wheels synchronous and precise. These qualities of the gears are reflected by the movement of the 3 depicted contacts between cogs along the diagonal of the contact matrix (figure 5-03b-d).

Of course, much more complex timings and timers operate inside living mosaics. Their dynamic changes as reflected by the contact matrix will be precise and patterned, if the operation of the mosaic is.

Fig. 5-02 Schematic operation of a gear, where the cogs of wheels A and B lock precisely into each other and prohibit each other's movements, unless both move synchronously out of each other's way (see white arrows). If the cogs are properly shaped, namely as cycloids, then their surfaces roll rather than slide over each other.

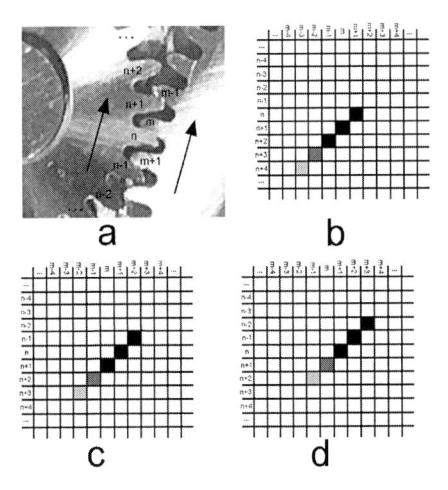

Fig. 5-03 The movement of a gear as reflected by the contact matrix. (a) The locking between the cogs of adjacent wheels of a gear. The cogs are numbered consecutively. (b)-(d) Consecutive locations of the contacts between the cogs as the wheels move them synchronously out of each others' way. The contacts yield precise and moving patterns. In this simple example they move in groups of 3 along the diagonal. The gray squares indicate their previous positions.

Fluid Mosaics

If solid tiles are numerous, very small compared to the frame, and able to slip past each other, the mosaics may appear fluid. Think of the seemingly fluid movements of a large amount of crystalline sugar as you pour it out of a bag!

Take the example of people or cars within the frame of a city. Naturally, people and cars are actually very large solid objects. However, if tens of thousands of cars are joined in 'traffic', or if thousands of people form 'crowds of sports fans', and are seen from a large distance within the much larger confines of a city or a football stadium, they too, appear to move like fluid mosaics [11].

Fluid mosaics combine properties of fluids with properties of solids. For example, the surfaces of solids are able to absorb molecules and exchange ions with their environment. However, fluid mosaics can do that, too; their solid particles can capture and adsorb ions from a reaction space, which true liquids could not do.

Alternatively, one can stream a real liquid past the particles, while keeping them fixed in place by mechanical or magnetic forces. After they have captured certain components from the liquid, one can collect, remove, and regenerate the particles. Ion exchange columns, water softeners, or magnetospheres operate based on this principle.

Another example may be the remarkable optical properties of colloidal suspensions. Since it takes the spatial order of crystal lattices to create birefringence, it may seem that only solid objects can be birefringent. However, if the solid particles are rod-shaped and in colloidal suspensions, they can orient parallel to the flow and each other, and thus create dynamically a lattice-like order. As a result, the flowing suspension becomes birefringent in spite of being liquid.

In many of the mentioned examples, the properties of the small, individual tiles are rather simple, e.g. they transmit light depending on their orientation, exchange ions, or adsorb contaminants.

In contrast, the tiles of liquid living mosaics may have almost unlimited complexity. For example, they may be as complex as entire living cells in the fluid mosaic of the blood.

Some of the cells in the blood are members of the immune system. These cells are free to move past each other, contact and communicate with each other intermittently, adsorb and process macromolecules, exit the blood stream and invade surrounding tissue, while cooperating in powerful ways as defenders and aggressors.

In the cases of city traffic or crowds of people the individual tiles are, of course, quite complex intelligent organisms. They can choose to cooperate in simple ways, like in the intermittent stop-and-go movements that chop the streams of cars into packets that interlace and cross each other periodically at intersections [11]. On the other hand, they can cooperate in hugely complex 'fluid mosaics' such as fire fighter squads, armies, the workforces of factories, or the students, staff, and faculty of hospitals and universities.

The 'Temperature' of Fluid Mosaics

The most natural definition of the temperature of a mosaic is the physical temperature of its tiles. It measures the average kinetic energy of the molecules in the material of the tiles.

On the other hand, we should not overlook the energies related to the movements of entire tiles of fluid mosaics. Unfortunately, measuring simply their total kinetic energy would not be sufficient, especially not, if there are living mosaics involved.

Temperature is measured on a scale beginning at $0°K = -273.15° C$. Therefore, it may seem that changes of temperature that are small compared with the number 273 have no significant effect. Nothing could be further from the truth. We all know that small temperature changes such as the rise of body temperature from $310°K$ (= $37°C$) to $315°K$ (= $42°C$) may be fatal. Even smaller changes of temperature can denature proteins. Sometimes they threaten the lives of individuals and even the survival chances of entire species.

Consider the disaster of the 'Challenger'-shuttle on January 28, 1986. The ambient temperature on that day was $-2°C$. Compared to the freezing temperature of $0°C$, it was $2°/273° = 0.7\%$ below freezing. It seems very small. Yet, the lower temperature during the launch decreased the resilience of the sealing gaskets so much that they leaked, which ignited the rocket fuel.

It was the result of a fatal misjudgement, based on the common myth that temperature describes the entire energy content of the various material involved. However, this only is true if the collisions between the molecules inside the materials were entirely *elastic*.

If there are also *inelastic* collisions between the molecules, they become sinks of energy as they swallow up large amounts of it, and hide or store it in their deformations, their density of molecular packaging, and in their inner fields (electrical, spin, spring tensions, etc.). The normal expression of these energy storage mechanisms is the heat capacity of the material.

For example, it takes much more energy to heat 1 kg of water than 1 kg of aluminium from $10° C$ to $20°C$. Although afterwards, their temperatures are the same, i.e. the average kinetic energy of the water molecules and the Al-atoms are the same, their total energy contents are quite different: During the warm-up phase the water needs and stores away much more energy by changing the packing of its mutually attractive dipole molecules. In contrast, the packaging of the Al-atoms changes very little within their rigid metallic crystal lattice domains, as the lattice warms up.

All this also applies *mutatis mutandis* to mosaics and especially to fluid mosaics. In addition to the kinetic energy of wiggling tiles, there are sinks of energy swallowed up by the changing distances between tiles and their inelastic deformations. Hence, it is important, in analogy to the heat capacity of materials, to define a suitable 'capacity of the heat sinks' that are the result of attractions, repulsions and inelastic deformations of the tiles of a fluid mosaic.

Adaptive Fitting: The Possibility of 'Learning'

The above text used the term 'fitting' predominantly in order to describe a property of *finished* mosaics. However, 'fitting' may also mean the *process of placing tiles* successfully inside frames. One of the most intriguing interpretations of 'fitting' as a process may be described as 'learning'.

All learning is about acquiring skills. It proceeds by a series of repeated actions, which contain a great many discrete component functions. Each of the component functions represents a basic ability or an earlier acquired skill of the 'learner'. During the learning process such components are selected, combined with each other, and their grouped function is tested. Eventually, a final selection and combination will emerge and constitute the newly acquired skill.

As mentioned earlier, the component functions of an action are the tiles of a functional mosaic. Therefore, the learning process is actually the selection of certain functional tiles and the honing of their functions and interactions.

Let us examine it a bit more closely. Learning proceeds by the selection and linking of the tiles towards a particular goal. At every step, the process determines how similar to the desired, ideal result each temporary mosaic of the series has become.

This very requirement of evaluation links the entire process of learning to the world of symbols: Since it relies on values, and since values are products of the mind, the evaluation of the learning progress must necessarily use symbolic quantities.

Thus, the learning process needs two kinds of symbolic, functional mosaics. We will call the first kind of symbolic mosaics, which depicts the successive results of the fitting attempts, 'achievement-mosaics'. The other kind, which reflects its idealized, intended result, will be called the 'target-mosaic'.

Comparing the two mosaics requires a third symbolic mosaic, which compares tile-by-tile the most recent achievement-mosaic with the projected target-mosaic. It also initiates the assembly of the next, altered achievement-mosaic, depending on how well the previous mosaic matched the target. We will call it the 'comparator-mosaic'. Its functions of comparisons and initiations make it a symbolic and functional mosaic.

In order to illustrate the principal players in the process, we turn again to pentomino tiles and the model-frame. The student may remember that we had shown how increasing numbers of the same 'P'-pentomino replaced the 12 different tiles of a pentomino mosaic (figure 3-05).

In some sense, the process of learning takes the reverse approach. The intermediary stages with their – still erroneously - placed 'P's turn systematically into the correct target-mosaic, which contains all 12 different pentominoes in their perfect fitting (figure 5-04). At no stage is the achievement a random, nonsense product, but always a fitted mosaic, albeit not the right one.

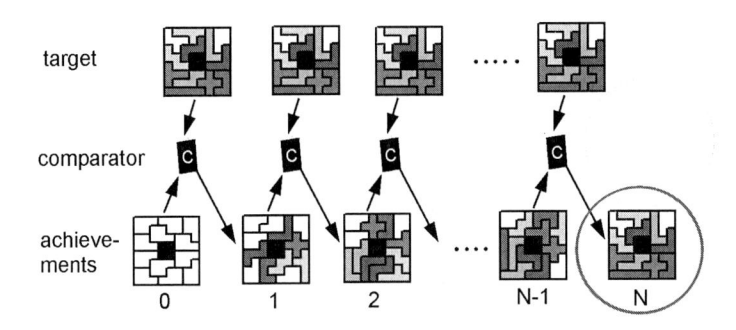

Fig. 5-04. Illustration of the principle of the interactions between symbolic and functional mosaics involved in the 'learning' process. Starting with the symbolic and initially quite dissimilar 'achievement-mosaic', it advances in 'N' stages towards the replication of the likewise symbolic 'target-mosaic'. The progress is driven by the 'comparator-mosaic', a functional mosaic, which evaluates the similarity between the successive mosaic stages and the (normally) unchanging target.

Figure 5-04 assumes that the target remains constant. In the reality of learning, this is not always the case. In fact, the comparator may select different targets while the learning proceeds.

The illustration does not answer the most urgent question the student may have at this stage, namely how the comparator puts together the next combination of tiles in the series of test them. Most likely, there is no general answer to the question, because the mechanism of learning will depend on its available component functions ('tools') and its particular targets. Nevertheless, as suggested by the success of neural networks, these mechanisms may have an element in common, namely a 'reward' mechanism that 'strengthens' groupings, it they received favourable comparisons.

Regardless of these and further details, I submit that the core of learning contains two symbolic mosaics and their – likewise symbolic - evaluation mosaic. The achievement- and target-mosaics are technicalities, which contain information that varies from learning-case to learning-case. Hence, they are case-specific, but not learning-specific. In contrast, the evaluation mosaic contains all the challenges and puzzling mechanisms of learning.

Naturally, all this only applies to living mosaics. Whatever the changes of inanimate mosaics over time may be, they cannot possibly be mistaken for 'learning'. (Note: Neural networks can learn and can be inanimate mosaics. However, humans design and drive them, and assign their comparator weights. As parts of living mosaics, they are living mosaics themselves).

Delayed, Deferred, or Remote Fitting of Tiles in Living Mosaics

We had mentioned earlier the role of scaffolds, i.e. exquisitely fitted mosaics that are indispensable for the assembly of a mosaic, but become obsolete at a later stage and are discarded.

This situation occurs quite frequently in living mosaics. However, living mosaics may turn the rationale of scaffolds on its head, when the scaffold does not protect the mosaic from some exogenous danger, but when it needs to protect the mosaic from the scaffold itself. The situation occurs when the mosaic has to assemble precursors of its structures or functions.

In this case, the mosaics build unused and unusable components. They may even leave them as temporary misfits in their local environment. However, later on, specific alterations may render them useful and exquisitely fitting. In this sense, the mosaic delays or defers their incorporation until the tile/mosaic has had the opportunity to develop into a different form.

Take the example of the embryonic lungs of land living animals! During embryogenesis, the lungs-to-be are immensely complex organs with no apparent respiratory use or function. Yet, if they would begin their normal breathing function before birth, or before hatching, they would drown themselves and the embryo along with it. Only after the embryo enters the air, they almost instantaneously acquire their vital function. Here their common dysfunction during early development is their protective scaffold.

Another example may be the formation and migration of the neural crest cells of the neural fold of vertebrates. Initially, they seem to be supernumerary, dysfunctional cells. However, after the closure of the neural tube, they begin their migration away from their place of origin to quite remote locations. There they become the seeds of vital tissues, such as bone, smooth muscle, and peripheral neurons.

Even biological molecules such as pro-insulin or angiotensin are generated initially in an inactive form, where they are safe and do not endanger other components. Yet, a certain modification makes them almost instantaneously active at a later stage and in a different location.

Another important use of building temporarily dysfunctional tiles/mosaics is the storage of raw materials and spare parts, such as the storage of fat and starch, or the 'assembly line' formation of replacement teeth in sharks.

The principal of deferred fitting is most important for the reproduction of animals. The development of animals and plants involves elaborate programs of cell differentiation, which leave substantially altered genomes behind. Yet, the germ cells of animals and plants need to remain undifferentiated and 'pristine'.

Plant development found elegant ways to circumvent the problem. After having differentiated their cells and genome expressions up to the level of flowers with styles and stamens, they are able to return the genomes of some of these cells back to the germinal state.

In contrast, animal cells are not able to do the same. They cannot reverse their differentiated state, without turning on a pathological or even malignant process. Therefore, their embryos are forced to set aside the so-called primordial germ cells in the earliest stages of development, during which time their genomes are still in the germinal state. Throughout the rest of development, they remain undifferentiated and dormant. Much later, after the animal has developed its reproductive organs, the primordial germ cells begin to migrate towards them and eventually populate their germ layers. In this way, the functional fitting of the primordial germ cells is deferred until the 'shells' of the reproductive organs are completed.

Perhaps the most spectacular application of temporarily dysfunctional tiles in living mosaics is the function of the imaginal disks in insects. During the larval stage they are dormant tissues. However, after the break-down of numerous tissues in the pupal stage, they become the 'seeds' of antennae, legs, wings, and many other structures of the adult insect.

There are even horrifying variations on the theme of using dysfunctional mosaics for temporary storage. For example, the sting of the *venomous emerald cockroach wasp forces* cockroaches into a state of suspended animation. During this state, the cockroach serves as future 'fresh' food storage for the wasp's larva, which develops inside the prey, and literally eats it up from the inside. Another such example is the fungus *ophiocordyceps unilateralis*. It alters the behaviour of an infected ant in such a way the animal seeks places that are optimal for fungus growth, and then turns its body into food supply and a protecting mechanical frame for the fungus and its sporangiophores.

In the above examples, the cells themselves control the developments towards functional fitting from within. There are also numerous cases, where the control is exogenous.

An obvious example is seed germination, where the seed remains dormant until gravity, humidity, special nutrients, duration of daylight, or other exogenous factors trigger the onset of the 'delayed fitting' processes. In other cases, exocrine or endocrine glands control the fitting of cells and organs. In other words, remote sources elsewhere in the same organism act as quasi-exogenous sources for the changes of functionality.

In the cases of symbolic mosaics, the use of delayed and remote fitting tiles is often a major instrument of art. A writer may create, yet temporarily put on hold a number of situations and characters that will play important roles much later in the work. In musical compositions, especially in polyphonic works, individual motifs fulfil the demands of musical harmony at their present place, as well as several measures downstream. The immediate *and* deferred fitting of the musical harmonies elevates canons and fugues to the most complex examples of such exquisitely woven compositions.

Information processing is another field where delayed, deferred and remote fitting are indispensible. Especially, in the cases of computer programs, most functions and databases are routinely 'dormant', until they are called by the program flow. After executing their code they return to 'dormancy'. Other examples are the widely used transmission protocols of information over the internet. They chop up information strings into dysfunctional packets. Subsequently, each packet is supplied with labels and addresses, which alert remote locations of the net to receive and combine these temporarily dysfunctional components at later times and remote locations of the transmitted code to restore their original meaning.

In most cases of living mosaics, the delayed, deferred, or remote fitting prevents damage, which certain tiles/mosaics would inflict on the organism, if they were fully functional at the wrong time or at the wrong location. Therefore, their mature functions remain suppressed, until they pose no longer a danger to themselves or others. Afterwards, they may literally leap into action, without having to go through lengthy maturation stages.

A special kind of damage may occur, if tiles and mosaics 'fall in the wrong hands, i.e. if they were inserted into unintended mosaics. In order to prevent it, the delayed, deferred and remote fitting of some symbolic mosaics may be reversible or even cyclical, as in the case of encryption. Here, the encryption renders individual tiles, such as words, sentences, and packets temporarily dysfunctional. After their transmission to the intended recipient, their functionality is restored. Subsequently, they may be re-encrypted and re-sent.

However, encryption not only is a tool for human needs of secrecy. Among animals, we find several cases of reversible and even cyclical 'encryption' as in the examples of the camouflage of the stout or the arctic foxes. Depending on the seasons, they change the colour of their fur, thus 'encrypting' their body surface for either snow-covered or normal plant-rich environments.

In the case of these animals the encryption and re-encryption is very slow, and may take weeks. In contrast, the amazingly effective changes of the camouflaging pigment patterns of octopus and squids require mere seconds to accomplish their kind of surface 'encryption'.

Chapter Six

Movements and Growth of Living Mosaics

Living Mosaics Can Only Grow by Adding More Tiles, or Opening Up More Empty Spaces; But Either Way, They Must Change Their Frames (=Constraints)

Whenever living mosaics move, they change the locations and shapes of their tiles, and when they grow, they also change their sizes and numbers. Yet, unless they are insects, amphibians, or certain kinds of fish that undergo metamorphosis, the individual mosaics seem to leave the 'essence' of their shapes intact. This kind of essential invariance of shape is not restricted to the body transformations of individuals. As argued by D'Arcy Thompson [23], the 'essential' body shapes of related species seem to remain preserved.

What exactly are 'essential' shapes? The mathematical discipline that determines the essence of shapes is the field of 'topology'. Its method consists of subjecting a shape to continuous deformations, while defining all the resulting intermediary forms as 'essentially' the same. Two shapes only belong to different classes, if no continuous deformations can turn one shape into the other.

The often-cited example of this kind of shape-test is the continuous transformation of a doughnut into a cup with a handle: By continuously deforming either one, one turns it gradually into the other. Therefore, both belong to the same topological class.

In contrast, continuous deformations of a doughnut can never create a sphere or a cube, because they cannot remove the hole in the middle. Hence, doughnuts and spheres belong to different topological classes.

The study of topology has discovered fascinating and even 'pathological' shapes such as Moebius strips or Klein-bottles, which are 3-dimentional objects that only have one surface and/or no volume.

There are elegant animations of topological shape transitions available on the internet, including, of course, the popular transition of doughnuts into cups with a handle, and the quite surprising fate of Moebius strips or Klein-bottles, if one cuts them in half.

In these and similar cases, the topology of living mosaics can be quite revealing without being pathological.

'Quilt' Mosaics

Let us first consider living mosaics with tightly connected tiles, such as the bodies of animals, plants, free living cells, and polymeric molecules. Like quilts, they leave no space between their tiles. Therefore, we will call them 'quilt' mosaics. Yet, they can be quite deformable, and are not necessarily symmetrical. There are continuous changes to describe the movements of such biological objects and most of their growth, provided we observe them at low enough spatial resolution. Of course, they remain mosaics at all times. In particular, the neighbours of each tile remain the same during such movements. For example, the adjacent amino acids in a protein molecule remain connected to the same neighbour amino acids during the biologically so important folding of the protein.

Continuous deformations apply even to metamorphosis. For example, the growth of four limbs, the resorption of the tail, atrophy of the gills, and the expansion of lungs of a tadpole turning into a frog, appear largely continuous. Naturally, they result from the movements, replication, and differentiation of individual cells, which are discrete objects. One may even skip intermediary shapes at regular intervals. The only effect of such time-lapse observations is an apparent speeding up of the seemingly continuous deformations.

'Swarm' Mosaics

Entirely different movement and growth phenomena may occur, if the individual tiles of deforming mosaics are not tightly connected with each other, and if there are spaces between the individual 'tiles'. We will call them 'swarm' mosaics.

Consider an ant nest, a school of fish, or a herd of caribou! Here the living mosaics, i.e. the 'swarms', are represented by the entire nest, the school, and the entire herd. The individual ants, fish, or caribou function as their tiles.

Swarm mosaics can grow by reproduction of their individual tiles, but also by expanding or compressing their internal spaces. Their movements can cause either dramatic shape changes or else the migrations of the entire swarm. They can be highly coordinated as in the schooling of fish, or become seemingly chaotic, as in the foraging of ants.

If one observes swarm mosaics at spatial resolutions that are low enough to hide the distances between individual tiles, their deformations may appear continuous. In contrast, if observed at higher spatial resolutions, one can often see their individual tiles change location or re-group independent of the movements of their neighbours in chaotic ways.

This impression of discontinuous re-groupings becomes much stronger, if we use time-lapse observations. Unlike the deformation of quilt-mosaics, which time-lapse merely speeds up, the swarm mosaics under similar conditions appear to leap into different patterns of their tiles in a discontinuous fashion. Invoking our more formal terminology, we may say that these mosaics seem to transition frequently from one sibling-solution to another.

Deformations of Mosaics at Constant Size (Body Movement)

Take a colourful quilt and mark a series of patches that touch each other like the links of a chain! No matter how much and how complicated you move, stretch, or fold the quilt, none of these deformations will change the relative order of the patches, even if the material is elastic. The only way to change the order would be to tear the quilt.

The same is true, if an animal moves its body. The relative order of its body parts will not change, as long as we follow them up along their natural connections. For example, the order of the organs along an internal path from heart→lungs→stomach→kidney will not change, no matter how fast and forceful the animal moves and distorts its body. The only way to change the order of the organs would be to injure the animal.

In a thought experiment, one may mark more and more, yet smaller and smaller patches, until they form continuous lines from one point of the quilt or the animal's body to another point. All body movements that do not disrupt the quilt can be described by a continuous mathematical function. Therefore, speaking in terms of topology, during body movements of living mosaics, all the body shapes cannot change to other classes, but remain essentially the same.

Let me repeat, though all this requires low spatial resolution, i.e. that the observer of the movements 'squints' quite a bit. Indeed, under a microscope, the movements look considerably more complicated, because the actual molecules, blood cells, etc. may actually leap around chaotically during seemingly smooth body movements. Still, it may be practical to choose a low spatial resolution, describe all shape changes during body movements as continuous and, therefore, as topologically identical. Even if a body seems to open a hole, topologically speaking, its centre is a discontinuity, which had been there all along, even if it was not easily detectable.

Needless to say, no matter how continuous we may describe them, the 'deformations' of living mosaics can be 'explosive', violent, with dramatic, life-threatening consequences. Just watch an attacking cobra, a pulsating jellyfish, or an 'ruffling' or 'blebbing' macrophage in action! In order to deal with the danger of the powerful and often fatal consequences of such movements, many animals including humans have evolved an exquisitely auxiliary detection system such as peripheral vision and other crude, yet fast sensory methods that detect their threats.

Not surprisingly, predators followed suit and evolved making their detection increasingly difficult. In most cases they resorted to natural camouflage, which depends on destroying the outlines of the animals' body.

The result can be quite counter-intuitive. Consider the striking patterns on a tiger's body! How can black vertical stripes on a brightly yellow body serve as camouflage? Nevertheless, it is effective as long as the tiger does not move, because a few vertical stripes in a forest do not look remotely like a tiger to the eyes of the casually passing prey. Furthermore, in the forest many bushes with bright yellow leaves or flowers may cast dark, long, vertical shadows.

Changing the Size of Mosaics (Allometry)

The same invariance of the order of tiles applies to the uniform growth of living mosaics. If the sizes of a quilt's patches or of an animal's paws increase, the order of every marked series of parts remains the same. Hence, the growth of such mosaics leaves them in the same topological class.

Nevertheless, growth may subject living mosaics to dramatic and sometimes counter-intuitive forces that arise from nothing more than their sizes. To illustrate this, let us look at the simplest shape a mosaic can have, namely a sphere. If a sphere doubles its radius R to $R' = 2R$, the laws of geometry dictate that its volume $V = 4\pi/3 \cdot R^3$ increases 8 fold to $V' = 4\pi/3 \cdot R'^3 = 4\pi/3 \cdot R^3 \times 8$. In contrast, its surface area $O = 4\pi \cdot R^2$ only increases by a factor of $2^2 = 4$.

The discrepancy between the increases of volume and surface area of a sphere may have surprising consequences. Let us begin with an example from the inanimate world. Assume that this sphere is a planet whose interior is large enough to produce heat by radioactive decay. As its radius grows by incorporating more and more asteroids, its internal heat production grows by volume, i.e. by the third power of the growing radius. In contrast, its heat dissipation, which is proportional to its surface area, only grows with the second power of size. Inevitably, after reaching a critical size the surface area of the growing planet will no longer be able to dissipate all the heat it produces in its interior: The surplus heat will begin to melt the core of the planet, as it did in the case of planet Earth.

Of course, this kind of inevitable linkage between geometric and physical properties applies to every tile and frame of every living mosaic, too. For example, consider the body of a 2 ft large otter as a tessellation of bones, muscles, organs, etc. It weighs about 2 pounds.

Now assume that the linear dimensions of the otter were doubled. For the mentioned simple geometric reasons all surfaces of the otter increase by a factor of 4, while all volumes increase by a factor of 8. Consequently, the weight, being proportional to the volume, also increases 8-fold, whereas the cross sections of the bones, being surfaces by definition and, therefore, 2-dimensional objects only increase 4-fold.

Likewise, the cross sections of the muscles only increase by a factor of 4, as well. As the force of a muscle is proportional to its largest cross-section, this means that the force of each muscle has only increased 4-fold. Compared to the small, original animal the 4-fold stronger muscles and bones of the enlarged one now have to move and carry 8-fold more weight. Hence, the movements of the scaled-up organism must be weaker and slower by comparison.

Let us make the animal even bigger by a factor of 12.5 until it reaches the size of a walrus. Now the cross sections of muscles and bones have increased by a factor of 159 while their weight has increased 2000-fold. By now the animal's muscles and bones have to carry and move 12.5 times their own weight. Obviously, the animal will be much slower and is in danger of breaking its own bones and joints if it jumps too daringly.

Finally, let us increase the animal by a factor of 57.5 until it reaches the size of a blue whale. The resulting animal would be enormously strong, because the cross sections of its muscles and bones have increased by a factor of 3,300. Yet, their strength is no match for its horrendous weight increase of 190,000 times. In spite of all its strength, an animal of this size cannot move even its own weight on land any more.

Animal sizes range between gnats and whales. It is easy to observe the described consequences of scaling. For example, a stranded whale cannot move its huge body back into the water, whereas an ant can easily balance the bodies of several other ants on one of its legs.

Of course, evolution has compensated the effects of size. For example, if two mammals differ in size by (say) a factor of 4, but are otherwise rather similar, it turns out that the bones and muscles of the larger animal are not merely 16 times larger, as geometry would predict. Instead, evolution has made them closer to 64 times larger to handle the 64 times larger weight and inertia.

Other properties, such as intestinal resorption area, total length of capillaries etc. may change in ways other than the geometry of scaling would predict. In some cases, the animal even has to change its habitat, as was the case with whales, which evolved to live in water where buoyancy renders their effective weight manageable.

Biologists have formulated the observed adjustments to size changes by mathematical formulas called 'allometry'. They express how evolution managed to cope with the physical consequences of size changes. The deviations from simple geometric increases of size can tell us a great deal about the habitats, physical properties and interactions between the inner components, i.e. the 'tiles' and the environment of the biological object, even if they are extinct and only known to us as fossils.

Let us look at the quantitative expression of allometric relationships between 2 quantities Q_1 and Q_2 of a living mosaic. They are log-linear (or at least approximately log-linear) laws of the type

equ. 6-01
$$\log[Q_2] = \alpha \cdot \log[Q_1] + \beta.$$

Let us derive this formula. Given a scale factor R, then Q_1 and Q_1 depend on it by certain power laws like the ones above, where we discussed the effects of size on volumes and surface areas. They contain constants A_1 and A_2, and exponents e and f.

$Q_1 = A_1 R^e$ (e.g. $V = 4\pi/3 \; R^3$),   and
$Q_2 = A_2 R^f$ (e.g. $O = 4\pi \; R^2$),

Hence,
equ. 6-02
$$\log[Q_1] = \log[A_1] + e \cdot \log[R], \quad \text{and}$$

$\log[Q_2] = \log[A_2] + f \cdot \log[R]$.

Eliminating the variable $\log[R]$ from these equations will yield equation 6-01 with $\alpha = f/e$ and $\beta = \log[A_2] - \log[A_1^\alpha]$.

As mentioned before, there are two principal kinds of size changes of mosaics, 'scaling' and 'growth'. 'Scaling' means that the linear dimensions of tiles and frames increase by the same scale factor R. 'Growth' means that the tiles remain the same size, whereas their numbers increase. In the latter cases, the frame increases and makes room for additional tiles to fit in. 'Negative growth' means that the frame shrinks and that the mosaic turns over or discards tiles that no longer fit in.

For most biological objects, scaling is the result of growth. Since the living mosaics of animals, plants, bacterial colonies, etc. are composed of cells, the sizes of their tiles are essentially fixed. Similarly, an ant nest or a herd of animals only can grow or shrink in unit 'tiles' of single animals. Therefore, the only realistic way of shrinking or growing living mosaics is either to leave out some tiles or else to add new ones.

Although this sounds simple enough, the functional consequences may be quite complex. If a mosaic shrinks by deleting tiles, the remaining tiles are forced to do the job of formerly many: Hence, the function of each tile has to rise to a higher level of universality and complexity. For example, the few employees of a small business handle daily much more diverse tasks than the individual employees of a large corporation with thousands of workers.

In contrast, if the mosaic grows by adding new tiles, it may experience two very different fates. Either the increased number of tiles will improve pre-existing tasks to more advanced levels, or else they handle entirely novel tasks. Using again our example of the above 'business-mosaics', it is obvious that large corporations can handle much larger numbers of tasks with a much broader scope than small businesses. In addition, large corporation can expand into novel areas of trade and technology that exceed the capacities of small businesses.

In other words, growth of living mosaics is not merely an increase of the numbers of tiles, but it requires changing the functions of and interactions between the tiles. If they fail to do so, loss of function, 'wild' tissue growth, or atrophy can create severe disabilities and diseases.

In summary, living mosaics grow and scale in numerous and surprising ways that obey geometric rules and/or compensate for them. Recently, Geoffrey West [27] published a veritable treasure trove of the data and the commonalities of the growth and scaling behaviour of living mosaics that range between organisms, cities, economies and companies.

At this point, the student may wonder, whether the described size effects on mosaics contradicts our initial claims that the concept of mosaics is size independent. I do not believe that it does. Allometry deals with the need to adjust the physical and geometric properties of mosaics as they change their sizes. However, the adjustments do not change their 'mosaic-ness'. Regardless of size, they retain frames, discrete and semi-autonomous modules, and their fitted quality. In fact, their very 'mosaic-ness' is the key to the successful adaptations of every tile to the new sizes.

The Relatedness of Sibling-Solutions in spite of Their Mandatory Differences

Whenever the tiles of fitted mosaics are free to regroup, the result must be either the loss of fitting or else the transition to another sibling-solution. We have mentioned already the example of swarm-mosaics. Their tiles are able to regroup rather easily, although in spite of the spatial freedoms of the tiles, the requirements of functional fitting impose very demanding constraints on the tiles, as in cases of a wolf pack encircling an elk, or a school of fish evading a predator.

The transitions between sibling-solutions may be even more complex in cases of symbolic mosaics. Examples are the transposition of letters in a verbal composition, or of notes in a musical composition, or during the composition of variations on a theme.

Surprisingly, in most cases the re-grouping of the tiles is not enough to switch from one solution to another. Closer inspection shows that they are usually rebuilt from scratch. Compared to the method of selecting tiles, excising them, regrouping them, and returning the modified group to the mosaic, it seems to be the much simpler and safer method of building a new one.

At least, one would think, making a copy of a mosaic should be a straightforward, templated process. However, most surprisingly, *living mosaics do not copy, but rather rebuild from scratch, even if the object is the same mosaic.*

In chapters 14 to 19 we will discuss the replication and cycling of mosaics in more detail, but at this point it may suffice to remind the reader that repeating a word, playing music, or reciting a poem is never the result of memorized, identical repetition of words, sounds, or notes. Be it a concert pianist or a stand-up comedian, as every performing artist will confirm, each of their performances, although guided by memory, is always a *de novo* recreation from their – equally memorized – components.

The obvious evidence for the mandatory re-creation of repeated living mosaics is the inability of living mosaics to produce an *exact* replica of them or other living mosaics.

For example, it is impossible to sign one's name twice in the exact same way. In fact, if two signatures are identical, the courts accept it as legal proof that one of them is a forgery. Obviously, every time we sign our name, our motor controls must recreate anew every move of the hands holding the pen.

Similarly, a close examination of their sound tracks proves that the acoustics of two identical words, songs, dances, or other acoustic communications between organisms are not exact copies of each other. They are all re-created sibling-solutions for their particular type of mosaics. The student, who wishes to see for him/her-self a simple demonstration of this kind of non-repetition in nature, should cut out the pits of several peaches and examine their large, complex ridges!

Perhaps, more amazing than the reassembly of tiles into a seemingly endless series of 'almost'-identical sibling-solutions, is our ability to recognize their close relationship.

How do we know that a never seen before painting is an authentic Rembrandt, Cezanne, or Klee? How do we know that we listen to the recorded voice of a loved one? How do we learn that the person who returned after leaving the room is the same person after s/he returned? As shown by Piaget [116], early in childhood we learn this remarkable skill. Eventually, we are able to recognize when two 'almost'-identical images indicate the same object, even if they changed shapes, illumination, or distances, or became undetectable for other reasons for a while.

To be sure, these skills are not entirely safe. Impersonators, magicians, forgers, and camouflage can fool us all.

Recognizing common features among never repeating sibling-solutions of a mosaic is undoubtedly an important survival skill. Without the ability to identify tigers, snakes, poisonous insects and mushrooms from partial or distorted images or sounds, we would not have survived. All too often, our lives depend on the recognition of enemies, relatives, predators, and food, even if we had not seen the exact same shape or configuration at an unusual angle or a particular illumination before.

How did evolution teach us these skills? Trial and error alone could be much too costly. Mistakes could be meaningless, irreversible, or even fatal. It helps, of course, that a mistake in judgement does not always have to be fatal. If it is not, we have defences that may get us out of trouble – at least once. Yet, I believe, the ability to re-create in our minds a rapid succession of images of suspected mosaics or of their 'seeds' (see chapter 11) may play equally important roles. Perhaps, even watching the mistakes of others may 'educate' our so-called 'mirror neurons' [19], and help us recognize and avoid similar situations in the future, even if we ourselves never experienced them in the past.

FRACTAL MOSAICS

CHAPTER SEVEN

A MOST IMPORTANT FUNCTIONAL MOSAIC: THE ITERATION

LIVING MOSAICS HAVE INPUTS AND OUTPUTS (PRODUCTS), WHICH BECOME THE INPUTS OF OTHER LIVING MOSAICS. SOME OF THE MOST STARTLING PHENOMENA OF LIFE RESULT FROM THE SPECIAL CASES WHERE THE OUTPUTS QUALIFY AS INPUTS OF THE SAME FUNCTIONAL MOSAIC

Iteration, an Operator Mosaic

Considering that living mosaics are not merely *being something*, but primarily *doing something*, they are all functional mosaics, whatever else they may be. Hence, some of their tiles are operators that may be described as 'black boxes O' that turn certain 'inputs i' into 'outputs o'. For example, the inputs could be salt concentrations, voltages, speeds, temperature, masses, etc. Usually, the outputs differ from the inputs, and may be currents, accelerations, molecular weights, or whatever.

As living mosaics, they also carry out tasks. Hence, some of their tiles must operate as symbolic mosaics. Examples are mosaics that compute distances and speeds, or map certain numbers onto others, encrypt texts, or translate them into other languages.

One type of functional mosaics stands out, namely the 'iteration-mosaics, '**IT**'. Like all other functional mosaics, iteration-mosaics turn inputs into outputs. However, unlike most other functional mosaics, their inputs and outputs have the same quality. In other words, if the input is a voltage, so is the output; if the input is a complex number, so must be the output. This very special property enables iterators to feed back on themselves by using their outputs as their next inputs. Since a hallmark of life is to feed on itself, iterator mosaics are able to act in many life-like ways. They oscillate, run asymptotically toward stable states, or drive themselves into catastrophes; they create fractals and chaos.

One of the immediate implications of this self-feeding action is a natural link to integers that results from counting how many times the operator **IT** has fed back its own output.

equ. 7-01

$o_n = \mathbf{IT}(i_n)$.
Using the output as the next input
$i_{n+1} = o_n$.   ($n = 0, 1, 2, ...$)

Or, in a shorter formulation,

equ. 7-02

$i_{n+1} = \mathbf{IT}(i_n)$.   ($n = 0, 1, 2, ...$)

Each round of iteration is a self-contained, discrete action, and hence, iterations are functional mosaics.

They also have important conceptual implications. Not only do they project self-reliance, they also may appear paradoxical, as they seem to 'pull themselves up by their own boot straps'.

Most importantly, the states of an iterative system are the immediate or delayed consequences of its very first input, and of all subsequent outputs, i.e. they expressing explicitly the **history** of their past stages, including their origin.

All the mentioned implications, especially the role of history, make iterations attractive algorithms for biology and its evolutionary past. Of course, they are not foolproof. For example, iterations do not automatically incorporate the linkages to the history of *parallel* relevant biological and geological events. Yet, they are able to expand into networks of historical events by linking the corresponding iterators with each other.

Creating iterators is relatively easy to do. For example, let the input of an operator $\mathbf{IT}_1$ be a voltage and its output a salt concentration. They do not have the same abstract quality and, therefore, $\mathbf{IT}_1$ is not an iterator. However, one can let a second operator $\mathbf{IT}_2$ feed the salt concentration into a galvanic element that turns it into a voltage. As a result, the composite operator $\mathbf{IT} =$

$IT_2(IT_1)$ [also written as $IT = IT_2 \circ IT_1$] has the same abstract quality input as output and can be used for an iteration. In other words, once one of the links in a series of operators and sub-operators turns out a product that can serve as the input for another link of the series, an iterator is born.

Correspondingly, the sub-operators could be composed of sub-sub-operators, and so forth. In short, the operator of iteration could be a fractal mosaic of sub-operators (see chapter 8).

To guard against easy misunderstandings, the time intervals between rounds of iteration may depend on variable parameters, and hence be quite variable themselves. There is no guarantee that iterations are strictly periodic.

Furthermore, iterations require sources of energy and of raw materials that power the repetition of their rounds. The energy may be contained in hard-fought raw materials, or else the environment may supply it freely. Either way, besides their logical circularity, a central requirement of iterators is the access to large supplies of energy and raw materials.

Iteration as Expression of Causality

Since each step of iteration is a discrete event that was the consequence of the immediately preceding one, it is also a tool to express the workings of causality. After all, the previous output '*caused*' the next output, by becoming its input.

Hence, iterations are a way to formulate serial and branching causal chains of inputs and outputs. It is even possible to differentiate between various branches of causal chains by introducing weighted outputs for the iterators, which determine the likelihood of alternative outcomes.

Cycling Through the Iteration of Mosaics

One of the most ubiquitous properties of living mosaics is their cycling, i.e. their periodic or 'almost'-periodic re-appearance in time and space. Chapters 17 to 20 will deal with the question how iterations yield strictly cycling and irregularly cycling, i.e. 'almost' cycling mosaics.

Chapter Eight

Nested Mosaics

Tiles Can Be Mosaics Themselves; They Can Even Contain Smaller Versions of Themselves

Creating Fractal Mosaics by Recursion

The next chapter is an application of the concept of iteration. Instead of placing tiles into frames, we will replace tiles recursively with entire mosaics that consist of smaller tiles. The result will be fractal mosaics.

Hausdorff (Fractal) Dimensions

Whenever some or all of the tiles of a mosaic are mosaics themselves, fractal structures may result from the infinite regress of replacing tiles with their smaller component tiles. Everybody has seen the extraordinary beauty of mathematical fractals such as the Julia-set or the Mandelbrot-set. One of the less visible, but most intriguing properties of fractals is the possibility to ascribe to them 'spatial' dimensions 'η' that are no longer integers 1, 2, or 3. These non-integral dimensions give rise to important physical and biochemical properties of living mosaics, such as surface adsorption, metabolic rates, heart rates, and others.

About 100 years ago, Felix Hausdorff discovered these non-integral dimensions of certain sets. More recently, Benoit Mandelbrot renamed them as 'fractal dimensions' and applied them to numerous natural phenomena. What are fractal dimensions?

Let us first recap what the familiar integral dimensions are. Ascribing integral dimensions 'p' to an object means that the object fulfils a number of conditions.

1. One can always find an integer 'p' that counts the mutually vertical directions into which the objects expands.
2. Walking along any one of the p directions, one finds continuous sets of points in space that all belong to the object.
3. Each such domain of the object is larger than a certain minimal 'grain size' > 0.

Fractal objects are very different. Walking along a fractal object, we find in infinitely many places that their points are densely interspersed with 'foreign' others that do not belong to the set. Any attempt to measure their expanse into such directions by placing a straight yard stick alongside must fail, because it cannot avoid covering these foreign points, too.

Hence, mathematicians resort to a recursive protocol, which applies smaller and smaller yardsticks in order to avoid the many points that do not belong. The result is an infinite process by which the needed number of yard sticks grows beyond bounds, while the sizes of the yardsticks decrease correspondingly to zero. The fractal dimension 'η' measures how fast the number of required yardsticks increases with their decreasing sizes. Strange, as this quantity sounds, a little further below, we will show how it is a generalization of the normal, integral dimensions.

Self-Similarity and Resolution Units

Reducing the sizes of the measuring stick does not occur in a continuous fashion. In many examples, but certainly in the following cases of fractal mosaics it occurs in distinct steps. This has to do with the notion of self-similarity of fractal sets.

In the 'classical' studies, such as the Koch-curve, the Sierpinski gasket, or the Barnsley fern, the fractals were strictly **self-similar** in the following sense. After magnifying a portion by a certain factor 'r', the enlarged portion looked identical to the original. However, for this to be true, the factor 'r' could not be just any number, but had to have a specific value.

This kind of self-similarity has a peculiar consequence. After the first such r-fold magnification, one could apply the same procedure again to the enlarged set, which looks exactly like the starting set. Inevitably, after enlarging it a second time r-fold, one would yield the same identically looking set, and so on. In other words, if 'r' was able to reproduce the initial set identically, then $r^2, r^3, \ldots, r^k$, etc. must be able to do the same.

Naturally, a magnification by r^k, meant that any unit of length δ_0 had been reduced to the much smaller length of $\delta_k = \delta_0/r^k$ in order to appear as large as the original unit length. Since k is an integer, the series of length units δ_k was necessarily discontinuous.

We will refer to δ_n as the **resolution unit**, to $1/\delta_n$ as the **resolution at n^{th} level**, to 'r' as the **reduction factor**, and to 'n' as the **n^{th} level of refinement**. The smaller the resolution units δ_n of the measuring rods become, the larger the number $N_n(\delta_n)$ of 'rods' that are needed to cover the set of points. For true fractal sets, there is no limit to the process.

equ. 8-01
Given a series of resolution units $\delta_n > 0$; (n = 0,1,2,…)
with $\delta_{n+1} = \delta_n/r$, and r an integer >1, it follows that
$\delta_n = \delta_0/r^n$.

Size Measurements of Fractal Sets by Recursion

Assume we wish to measure the perimeter of a fractal set. As mentioned earlier, we need to use a recursive protocol that applies smaller and smaller resolution units. Figure 8-01 illustrates two steps in such iteration. The left side of panel (a) shows a portion of the set. Obviously, it is very difficult to decide, where the set's perimeter runs, and how to line up a measuring rod along its length.

The right-hand side of panel (a) illustrates how to solve this problem. Instead of using a single yardstick of length δ_n that is pointing in a particular direction, one uses a circular patch with diameter δ_n that represents all possible directions. Nevertheless, each patch counts the length of δ_n as one length unit for the measurement. It covers crudely, if not the perimeter, so at least the area where the set ends. True, it covers additional points, but that does not matter. As δ_n shrinks, the increasing number of patches will always include all the points of the perimeter, but they will also exclude successfully more and more points that do not belong. Eventually, the measurement must exclude all the 'wrong' points, while still covering the entire perimeter of the set.

Panel (b), which shows how 7 times smaller patches cover a detail of panel (a) illustrates the procedure of asymptotic refinement.

The central findings of such iterative and asymptotic measurements of the size of fractal point sets were as follows:
1. The number of patches, namely $N_n(\delta_n)$ grows to infinity as the resolution unit δ_n shrinks to zero. For instance, consider the details of the object shown in Figure 8-01a. They are covered by

94 Chapter Eight

approximately 4 patches. Hence, they have a total length of $L_n = 4 \cdot \delta_n$. Using the 7 times smaller patches shown in panel b covers it more accurately, but we need 49 patches, i.e. $L_{n+1} = 49 \cdot \delta_{n+1} = 49 \cdot \delta_n/7 = 7 \cdot \delta_n$. In other words, *such sets have no finite size.* However, the measurement is far from useless, because...

2. A plot of the logarithm of the number of patches versus the logarithm of the resolution ($1/\delta_n$) at n^{th} level yields a **straight line** with good approximation.

equ. 8-02
$$\log[L_n] = \eta \cdot \log[1/\delta_n] + \gamma, \text{ with } (n = 0,1,2,...)$$

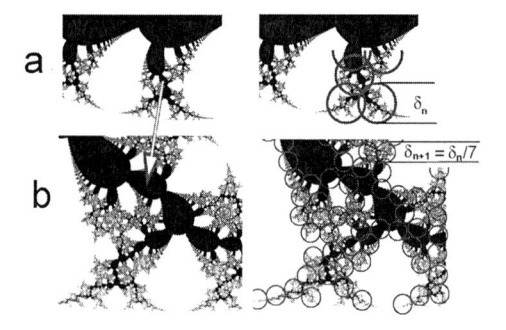

Fig. 8-01. Example of the method to measure the size of the perimeter of a fractal set by covering the perimeter with patches of decreasing sized units of resolution δ_n. (a) Crude coverage with a unit of δ_n. (b) better exclusion of points that do not belong to the perimeter by using a unit of $\delta_n/7$.

Now we can define the **fractal dimension**, or **Hausdorff dimension 'η'** [15]. It is defined as the slope η of the curve that links the shrinking sizes of yard sticks to the increasing number of yard sticks needed to cover the fractal object. Written more formally, ...

equ. 8-03
The fractal dimension η is the slope of the linear regression of '$\log[L_n]$ vs. $\log[1/\delta_n]$'.

Nested Mosaics 95

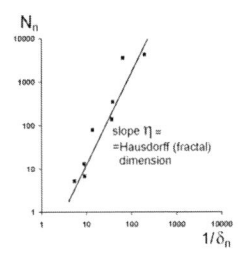

Fig. 8-02. Power law relationship between the measured size of the fractal set N_n and the resolution $(1/\delta_n)$: The log-log plot of the data yields a straight line with a slope of η.

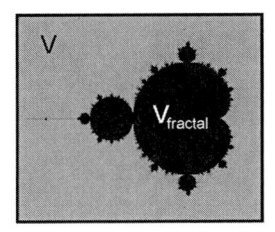

Fig. 8-03. Example that the 'volume' $V_{fractal}$ of a fractal set may be finite, although its 'surface' is infinite (The illustration places the figure of the well-known 'Mandelbrot-set', which is certainly a fractal set, inside a finite volume V.).

Consequently, the essence of measuring the fractal dimension of a fractal mosaics is to measure in a log-log plot the slope η of the number of patches needed to cover it as a function of resolution units. Figure 8-02 shows a typical log-log plot of such a measurement.

Measuring the Topological Dimensions of 'Normal' Objects

The above procedure of measuring the dimension of a fractal set does not sound like a natural way of determining what we normally call a 'dimension'. Yet, it would be a legitimate generalization if it would yield the correct dimensions after we apply it to the familiar cases of 1-, 2-, and 3-dimensional objects. So, let us treat normal objects as fractals and measure their fractal dimensions.

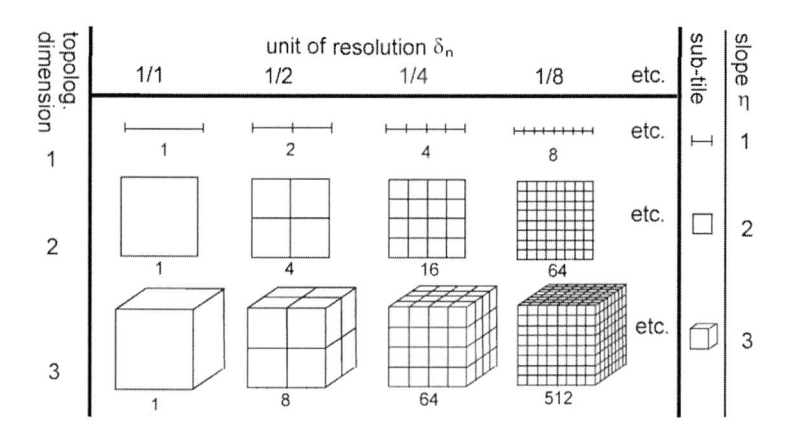

Fig. 8-04. Illustration that the method of measuring fractal dimensions yields the correct result, if one uses it to measure the topological dimensions of 'normal objects such as a cube (details see text).

We will take the examples of 'cubes'. The procedure requires that we choose a series of geometrically decreasing units of resolution, and then select patches of the size of these units, cover the cubes with the corresponding patches, and count the number of patches it takes to cover them completely. For convenience, we select the shapes of the patches as matching cubes (figure 8-04), and use a reduction factor r = 2. According to equation 8-01, the units of resolution become $\delta_n = \delta_0/2^n$ (n = 0,1,2,3,…).

1-dimensional cube. In the case of the unit interval (another word for a 1-dim cube) the numbers of patches needed are $N_n(\delta_n) = 2^n = \delta_0/\delta_n$, and $\log(N_n(\delta_n)) = 1 \cdot [\log(1/\delta_n) + \log(\delta_0)]$. Obviously, a plot yields a straight line with a slope of $\eta = 1$, as one would expect from a 1-dimensional cube.

2-dimensional cube. In the case of the square (= 2-dim cube) the numbers of patches needed are $N_n(\delta_n) = 4^n = (\delta_0/\delta_n)^2$, and $\log(N_n(\delta_n)) = 2 \cdot [\log(1/\delta_n) + \log(\delta_0)]$. Indeed, it yields a straight line with a slope of $\eta = 2$, as one would expect from a 2-dimensional cube.

3-dimensional cube. In the case of the cube (= 3-dim cube) the numbers of patches needed are $N_n(\delta_n) = 8^n = (\delta_0/\delta_n)^3$, and $\log(N_n(\delta_n)) =$
$3 \cdot [\log(1/\delta_n) \log(\delta_0)]$. Indeed, it is a straight line with a slope of $\eta = 3$, as one would expect from a 3-dimensional cube.

The deeper reason for the exact match between the fractal and the topological dimensions is not hard to see. Whether the p-dimensional object is a cube or any other familiar shape, after reducing the unit of resolution by the factor 'r', each of its p dimensions needs r-times more patches to cover it. Hence, we need exactly r^p-times more patches at every round, which generates a slope of 'p' in the log-log plot.

Therefore, the described method of measuring fractal dimensions is a valid generalization of a method to measure the familiar topological dimensions. The fractal dimension moves the closer to an integer, the more 'normal' its fractal set becomes.

Finite Properties of Fractal Sets

Not all properties of a fractal set are infinite. For example, a fractal set may have an infinite fractal surface, while its volume has a finite size. Consider the often-cited example of the fractal nature of the coastline of England. It is clear that the total area of England is finite, while the length of the coastline is not.

Or take the case of the well-known Mandelbrot set. It certainly is a fractal with an infinitely large circumference. Yet it fits easily into a finite area V (figure 8-03). Since it is smaller than the finite set V, the volume of the Mandelbrot figure has to be finite, as well.

Extended Definition of Fractals

All this applies also to fractal mosaics. Their tiles of the nth level of refinement are mosaics themselves. Although we can be sure that the tiles of the $(n+1)^{th}$ level are smaller than the tiles of the nth level, they do not necessarily have the same shapes.

In most cases, the tiles of the living mosaics of biology have different shapes at different levels of refinement. For example, if we consider the organs of an animal as the tiles of its body mosaic, and each organ as a mosaic made of the much smaller cells, it is obvious that the functions and shapes of the cell-level tiles are quite different from those of the organ-level tiles. Similarly, each cell is a mosaic of organelles and polymers, which are much smaller than cells, but have different shapes than the cells. And so forth.

This fact also forces us to accept a less rigid definition of fractals, just as mathematics and meteorology have done so long ago. If the shapes of the tiles change at different levels of refinement, then we can no longer call the set rigorously self-similar. In summary, *the definition of fractality of sets does not rest on self-similarity, but on the existence of infinitely many layers of layer-specific details that become visible, as one views them at smaller and smaller resolution units.*

A closer look at the biological examples reveals another practical aspect of measuring fractal dimensions. In reality, the levels 'n' of refinement of the natural living mosaics do not extend to infinity, but (say) only cover the sizes between meters and Ångstroms.

Yet, this does not preclude an estimate of their fractal dimension. We can pretend that the resolution units continue to shrink to zero, but we do not plot all of them. Instead, we plot in our graph only the number of points down to a smallest size.

Regardless, in graphs like Figure 8-02 one can still compute a slope η by linear regression. As always, the interpretation of such approximated fractal dimensions is not a matter of mathematics, but predominantly up to the judgement of the scientist.

Construction of Fractal Mosaics by Recursion

Now let us turn from *measuring* fractal mosaics to the methods of *constructing* them. Like the methods of measuring them, the method of constructing fractal mosaics will employ iteration as its main technique. We create them by replacing recursively each tile with a mosaic made of smaller tiles. These kinds of 'tiles-that-are-actually-mosaics-made-of-smaller-tiles' will be called in the following '**tile dissections**' and their component tiles will be called '**iterative sub-tiles**'.

The Conditions of Recursive Nesting

The recursive replacement of tiles does not mean that every tile is replaced with its dissections. Mosaics may contain several different types of tiles, but only one type is recursively replaced with its dissection. As a result, these other tiles appear after every round, exactly as they were at the first round. Simple examples of tiles that do not participate in the iteration are the frame of the mosaic, or a hole in the pattern. Neither changes during the recursion. Obviously, this aspect of recursive nesting is not particularly interesting.

A much more interesting situation arises, if it is not the mosaic, but the dissection, which contains 'dead' spaces that do not participate in the recursion. We will call them '**static sub-tiles**'.

As invariant parts of the dissection, the static sub-tiles re-appear in every round, but shrink in size together with the iterative tiles. However, they never appear inside the larger static sub-tiles of earlier rounds. The earlier ones simply stay where they are, but new and smaller ones appear after every round elsewhere in the mosaic.

For example, take the mosaic of the heart muscle. It contains iterative sub-tiles in the form of contractile muscle fibres. However, they terminate in connective tissue domains that are not muscle fibres. The muscle fibres, in turn, contain the contractive cardiac muscle cells as iterative sub-tiles. However, Interspersed into *their* arrangement are intercalated disks which are not cardiac muscle cells. Even each individual cardiac muscle cell contains as its iterative sub-tiles the contractive sarcomeres which are parallel arrangements of actin and myosin filaments. Yet, each sarcomere ends on either side by its own 'miniature intercalated disk', namely the so-called Z-lines, which are not sarcomeres. The mentioned connective tissue domains, the intercalated disks, and the Z-lines are static sub-tiles of the heart muscle.

Calling one kind of sub-tiles iterative while calling others static sub-tiles may seem arbitrary. Upon closer examination, the static sub-tiles of living mosaics may be fractal mosaics, too.

Using the above example of the heart muscle, the mentioned static sub-tiles such as the connective tissue domains, the intercalated disks and the Z-lines, are fractal mosaics themselves. However, there is an important difference.

The sarcomeres nest inside the cardiac muscle cells; the cardiac muscle cells, in turn, nest inside the muscle fibres; the muscle fibres, in turn, nest inside the heart muscle. In contrast, the Z-lines are *not* part of the intercalated disks, and the intercalated disks are *not* part of the connective tissue domains.

The basic rule is this. We consider mosaic components as iterative sub-tiles, if they are replaced at every round of a particular recursion. In contrast, we consider them as static sub-tiles, if they do not participate in a particular series of recursions. As noted before, it does not mean that static sub-tiles are never fractal mosaics. They just do not participate in the same recursion as the iterative sub-tiles, but may have their own recursive structure.

There is another important distinction between the different methods of replacing tiles with their dissections. Either the dissections are always the same, or they may vary from round to round of replacement. We call the first kind 'level independent dissections', and the other 'level dependent'. Each kind may have iterative sub-tiles with or without static sub-tiles

For a better understanding of the many different patterns that may arise from fractal mosaics, let us use the pentomino model mosaic to look into the various kinds of recursive replacements.

A. Dissections Containing No Static Sub-Tiles

A1. Level independent dissections

A1a. Different sub-tile shapes

We begin with the simple situation, where the smaller iterative sub-tiles have the same shapes as the larger ones. As we use pentominoes for an illustration, the category of 'level independent tile shapes' means that the infinite regress procedure interprets each pentomino tile as a mosaic of smaller pentominoes (figure 8-05).

Since each of the initial tiles fit tightly to each other, and since each of them was replaced by equally tightly packed tile dissections *ad infinitum*, the result has no chance of forming any loosely packed areas: The recursion leaves no point inside the initial frame unchanged. Every spot of the interior belongs to the surface or the volume of one of the ever shrinking tiles. The pattern of the total surface is like a spider web, whose openings (= volume) are covered by smaller spider webs, whose openings are covered by even smaller spider webs,… until it acts like an almost planar object.

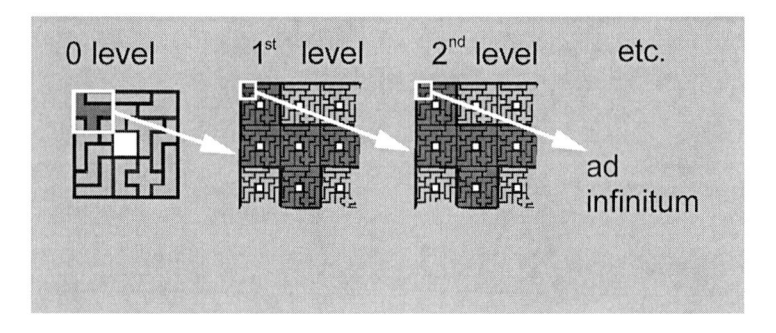

Fig. 8-05. Recursive construction of a fractal pentomino mosaic by the substitution of each pentomino with one of the level independent pentomino-dissections (= pentomino-shaped mosaic composed from nine 3-times smaller pentomino active sub-tiles (see figure 4-02)). The initial mosaic is a solution of the model-frame. The procedure is an example of the recursion using 'level independent dissections'

Obviously, the tiles, or iterative sub-tiles need not be pentominoes to create this kind of object. Still, whatever the tile shapes may be, their dissections replace all larger tiles with a mosaic of smaller ones. Hence, the rising numbers and shrinking sizes of the tiles at ever-higher levels define ever-smaller units of resolution δ_n.

A1b. Repeated sub-tile shapes

In contrast to the above example, the shapes of the tiles of the dissections may all be the same. As long as the dissections are well-fitted mosaics, the specific tile shapes do not really enter the recursive procedures of the construction algorithms. Therefore, it does not matter whether some or all iterative sub-tiles have the same shape, as long as the computation is able to tell them apart during the recursion.

An easy way to distinguish tiles, even if they have the same shape, is to give them individual names or by colouring them differently (figure 8-06). Obviously, the colour does not interfere with the fitting process. However, it renders the tiles of each dissection different from each other. These or other labelling methods that differentiate between identical tiles pose no problem, as long as they do not interfere with the fitting requirements of the mosaic.

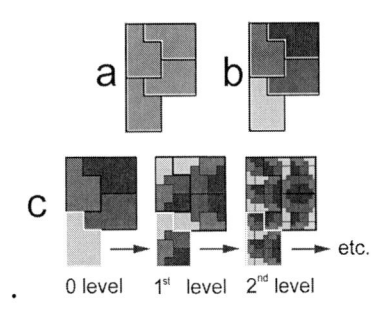

Fig. 8-06. Example of a colouring scheme that turns a level independent dissection of identical sub-tiles into one that contains different sub-tiles. (a) Initial mosaic composed entirely of 'P' pentominoes; (b) Tile dissection of the 'P' pentomino from smaller 'P' shaped tiles which are distinguished by their gray-level (=colour). (c) Progression of the infinite regress of replacing each tile with its coloured version, hence individualized, dissection.

A2. Level dependent dissections.

The construction of fractal mosaics with level independent tile shapes is rather easy to compute, but unfortunately not quite typical for living mosaics. Although living mosaics often display a hierarchy of tile shapes nested inside each other, as illustrated earlier by the example of the heart muscle, they are usually vastly different at the different levels.

The most obvious example is the size hierarchy of the entire biosphere. Here the dissections are as different from each other as ecologies are different from populations, organisms different from organs, tissues different from cells, organelles different from molecular complexes, polyamino acids different from water structures and ions, and so forth.

Sometimes, the tiles of different size levels share at least a certain similarity, as in the category of muscle, where the tile hierarchy of whole muscle, muscle fibres, myofibrils, thick filaments, and thin filaments are all more or less rod-shaped objects. Or consider the 'mosaic' of cartilage which is composed of proteoglycan aggregates, each of which appears like a brush with 'hairs' of proteoglycan subunits, each of which appears like a brush with 'hairs' of chondroitin sulfate molecules. Still, even in these categories we find pronounced differences between the 'tiles' of different levels.

Again we turn to pentominoes for a simple illustration of level-dependent dissections. Assume that there are 7 different pentomino-shaped tiles at the (n-1)th level ($\tau^{(n-1)} = 7$), but only 2 tile shapes at the nth level ($\tau^{(n)} = 2$). (Inevitably, the 2 tiles of the nth level are not pentominoes, because one cannot compose 7 pentominoes from 2 smaller ones (figure 8-07)). Furthermore, one can combine 1 or 2 tiles of the nth level to create each tile of the (n-1)th level.

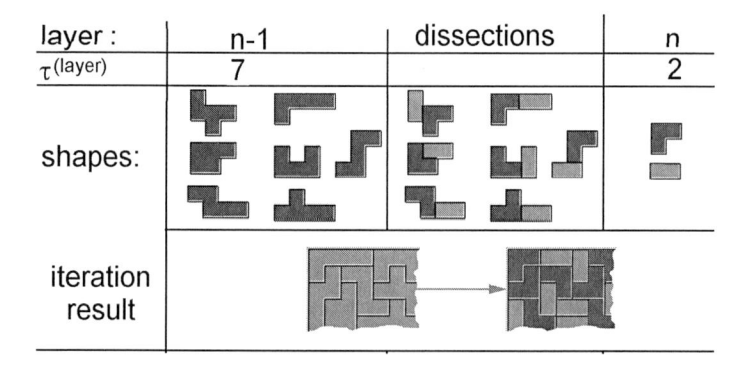

Fig. 8-07. Illustration of 2 levels of refinement ('n-1' and 'n') of the recursion with level dependent tile shapes. In this example, the number of different tile shapes changes from 7 at the (n-1)th level to only 2 at the nth level.

The next step in the iteration is to replace all 7 tiles of a (n-1)th level mosaic with the 2 iterative sub-tiles of the nth level. The tile dissections shown in the left column of Figure 8-07 demonstrates this step. The last row shows the result.

Similarly, the iterative sub-tiles on the next, namely the (n+1)th level may have very different shapes than all of the previous ones, but must be able to compose the 2 tiles of the nth level. And so forth.

The situation raises a number of questions. (a) What effect does this have on the resulting fractal mosaic, and (b) specifically what effect does it have on the plot of the log[total surface] vs. the log[resolution]?

Question (a): In spite of the variable number of iterative sub-tiles of the dissections at different levels, the resulting fractal mosaic will eventually consist of densely packed infinitely small tiles. As we mentioned the argument before, there cannot be a loosely packed or empty spot, because all dissections are perfectly fitting into the shapes of their predecessor tiles. As they are dissections, they are also perfectly fitted mosaics composed of the next level tiles. Hence, there cannot be a spot without tiles. If one existed, there would be an appropriate dissection to fill it. Consequently, the resulting fractal mosaic will be completely compact.

Question (b): One might think that the log-log-plot displays flat or steep increases, depending on the changing numbers of iterative sub-tiles at their different levels. However, the log[*resolution*] *also changes concomitantly depending on the changing numbers of iterative sub-tiles*. Hence, the slope of the log-log plot stays approximately constant.

B. Dissections Containing Static Sub-Tiles

B1. The major effects of static sub-tiles

Regardless of their shape and locations inside their dissections, if used in a recursive replacement scheme, static sub-tiles have one dominant effect: **Eventually, they replace every iterative sub-tile and fill the entire mosaic frame with their own kind.**

The reason is simple. As level follows level of the recursive replacements, every static tile stays exactly where the previous round had placed it. In contrast, every active sub-tile continues to be replaced by a combination of smaller active and more static sub-tiles. Again, these new static ones will remain for the rest of the recursion, whereas the iterative sub-tiles will not. Hence, eventually the active tiles disappear altogether, whereas the static sub-tiles stay behind and fill the entire mosaic frame. We show examples for three very different cases.

Example 1: (figure 8-08 and figure 8-09a). We use here a modification of Figure 8-06 by declaring one of the tiles as a static one (marked in white).

104   Chapter Eight

Already after the second round of replacement, static sub-tiles occupy most of the area of the mosaic frame, while the remaining iterative sub-tiles are crowding the left edges of the frame.

Assume that the initial frame in the shape of a 'P' has the area A_0 (figure 8-08 and figure 8-09a). Then the sizes of all sub-tiles, including the static tile occupy $1/4^{th}$ of the area. Hence, at the starting round 0, the total area of the static tiles $A^{ST}_0 = A_0/4$. At the first round, all 3 iterative sub-tiles, and only those, are replaced by a dissection with sub-tiles that are $1/16^{th}$ of the original size. At the second round, the iterative sub-tiles will be replaced with 9 sub-tiles of $1/64^{th}$ of the original size, and so forth. Hence, cumulative size of all static tiles at round n is

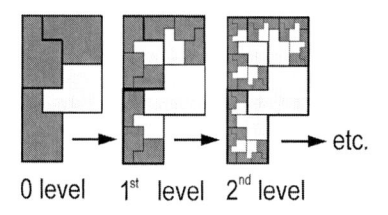

0 level   1st level   2nd level

Fig. 8-08. Recursive construction of a fractal pentomino mosaic using exclusively 'P' tiles with one static sub-tile (marked white)

$A^{ST}_n = A_0/4 \cdot (1 + 3/4 + 9/16 + \ldots + 3^n/4^n)$. For $n \to \infty$,

$A^{ST} = A_0/4 \cdot [1/(1-3/4)] = A_0$. In other words, the infinite sum of all static files fills the entire frame. No iterative sub-tiles remain. As the fractality increases, the static sub-tiles dominate, and **the active ones become 'extinct'**.

Example 2 (figure 8-09b): It does not matter where the static sub-files are located. In the present example the static sub-file is located at the centre of the square frame which has an area of A_0. Hence, the static sub-tile has the area of $A^{ST}_0 = A_0 \cdot 4/64$. Every round of the recursion places one 64-times smaller square inside every one of the 60 squares. In this manner it adds 60 static sub-tiles, each of which has an area of $A_0/64^{th}$. Therefore after n rounds, the cumulative area of the static sub-tiles is

$A^{ST}_n = A_0 \cdot (4/64) \cdot (1 + 60/64 + (60/64)^2 + \ldots + 60^n/64^n)$.

As in the previous case, for $n \to \infty$,

$A^{ST} = A_0 \cdot 4/64 \cdot [1/(1-60/64)] = A_0$. Again, the cumulative static sub-tiles fill the entire initial frame. No iterative sub-tiles remain.

Nested Mosaics 105

Example 3 (figure 8-09c). Both previous examples used level-independent dissections. In contrast, the final example uses level-dependent dissections, where $\tau_0 = 2$, $\tau_1 = 4$, $\tau_2 = 3$, etc. More specifically, the subsequent stages are as follows (gray shapes = active shapes; white shapes = static shapes).

Stage 1: Each active square splits into 2 active rectangles.
Stage 2: Each active rectangle receives 3 small static squares.
Stage 3: Each residual active shape becomes all static, except for 4 diagonally placed active circles.
Stage 4: Each active circle reduces to 3 small active triangles.
……

Simple inspection of the advancing white area of the static sub-tiles shows the area of the iterative sub-tiles only can shrink and eventually disappear.

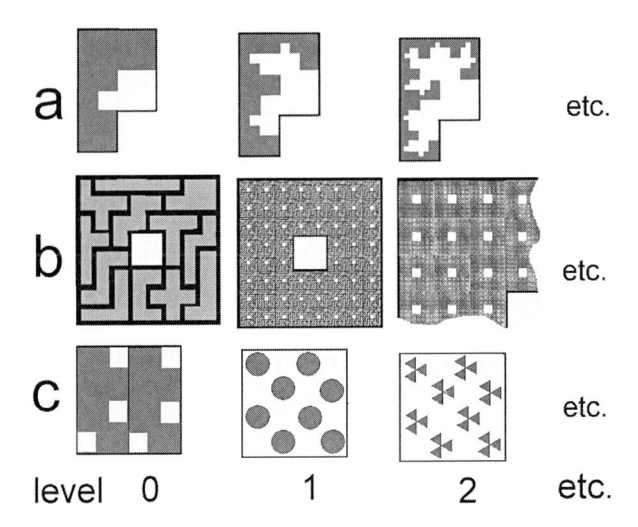

Fig. 8-09. 3 stages of the recursive replacement of dissections containing static sub-tiles (marked white). (a) Level-independent dissection; single static sub-tile at the edge of the frame. (b) Level-independent dissection; single static sub-tile in the centre of the frame. (c) Level-dependent dissections; variable numbers of static sub-tiles (see text).

The three examples illustrate the mentioned reason that the static sub-tiles will eventually fill the entire mosaic frame. Moreover, if the frame has a topological dimension 'k', so will the total sum of all static sub-tiles. In other words, *all fractal mosaics, which result from static sub-tiles, but which have a non-fractal frame, have as fractal dimensions an integer that indicates the topological dimensions of their frame.*

However, it does not mean the resulting mosaic does not *contain* a fractal one with non-integral fractal dimensions. On the contrary, since each static tile has its own surface, the filling of the frame with ever increasing numbers of smaller and smaller static sub-tiles, produces an infinitely growing size of their total inner surface. In fact, these moieties of static sub-tiles are good examples of *fractal mosaics, whose fractal dimension is always an integer, although they differ substantially from a 'normal', non-fractal mosaic.*

It is important to keep in mind that a non-integral fractal dimension indicates a fractal, whereas an integral fractal dimension does not indicate a 'normal' set. The next section will show an example based on a particular static tile, namely the central hole in the model-frame.

Another caveat to keep in mind is the infinity of the recursion. The static sub-tiles in the dissections only can remove all iterative sub-tiles from the resulting fractal mosaic, *provided the dissections contain such tiles infinitely often*. If only (say) the first 2 rounds of replacement used dissections with static tiles, but afterwards all dissections would only contain iterative sub-tiles, they would have effectively have done nothing more than deformed the mosaic frame.

For example, consider the first 3 stages of Figure 8-08. Assume that the static tile would turn into a normal 'P'-shaped sub-tile for all following replacements. Obviously, it would only affect the gray, 'normal' areas of Figure 8-08, because the static tiles, which reside already inside the frame will not be changed any more. At this stage, the (gray) area of the iterative sub-tiles still only has a finite circumference, which makes it effectively a new frame. Continuing the recursion with only iterative sub-tiles would be equivalent to generating a fractal mosaic with it as its frame.

B2. The surface fractal of the static sub-tiles (level independent dissections)

As mentioned, a fractal mosaic containing static sub-tiles may have the integral fractal dimension of its frame. Yet, it may yet contain a fractal set with non-integral fractal dimension. As an example, let us examine the fractal mosaics that result from the recursive replacement of the model-frame with smaller model-frames as in Figure 8-09b. They display prominently a static tile in the centre, namely the middle hole. Furthermore, since each such mosaic has the overall shape of a square, and since every pentomino contains 5 squares, we can arrange 5 sufficiently small copies of these mosaics into the shape of one pentomino.

Figure 8-10 shows the example of an 'F' pentomino dissection constructed this way. There are countless ways of doing this. One can either use 5 times the same solution of the model-frame (figure 8-10a), use 5 different siblings (figure 8-10b), or apply a combination of 'same' and 'sibling'.

Either way, we create dissection of every pentomino made from 8-times smaller pentomino shaped iterative sub-tiles. More importantly, all these dissections also contain 5 static sub-tiles.

During the recursion (figure 8-11), as the tiles shrink to infinitely small sizes, so do the static sub-tiles. However, according to the definition of static sub-tiles there is an important difference: *None of the static sub-tiles, regardless whether large or small, which were placed at a particular round of the recursion, will ever be replaced by anything at a later round.* Therefore, the resulting fractal mosaic will contain more and more 'pores', 'foreign objects', 'associated components', 'connectors', 'binding compounds', or whatever the biological interpretation of the static sub-tiles may be, which become smaller and smaller as the recursion proceeds.

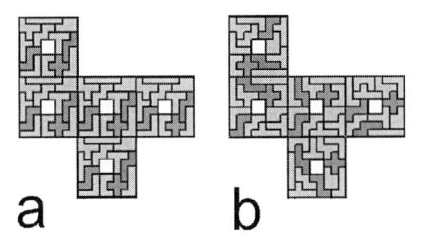

Fig. 8-10. Example of the construction of level independent pentomino dissections with static sub-tiles using solutions of the model-frame. (a) Construction of the 'F' pentomino using the same solution. (b) Construction using multiple sibling-solutions.

As mentioned before, the total area of all these static sub-tiles will eventually be equal to the area of the mosaic frame. Therefore, it is certainly finite. However, the same is not true for the total surface (= perimeter) of the static sub-tiles. Let us calculate it.

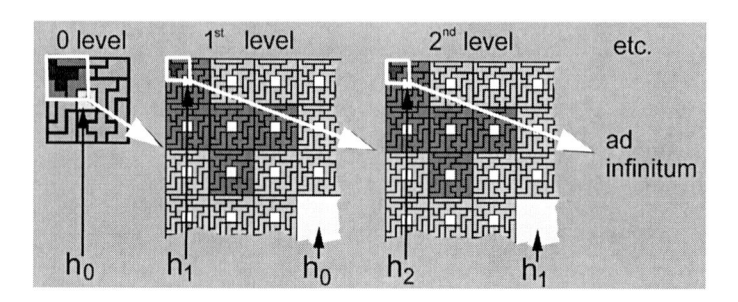

Fig. 8-11. Recursive construction of a fractal pentomino mosaic from the level-independent dissections of figure 8-10 using tiles with static sub-tiles. The static sub-tiles (marked in white and labeled successively as $h_0, h_1, h_2,...$) will never be replaced by anything during the recursive replacement of the tiles.

The static tile at the 0 level has a perimeter $hp_0 = 8$. Each subsequent level shrinks pentominoes and static sub-tiles alike by a factor of 8. Each of the 12 pentomino-shaped tiles contains 5 miniature model-frame solutions. Hence, level 1 adds the 60 combined perimeters of their 8x smaller holes, each having a perimeters of $8/8 = 1$. Hence, $hp_1 = 8 \cdot (1 + 60/8)$. Level 2 adds to them 60·60 static sub-tiles with perimeters of $8/(8 \cdot 8)$ each. Hence, $hp_2 = 8 \cdot 1 + (60/8) + (60/8)^2)$. It is easy to see the nth level total perimeter of the static sub-tiles is

$hp_n = 8 \cdot ((60/8)^0 + (60/8)^1 + (60/8)^2 + ... + (60/8)^n)$.

Since $60/8 = 7.5 > 1$, the series diverges rapidly with increasing 'n' to infinity. What is the fractal dimension of this fractal mosaic, which only consists of static sub-files?

Using the well-known formula for the sum of a geometric series yields $hp_n = 8 \cdot ((60/8)^n - 1)/((60/8) - 1)$. Since $(60/8)^n >> 1$, we can neglect the '1' in the nominator and obtain

$hp_n = (64/52) \cdot (60/8)^n$. Hence, $\log[hp_n] = n \cdot \log[60/8] + \log[64/52]$.

Concomitantly, every round of the recursive replacement reduces the unit of resolution δ_n by a factor of 8: Hence, $\delta_n = \delta_0/8^n$, or $\log[1/\delta_n] = n \cdot \log[11] + \log[1/\delta_0]$. Eliminating 'n' yields

equ. 8-04
$\log[hp_n] = \eta \cdot \log[1/\delta_n] + \beta$, with
$\eta = (\log[60]/\log[11] - 1)^{-1} = 1.03203...$ and
$\beta = \log[64/52] - \eta \cdot \log[1/\delta_0]$.
Hence, the fractal dimension $\eta = 1.03203...$

C. Quantitative Aspects of Recursive Replacements

Independent of the particular dissections, there are some universal consequences of the recursive replacements.

1. Every time, dissections have replaced all the tiles, the resulting mosaic will no longer contain any tiles of the previous size. Instead, it is now filled entirely with the iterative sub-tiles of the dissections.
2. Even if all tiles of the preceding level had different shapes, the replacement steps must duplicate many of the iterative sub-tiles, because the different dissections contain the same set of iterative sub-tiles.
3. The total number of tiles cannot decrease, as each dissection that replaces a tile, contains at least one sub-tile. Since the frame volume V_Φ of the mosaic has a finite size, the ever increasing numbers of iterative sub-tiles forces them to shrink in size. Concomitantly, the units of resolution only can remain constant or shrink. They cannot increase as the iteration proceeds.
4. If there are static sub-tiles, then they will never be filled with anything (see figure 8-09).

The most obvious effect of the recursive replacements is the ever increasing number of tiles N_T in the mosaic. After 'n' rounds of replacement, their number will be $N^{(n)}{}_T$.

Appendix D gives an explicit calculation of the fractal dimensions.

The 'Interesting' Sub-Sets of Fractals Mosaics

As mentioned earlier, the physical, chemical, and biochemical properties of biological and other real objects link to their fractal dimensions. For example, the diffusion rates, thermal conductivity, resorption surface, etc of a real object with a fractal dimension of (say) $\eta = 1.4$ may act as if it was 'a little' 2-dimensional, even though it only consists of basically 1-dimensional components. It acts as if its components weave a dense fractal structure of such density it acquires properties, which only the objects of higher topological dimension possess.

Is the same true for fractal mosaics with static sub-tiles? They have as fractal dimensions the integral topological dimensions of their frame. Does this mean that we should expect these kinds of fractal mosaics to generate diffusion rates, thermal conductivity, etc., as if they were no fractals at all?

Of course, no! As mentioned before, such fractal mosaics contain real fractals, such as their total inner surface with non-integral fractal dimensions. These lower-dimensional sub-mosaics create the counter-intuitive values of their thermal conductivity, resorption surface, and other physical qualities.

For example, the body of a tiger is, of course, a 3-dimensional object. Inside, however, it is an immensely complex fractal mosaic, whose structure results from its dissections of organs, tissues, cells, cell organelles, macromolecules, etc. Still, when the tiger leaps forward to capture a prey, its impact is the same, as if it was a single macroscopic boulder of the same mass and speed. Nevertheless, when he fills its lungs before it leaps, the gas exchange in its lungs remains that of a fractal mosaic consisting of trillions of alveoli.

A Paradox of Continuity of Fractal Mosaics?

In the beginning of this book, we emphasized that mosaics are quite different from the 'favourite' objects of physics, such as differentiable fields and functions, as all mosaics are finite objects with no gradual transitions between their tiles.

Yet, after having studied fractal mosaics, we seem to arrive at a contradiction. After all, in the limit of infinitely many recursions, the fractal mosaics contain no more tiles. Instead, infinitely many, infinitely small point-like objects fill them infinitely dense. What right do we have to call them mosaics?

The simple answer is that all biological objects are finite. No recursion is ever repeated infinitely often in biology. Hence, living mosaics are never actual fractals.

Yet, even if we would consider finite, although very large numbers of recursions, it does not mean that biological mosaics would approach asymptotically the differentiable fields and functions of physics. On the contrary, they move ever further away from them: The reason is that *these objects are nowhere differentiable.*

The proof of this strange property offers little for the student of biology, but it is of great historical interest, because it was the first time that a great mathematician deliberately constructed a fractal to demonstrate this strange property of fractals. Therefore, we add a brief detour to this chapter, in order to demonstrate this fascinating property of such fractals.

Riemann's First Constructed Fractal

Around 1850 the great mathematician Bernhard Riemann wanted to disprove the wide-spread opinion that a continuous function was always differentiable, at least somewhere in a few small intervals.

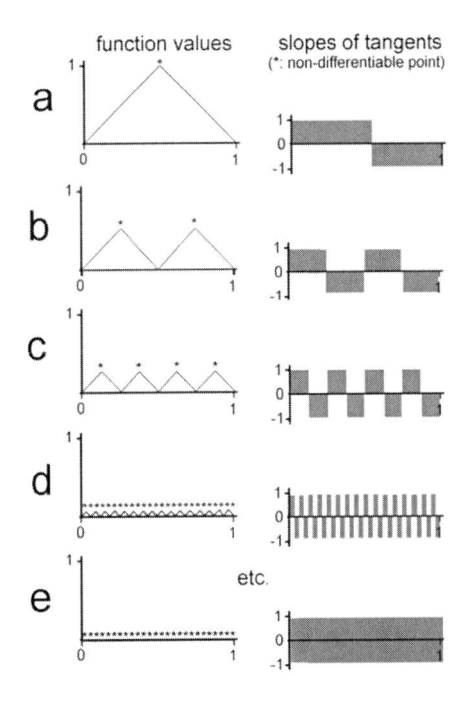

Fig. 8-12. The historically first fractal set: Riemann's construction of a real function that is defined over the interval [0,1] and is continuous everywhere while being differentiable nowhere. The function is the limit of an infinite series of saw-tooth functions that oscillate faster and faster while their amplitude converges to zero.(details see text).

So he constructed a continuous function that looked like a saw-tooth with a height of a = 1 and was differentiable throughout 2 whole intervals, namely within either flanks of the saw-tooth (figure 8-12 (a), left column). At the tip of the saw-tooth, the function was still continuous, because it did not leap to another value. Yet, it was not differentiable, because the point at the tip had two different slopes (figure 8-12 (a), right column).

The next function in the infinite series had twice as many saw-teeth, but only half the height (figure 8-12 (b)). The third function had 4 saw-teeth and 1/4 of the height (figure 8-12 (c)), and so forth. No matter how small the height of the saw-teeth became, the slopes of either flanks remained 1 or -1, and their tips remained non-differentiable.

As shown in figure 8-12 (e), continuing this procedure *ad infinitum* leads to a continuous function that had everywhere a height of zero. Yet, there was not even the smallest interval left where it was still differentiable: In a sense, it consisted of nothing but tips of saw-teeth, whose slopes oscillated with infinite frequency between the values of 1 and -1 (figure 8-12 (e), right column).

Today the limiting function of this series would be called a fractal. The proof for the aforementioned reason that fractal mosaics are continuous everywhere, and differentiable nowhere follows a similar rationale.

A Continuous, Yet Nowhere Differentiable Fractal Mosaic

In order to illustrate how fractal mosaics resemble Riemann's function, namely to be continuous everywhere and differentiable nowhere, we apply Riemann's approach to the pentomino model-frame. As shown in figure 8-13, we depict the model-frame as a real function $f_I(x,y)$ which depicts the body of every pentomino tile by $f_I(x,y) = 0$ and its edge as $f_I(x,y) = 1$.

Nested Mosaics 113

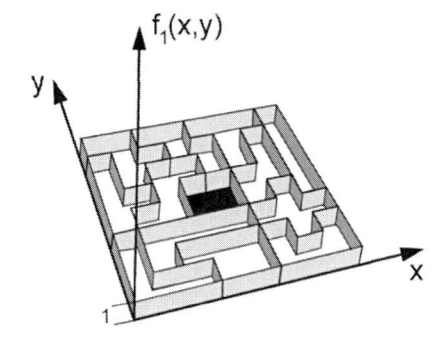

Fig. 8-13. Definition of a 2-dimensional mosaic as a real function $f_1(x,y)$ with $f_1(x,y) = 1$, if (x,y) is a point of the surface of a tile, and $f(x,y) = 0$, elsewhere.

The result is a function that is continuous and differentiable inside every pentomino, but discontinuous at the edge of each pentomino. Now let us subject this mosaic to the infinite recursion of replacing each pentomino tile with nine smaller pentominoes. There are many ways of dissecting each pentomino into 9 smaller ones. We shall use the examples of Figure 4-02. Each of the smaller sub-tiles of the nth round of recursion is depicted in a similar way as the initial ones, namely as $f_n(x,y) = 0$ inside and $f_n(x,y) = 1/n$ at the edge.

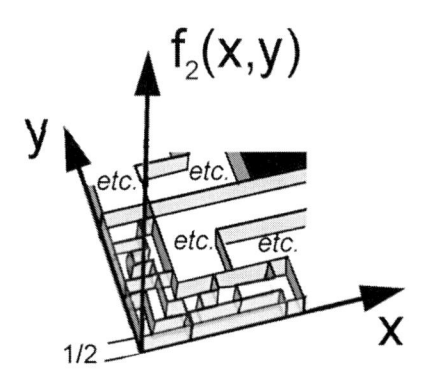

Fig. 8-14. The next stage of the infinite recursion. It replaces every pentomino tile with 9 sub-pentominoes as in figure 4-02, and depict each of them as the real function $f_2(x,y)$ with $f_2(x,y) = 1/2$, if (x,y) is a point of the surface of each sub-pentomino tile, and $f_2(x,y) = 0$, elsewhere. (For the sake of clarity, only the 'V'-pentomino has been replaced in the illustration.)

As the recursion continues to infinity, the inside of each sub-pentomino retains always the value zero, while the height of the 'wall' of each pentomino tile shrinks to zero.

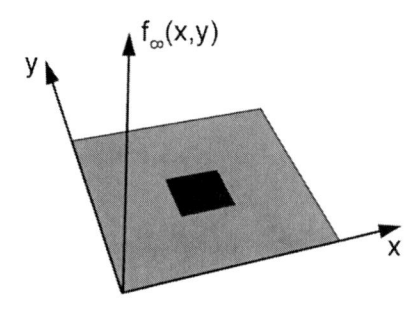

Fig. 8-15. The limit stage of the infinite recursion. The limit function $f_\infty(x,y)$ is equal to zero everywhere and, therefore it is continuous. Yet it is differentiable nowhere, as it oscillates infinitely often between the height of each pentomino's 'wall', and the constant value of zero of its inside volume.

In the limit, the function $f_\infty(x,y) = \lim_{n \to \infty} f_n(x,y)$, which depicts the 'final' fractal mosaic, has the value zero everywhere and, therefore, is a continuous function. On the other hand it is differentiable nowhere as the function at each point (x,y) is surrounded on all sides with oscillating values, which create slopes that alternate between $\pm\infty$.

A ROLE FOR TELEOLOGY IN LIVING MOSAICS?

CHAPTER NINE

THE TASKS AND INTERACTIONS OF LIVING MOSAICS

LIVING MOSAICS FORCE BIOLOGIST TO ADOPT A MORE TOLERANT VIEW OF TELEOLOGY; INANIMATE MOSAICS LIKE THE RINGS OF SATURN DEFINITELY HAVE NO PURPOSE, BUT LIVING MOSAICS LIKE THE LIVER DEFINITELY DO

Humans have an uncanny ability to invent, engineer, and, thus, interpret tools and machines. As mentioned earlier, we may consider all life forms as manifestations of a specific technology. I hasten to stress that this interpretation does not claim the existence of a supreme 'engineer' who designed all these objects. Nevertheless, it can be quite useful to pretend that living mosaics are engineered objects. By applying categories, usually reserve for human-engineered objects, the analysis and classification of living mosaics may seem more familiar.

The most significant of these categories is that of a **'purpose'**. Only if humans with specific uses in mind engineered an object, it can have purposes. In other words, they fulfil **'tasks'**. Tasks imply 'intent', and therefore, only living mosaics can have tasks. The planet Jupiter or water molecules *per se* have functions and effects, but never a task.

Admitting concepts like tasks opens up a wealth of related concepts. For example, we can classify the tiles of mosaics as 'starting materials', 'tools', 'processed materials', 'product storage', 'controls', 'scaffolds', 'waste', etc.. Among other applications, they can aid us in the formulation of a function-based taxonomy (See chapter 28).

As long as we never forget that living mosaics were not engineered by anyone, there should be no danger in using the same concepts as are used to explain truly engineered objects. More recently, with the onset of genetic engineering, there are now actually living mosaics that have at least partial purposes, because human engineers with specific tasks in mind had modified them.

Tasks are more than one of the properties of living mosaics. As we will argue below, their presence in mosaics is logically equivalent with their being *living* mosaics: In other words, every task implies the existence of a living mosaic composed of sub-tasks. Conversely, every living mosaic implies the performance of one or several tasks.

In order to show this equivalence in more detail, we need to begin with some definitions of the components of tasks that rely on the common-sense usage of the terms.

Basic Premises

We begin by calling one kind of variables of tasks by the name of **'materials'**. Materials have properties like amounts, compositions, time parameters, and shapes. One may think of them as portions (sub-sets) of the real world including the mental domains.

The second set of premises we need, will be called **'processes'**. One may think of them as transformations that change one kind of material into another.

Like all transformations, they may happen in time, physical space, or symbolic-informational space. They may operate on objects consisting of matter or information. Their action may be mechanical or informational. In reality, they take place to various degrees in all these dimensions, but some transformations may emphasize a particular dimension over others which are also involved in the task.

For example, when we mention the radioactive decay of radium, we mean predominantly a process in time. Of course, it also happens somewhere in space, but we take that for granted.

On the other hand, when we mention the precession of the Earth axis we mean predominantly a process in space. Naturally, it happens in time, too, but we find it too obvious to mention.

Likewise, when we mention the computer search for a fingerprint, we mean predominantly an event in the realm of information/logic. The fact that it happens at a certain time and location is usually not considered relevant.

There are three special kinds of materials linked to processes. One kind, called **'starting materials'** are the *initial* input materials of a process.

The second kind of materials, called **'processed materials (products)'**, is the *final* output materials of a process. In practice, one distinguishes between two varieties of products, namely **'desired products'** and **'waste'**. The distinction is arbitrary because each can function as the other.

For example, ATP (*adenosine-triphosphate*) is a product of glucose metabolism. From the standpoint of fast growing cells, which use glucose as a carbon source, it is an unwanted by-product, and may be considered a waste (see Warburg-effect of cancer cells). On the other hand, under normal conditions cells need it as their most important and useful metabolic energy source.

Similarly, bee wax is a waste secretion of bees, while also being a most important building material for the hive. *Guano* as the excrement of seabirds is a waste product of their metabolism, while also being a most desirable fertilizer for human agriculture.

The third kind of special materials will be called **'tools'**. They are materials that enter a process at various stages. They facilitate, support, or accelerate the process, but exit it again at a later stage, while their composition and shapes remain essentially unchanged by the process in which they participated. They leave the process before its termination and, therefore, constitute no part of any final product.

For example, enzymes or other catalysts facilitate biochemical reactions. Cellular transport like exocytosis or endocytosis uses vesicles as containers, but recycles them at a later stage. Claws, teeth, or spider webs are obvious tools of animal survival. Oceanic currents may be used as passageways for the migration of animal populations.

Definition of 'Tasks' and Their Basic Implications

Based on the above premises, we define a *'task' as an ordered set of processes to turn starting materials into products, while using specific tools.*

The definition has a number of important implications. We begin by showing that *every task is a living mosaic*. Subsequently, we will argue that the converse statement, namely that *every living mosaic performs a task, is also valid*. Together, these statements form a core thesis of this book, namely that living mosaics and tasks are equivalent concepts.

Every Task is a Living Mosaic

A task cannot be an isolated, singular item. It is always composed of a number of the fundamental sub-tasks that are listed below. These tasks are fitted to each other as the processed material of each sub-task becomes the starting material for its successor. Hence, the sub-task, being a task, contains sub-tasks of its own, which makes 'task' an iterative (and eventually a fractal) concept. However, the main task and each of its sub-tasks represent ordered sequences, which lead from a 'sub-starting material' to the particular 'sub-product' of that sequence. In other words, a task is a mosaic that, if carried out to its end, is a fractal mosaic consisting of ordered operational sequences, sub-sequences, sub-sub-sequences, and so forth.

Of course, the sequential sub-tasks are mosaics, as they change discontinuously into their successor task. There is no continuity between them; the next step in the sequence is an entirely different task, although their progress is a common parameter. However, since only living things carry out tasks, the tasks are also living mosaics.

The following is a list of sub-tasks that every task contains as its modules.

1. Detection of Starting Materials

In order to find its necessary starting material, the task must survey its environment and tell right materials from the wrong ones.

Examples of detection sub-tasks are detection of antigens by the immune system, the visual or olfactory search for signs of prey, or the search of a herd of wildebeests for a water hole.

In more abstract terms, the task of detection consists of the following predominantly informational sub-sequences.
1. Scan environment
2. Collect information about material properties
3. Compare material properties with the properties of in-built templates.
4. Assign a value for the desirability of each scanned materials.

2. Acquisition of starting materials

Once a desirable starting material has been detected, the next sub-task is to acquire it.

Examples of acquisition of starting materials are phagocytosis, the chewing of food, or the surrounding of an elk by a pack of wolves.

In more abstract terms, the task of acquisition consists of the following predominantly mechanical sub-sequences.
1. Move to target material
2. Select amount of target material.
3. Confine amount of target material.
4. Transport selected material into a controlled space.

3. Transitions of Materials

Once acquired, the starting materials undergo changes as dictated by their conversion into processed materials. Except in the trivial case where starting materials and processed materials are identical, the transition from starting material to processed material is not a quantum leap. Normally, the processed material must transition through one or several intermediary stages. (Note: the term 'transition' does not mean the mathematical continuity or even differentiability of the process, although it may appear so at a low enough resolution.)

Examples of such transitions may be the increasing mineralization of tissues, the intestinal digestion of food, or the different developmental stages of an organism.

In more abstract terms, the sub-sequence is
a. Identify the 'present' state of the material.
b. Energize a transition of the material.
c. Trigger the transition.
d. Control the extent and uniformity of the transition.

4. Selection between Pathways

Conceivably, different processes can turn the same set of starting materials into quite different products. Hence, a number of sub-tasks are required to decide which of the possible products the starting materials will become.

For example, specified proteins, which decorate the extra-cellular collagen matrix, influence the specific pathway of differentiation of the contacting cells. The gender-specific development of an embryo depends on signalling molecules such as estrogen and testosterone. The odour-tracks that other ants have left behind, determine the migratory paths of foraging ants.

In more abstract terms, the sub-sequence is
1. Encode the linkages between key signals and different pathways.
2. Deposit the key signals at cross-roads of migration and development.

3. Detect key signals among countless others.
4. Implement the programmed response to a key signal.

5. Protection of Products and Processes

There are a number of reasons to protect the conditions of the execution of a process or the qualities of a product.

α. *Protection from exogenous perturbations.* The mechanical, thermodynamical, electrical, or chemical forces that operate in the external environment may modify or inhibit some of the processes, or may destroy the products, if their influence exceeds certain boundaries. Therefore, the processes must be protected from their detrimental effects. The well-known, visible evidence for this kind of protection are walls, capsules, eggshells, foam layers, etc.

β. *Protection from endogenous perturbations.* Frequently, two vital sub-tasks exclude each other. In such cases, each must be insulated from the effects of the other. For example, some sub-tasks may require a gaseous medium, while others require physiological salt solutions. Some require near-freezing temperatures, while others need room temperature to function. Operating them at different times or inside insulated compartments, provides the protection against their unwanted interactions.

Examples: The high acidity of the stomach content is not compatible with the food tasting or the air breathing functions of throat and mouth. Hence, the stomach separates the *throat* and the mouth from the esophagus.

The various symbiotic protozoa that help cows to digest their cellulose-rich food, cannot co-exist and are, therefore, kept in several different stomachs of the animals.

γ. *Protection from excessive parameters.* The effects of the various physical agents involved must remain within certain limits. Excessive values of their strength lead to process failures.

Examples: The correct speed of epithelial and connective tissue growth is vital for the success of wound healing. Otherwise, excessive scar tissue or tumour tissue may jeopardize the healing. The correct ranges of the salinity, water acidity and temperature are vital for the hatching of fish eggs.

δ. *Protection from flawed products.* Certain production errors may occur during the process and generate dysfunctional products. They need to be weeded out, as they could damage the desired products, and compromise the execution of the subsequent sub-tasks.

Examples: Many proteins contain a tract of several glutamine residues (poly-Q proteins). A common error of the process of RNA-message translation leads to the extension of the size of the tract, which in turn creates misfolded, dysfunctional protein molecules. They cause diseased cells, and seem to lead to neuro-degenerative diseases such as Huntington's disease.

ε. *Protection from waste materials.* Especially detrimental processed materials are the waste products. In fact, their large damaging effect on the success of the task is the very reason for their identification and removal.

Examples: The insufficient removal of lactic acid from actively contracting skeletal muscles leads to their soreness and eventual paralysis. Pain warnings and extensive vascularization of the muscle tissue help protect the muscle.

Heat is a waste product of our data processing systems, which can jeopardize and ultimately limit their function. Hence, they always contain powerful cooling devices.

In the past, the insufficient removal of human waste has caused devastating epidemics of cholera and typhoid. Hence, modern human habitats are protected by elaborate and sophisticated sewage systems.

η. *Protection from immature or incompatible products.* In many cases, the materials undergoing processing are not compatible with the finished products or are not tolerated by the environment and need to be separated from each other as soon as they have been created. In other cases certain processed materials must not be functional until later when a trigger signal permits them to 'spring' into action.

Examples from human physiology are the maintenance of pro-insulin before a cleaving enzyme turns it into the mature form of insulin, or the renal secretion of angiotensin I remains in a dysfunctional state until it is modified by the lung into angiotensin II and influences the blood pressure.

In more abstract terms, the sub-sequence is
1. Build scaffolds and walls to protect processing centres and processed materials.
2. Remove waste materials.
3. Impose thresholds to stabilize processed materials.
4. Store processed materials safely.
5. Protect certain processed materials in their immature state.

6. Progress Control

Two very different actions drive the progress along a series of sub-tasks. One advances each sub-task from its beginning to its end The second is the selection of every next sub-task after the previous one is finished. Both actions need accurate timing. Hence, clock mechanisms have to be linked to the process and continuously read by it.

Another consideration concerns product errors. They are inevitable. In order to detect faulty products, the product qualities need to be monitored at regular intervals. Developmental processes, such as neurocrest migration, somite formation, epiboly, and many others present some of the most obvious examples of critical progress control. As another example, observe the actions of the limbs and wings of a fly in its 'landing approach' upside down on a ceiling. The limbs must time and execute their angles and position with great precision. The claws must grab the ceiling at the right moment, hold fast, and let the body pivot around them, thus turning it upside down.

In more abstract terms, the sub-sequence is
1. Time the beginning and end of each sub-task.
2. Select the required next sub-task.
3. Monitor the quality of each intermediary processed material.

7. Termination of Process

As everybody has experienced who climbed a mountain, raced downhill on a sled, or steered a sailboat to the pier, it can be more difficult stopping a process than starting it. Of course, the depletion of energy and starting materials will eventually stop the process, but often the processes have a certain inertia, and do not stop at the right moment and location (Imagine an airplane landing, whenever and wherever it runs out of fuel!). Unless the termination of the process is well planned and controlled, it may compromise or even destroy the success of the entire task.

Therefore, the successful termination of a task may actually require a full-fledged task of its own. In other words, the termination of a task needs a program and the means to execute it.

Typical the sub-sequences of the process termination are:
1. Establish the 'termination' parameters ("When, where, etc. should termination occur?").
2. Compare the 'running' parameters with 'termination' parameters ("Is it time to stop?").
3. Initiate termination task ("Put on the brakes.").
4. Carry out termination protocol ("Come to a complete halt.").

Of course, each of the listed sub-, sub-sub-, etc –tasks consist of their own further sub-tasks. The student can easily imagine and fill in their appropriate sub-steps.

All Living Mosaics Perform Tasks

The above section leads to a number of obvious observations about the tasks consisting of fitted (cooperating) and interacting sub-tasks.
1. Each of them is composed of many sub-tasks itself.
2. Each of them requires information processing, if only to decide when to stop and what to do next.
3. The individual sub-tasks follow one another as discontinuous objects, although their execution may seem to flow monotonously, if viewed at a low enough time resolution.

Hence, all tasks are not only mosaics; as their execution requires living mosaics, they are living mosaics, themselves.

However, the converse is also true: All living mosaics perform tasks. In fact, we recognized them as being alive by their performance of these very tasks. Examples are the maintenance of the body temperature, the maintenance of balance, the continuation of heartbeats, the actions of running, swimming, flying, searching, defence, etc.

The Irreversibility of Tasks

A task cannot successfully turn starting materials into its products without leaving marks in the environment. These marks cannot be removed by running the task backwards. Especially, it is not possible to return the products back into the starting materials without performing other, additional tasks. There are various reasons for this irreversibility.

1. Irreversibility by Reasons of Entropy

If any of the sub-tasks involve heat energy, it may dissipate in part during the task. In this case, thermodynamics proves that these changes that cannot be reversed without further consumption of energy and increase of entropy.

2. Irreversibility by Reasons of Practicality

The ultimate reason for the thermodynamic irreversibility is the impossibility of returning countless molecules that are engaged in chaotic thermal movements back to their precise initial state. Even if we knew all their initial states, it would be practically impossible to capture each molecule and return it to its initial state, without causing additional irreversible changes.

The same applies *mutatis mutandis* to the macroscopic tiles of living mosaics, although their number is, of course, much smaller than the number of thermally moving molecules in (say) a litre of gas. Nevertheless, the numbers of possible states, locations, interactions, etc. of the sub-tasks are frequently large enough to make it practically impossible to reverse them. The reason is that the number of data that are required to reverse the process grows much faster than the number of component tiles.

Take the simple example of a metal rod on a lathe and the task to cut it to a smaller diameter. Let us even ignore the thermodynamic irreversibility of the heat production by the cutting iron as it removes metal chips from the rod! It would still be practically impossible to find every drop of cutting oil that sprayed off the spinning rod, and return it in the proper order to its flask. Furthermore, how could one find in practice the proper order of the metal chips as they coiled under the knife and flew off the lathe, flatten them (i.e. reverse all deformations of the metal lattice inside every chip) and return them to the metal rod their original orientation?

This example only involves a very small number of macroscopic sub-tasks, such as the mounting of the metal rod in the chuck, aligning the cutting iron, tightening nuts, flipping switches to turn on the motor, and so forth. Usually, tasks performed by living mosaics contain much more elaborate and complex sub-tasks that cannot run in reverse without additional tasks, if at all.

3. Irreversibility by Reasons of Logic

The attempt of reversing the results of completed sub-tasks may encounter even logical obstacles if language or information processing are involved. As the saying goes, one cannot un-ring a bell. More specifically, transmitted and received information cannot be 'un-sent'. Of course, it may be possible to discard a copy of the received information, return it to the transmitter, or write a computer program with the codes in reverse order. However, this would merely add more information exchanges, or more program executions, but not return of the world to its state before the transmission or execution of the messages ever occurred.

Other cases of logical irreversibility arise if data compaction occurred. Consider a task that involves calculating the average of a set of data. In order to reverse the process, one would have to reconstruct its many input data from their average value. Obviously, after the data have been compacted into an average, countless other data sets would yield the same average. Hence, no exact reversion is possible.

Multiple Tasks

There are a number of important consequences, if a task is part of a set of multiple tasks $T_1, T_2, T_3,.$, Let us call their inputs by $J_1, J_2, J_3,..$, and their outputs $O_1, O_2, O_3,...$ Then performing one of tasks T_k means turning its input (starting materials) into its output (processed materials), i.e. $O_k = T_k(J_k)$.

(Please note: The inputs J_k and the outputs O_k are not necessarily numbers, and the tasks $T_k(J_k)$ are not necessarily continuous function of their inputs J_k, although some of them may be such functions. For example, the inputs may be centrioles, and the outputs may be transcription factors. We may describe them by logical symbols, or perhaps by tables of integers. The student familiar with programming languages may think of the variables as programming variables, and of the functions as lines of code.)

Assume next that task T_k is performed **after** task T_i. There are 2 possible situations.
1. The output O_i is stored away and the task T_k starts a new independent sub-task using its own input, or
2. The output O_i of T_i is fed into T_k as its input. In other words,

$$O_k = T_k(O_i) = T_k(T_i(J_i)).$$

Case (1) needs no special consideration, as T_k and T_i are simply 2 different tasks that need not have any relationship. In contrast, case (2) is typical for a great many processes, and needs to be discussed in more detail.

To simplify its notations we will abbreviate the sequential, dependent performance of tasks T_k and T_i in this order as

$$O_k = T_k \bullet T_i$$

Using these notations, we can formulate a number of important linkages between multiple sequential tasks.

1. Non-Association

If the tasks were simple physical events, the result of a series of dependent tasks $T_k \bullet T_i \bullet T_m$ is the same, regardless of the way we consider them as groups in our minds, as long as their sequence does not change.

equ. 9-01
$$T_k \bullet (T_i \bullet T_m) = (T_k \bullet T_i) \bullet T_m = T_k \bullet T_i \bullet T_m$$

Unfortunately, the same is no longer true for living mosaics, because their tasks depend on the meaning of information. Furthermore, the meaning of information is not associative.

Take the example of the following 3 tasks.
T_1 = 'identify conservative voters':
Variable = 'voter';
task = [if registered as conservative, attach '1'; else attach '0'].
T_2 = 'make a list of voters':
Variable = 'people';
task = [if qualified to vote, add to set of {voters}; else ignore]. ,
T_3 = 'count the ballots cast by the voters on a list'.
Variable = 'ballots';
task = [find listed voters; if a ballot cast, add 1 to sum, else add 0].

Each of the tasks is well defined. However, there are two ways of interpreting the meaning of the combined task $T_3 \bullet T_2 \bullet T_1$. One is to associate T_2 with T_1 and make a list of conservative voters before you count *their* ballots. The other is to first identify conservative voters but then associate T_3 with T_2 and count the ballots cast by *all* listed voters. Obviously, the results are quite different.

2. Non-Commutability

The order in which one carries out the tasks matters a great deal. Anybody, who ever attempted to assemble a piece of furniture that came as pieces in a box, has painfully learned that truth. Like in the case of most technical programs, the order of sub-tasks of living mosaics is most often critical and hence not interchangeable.

For example, consider the order of two sub-tasks during a eukaryotic cell division. Let T_i = replicate each pair of homologous chromosomes in S-phase, and T_k = separate each duplicated chromosome into its 2 sister chromatids during anaphase. Obviously, one cannot separate the sister chromatids before they have been produced, Hence, T_i must come before T_k.

$T_k \bullet T_i$ means that homologous chromosomes are at first replicated and subsequently separated into sister chromatids after forming the mitotic spindle successfully. The result is two diploid sets of chromosomes.

In contrast, $T_i \bullet T_k$ means that all chromosomes are supposed to be segregated into sister chromatids, before they duplicate. However, at this time there is no bi-polar spindle available for the mechanical separation.

Furthermore, the chromatids are still condensed. In short, the cell division will fail.

Another example may be the case of a protein that needs to be phosphorylated in order to be cleaved into 2 peptides. Here T_i is the task to interact the protein with a specific kinase, while T_k is the task to interact it with a protease that will cleave a properly phosphorylated protein at a specific site.

$T_k \bullet T_i$ means that the protein is first phosphorylated and subsequently cleaved. In contrast, $T_i \bullet T_k$ means that it is first cleaved and the resulting peptides are interacted with the kinase. The latter order of procedure will fail because both molecular interactions are usually quite specific. In other words, the protease will not be able to cleave the un-phosphorylated protein, and neither will the peptides be appropriate substrates for the kinase.

We express non-commutability as

equ. 9-02
$$T_k \bullet T_i \neq T_i \bullet T_k$$

It is not difficult to generalize the above examples. Non-commutability of two sub-tasks T_i and T_k occurs every time the subsequent task (T_k) requires a product of the preceding one (T_i) in order to work.

This requirement even applies to cyclical tasks, although it may seem that one should be able to interchange the order of identical repeats. Here the required 'product' of T_i is simply its completion. Otherwise, the next round, T_k cannot get started properly. For example, consider the cyclical repeats of your steps during walking. If you try to move your right foot forward, before the step of your left foot has firmly landed, you will most likely fall.

Two sub-tasks are commutable if either of them can work without input from the other. This may happen, if they have nothing to do with each other ('clearing your throat either before, or after taking a step forward') , or if the products of the two sub-tasks are held in storage and are always available for the other (production and storage of leucocytes in the bone marrow, or erythrocytes in the spleen).

Sub-tasks are also interchangeable as long as they are equivalent. For example, let T_i = 'a bee flying to a certain flower and collecting nectar and pollen', and T_k = 'the same bee flying to another flower, instead, but still collecting nectar and pollen'. Although not identical and not necessarily cyclical, the two sub-tasks are completely interchangeable, because their outcomes are essentially the same, and so $T_k \bullet T_i \approx T_i \bullet T_k$, because $T_k \approx T_i$.

3. Iterations

As mentioned, a very important and special form of multiple tasks arises, if a product qualifies as an input to its own task. Hence, after executing the task, the results can be subjected to the same task again and again. We encountered this procedure in earlier Chapters. It is called **iteration I.**

equ. 9-03
$I_2 = T_k \bullet T_k = T_k^2$

As we have seen above, the results of a series of iterated tasks such as $I_N = T_k \bullet T_k \bullet T_k \ldots = T_k^N$ can be quite complex as they lead to fractals and chaos [15].

Frames and Their Interaction with Other Mosaics

Based on the concept of tasks, we must finally bring up a subject about living mosaics that we have neglected so far, namely the role of the frames in the interactions between different mosaics. Of course, it was inevitable to talk about mosaic-mosaic interactions earlier and, indeed, we have done so whenever we mentioned interactions between tiles. After all, the tiles of living mosaics are usually living mosaics themselves. Hence, tile-tile interactions are actually mosaic-mosaic-interactions.

However, we have largely ignored the mosaic frames and their role in the mosaic-mosaic interactions. We treated frames as merely necessary, but otherwise inert components of mosaics. It was their task to limit the mosaic, to hold it together, to constrain its tiles, and to provide its mechanical, functional, or behavioural support.

Nothing could be further from the reality of living mosaics. It is correct that frames hold mosaics together, but very frequently, the tile-tile interactions are more important than frames for their cohesion. For example, the capsule of the kidney is not the main reason that keeps the nephrons in order and register. There is enough sufficient connective tissue between the nephrons that would hold the organ together even without the capsule.

The main task of the frames is to mediate the interactions with other mosaics or their isolation, whatever the case may be. In order to explore the true nature of the mosaic frames and their level of complexity, let us look at the example of a human kidney.

The student may skip the following detailed description of the tasks, subtasks, and topological complexity of the kidney. I added it for the student who finds it easier to trust abstract claims if the details of at least one realistic example accompany them.

Example: Kidney

The main task of the kidney is to filter the entire blood volume of 5 litres about 20 times every day, remove small-molecular toxic substances, and adjust the salt and water concentrations.

Curiously, the adult kidney also secretes erythropoietin, a hormone needed for our red blood cell production (erythropoesis). Kidney function is crucial. Its failure is rapidly fatal. Therefore, the body provides an extra copy of the kidney, although each of them can handle the task quite well alone.

The kidney achieves its main task with most elegant engineering. I hope to describe the required spatial organization and sub-tasks in sufficient detail to demonstrate it without overwhelming the student.

In addition, the example of the kidney illustrates the huge complexity of living mosaics, and raises further intriguing questions, such as 'How can such immensely complex objects be built during embryogenesis?' and 'How can they be repaired after an infection or injury?'

At first, however, we will have to present a relatively large number of details of the histology and physiology of the kidney. Afterwards, we can focus on the main subject, namely how the requirements of engineering shape the interface between the inside and outside world of the kidney, i.e. the frame of this living mosaic and its interaction with its own tiles.

Basic Task and Anatomy

The human kidney is a living mosaic containing about 1 million nephron units that operate in parallel. The nephrons are the main tiles of the mosaic. They fit together with great precision by keeping in exact register. Figure 9-01 only provides a crude schematic of the main structural and functional elements of two of them.

Each nephron not only is a tile, but also a mosaic, which contains 4 major 'tiles' of its own.

The glomerulus ('gl' in figure 9-01). Its task is to extract the so-called ultra-filtrate from the blood plasma (marked light gray in figure 9-01). It consists of a coil of capillaries whose wall cells (podocytes) leave countless slit-like openings between them. They hold back the blood cells, but let the plasma ooze out between them. After passing through a dense fibre-filter, the plasma becomes the ultra-filtrate', which is collected by a capsule around the coil and pushed forward into the proximal convoluted tubule for further processing.

The proximal convoluted tubule ('pct' in figure 9-01). Its task is to retrieve the nutrients and other valuable molecules from the ultra-filtrate. Specialized cells line its walls that are covered with myriads of tiny finger-like projections (microvilli) that create a much enlarged surface area, which allows them to extract and reclaim much of the salt, nutrients, amino acids etc. that the body needs to retain. At its end begins the loop of Henle.

The loop of Henle ('lH' in figure 9-01). Its task is to reclaim most of the water in the ultra-filtrate. That seems to be an impossible task! After all, the cells that make up the tubule walls consist themselves to 85% of water! How can water extract water from water?

The answer is one of the most elegant feats of biological engineering. It is accomplished by simply passing the tubule through the medulla of the kidney, which contains a very high concentration of salt (NaCl) and urea.

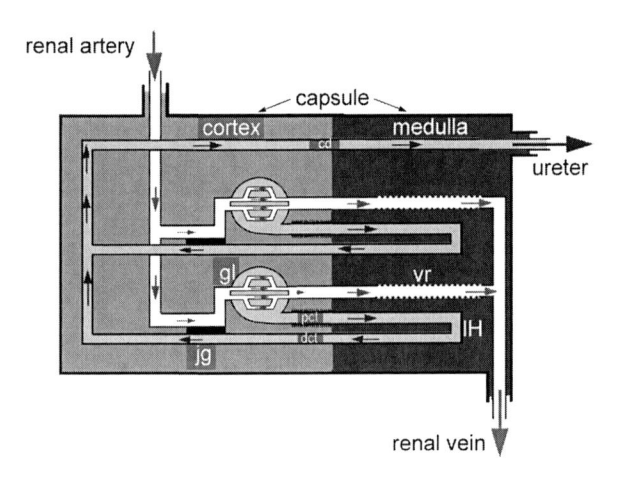

Fig. 9-01 The much simplified schematic of a human kidney as an example of the complexity and the mosaic character of the frames of living mosaics. It depicts only two of the approx. 1 million nephrons of the human kidney, and it shows only 2 branches of the many million branching vessels that allow all nephrons to operate in parallel.

The gray areas represent the actual kidney tissue (parenchyma). It is crudely divided into 2 domains, the isotonic cortex and the medulla with its very high concentration of salt (NaCl) and urea. The white and light gray areas belong to the outside world of the kidney, which consists of 3 major compartments, the body cavity (white area limited by the capsule surrounding the kidney), the circulating blood volume (white area containing arrows starting at the renal artery and leading to the renal vein) and the ultra-filtrate of the blood plasma (light gray areas with dark arrows leading into the ureter and eventually the bladder).

The black lines represent the interfaces between kidney mosaic and the 3 outside domains. Together, they form its frame. It is a mosaic in its own right. Some of its many specialized domains (tiles) are indicated by gl: glomerulus; jg: juxta-glomerular cells; vr: vasa recta; lH: loop of Henle; pct: proximal convoluted tubule; dct: distal convoluted tubule; cd: collecting duct.

(For graphical reasons, the illustration puts the renal artery in the wrong place. In reality, it enters through the medulla like the two other in- and outputs.)

The water of the ultra-filtrate – at this stage of the processing it has become a very dilute solution of salt and toxins - is literally sucked through the very thin wall cells of the tubule into the high salt environment. Again special cells are needed for this sub-task. After all, the cells of the loop of Henle have to live permanently surrounded by the very high salt concentration that would normally kill cells.

The distal convoluted tubule ('dct' in figure 9-01). At this point we have not yet explained where the high salt and urea concentrations of the medulla come from. The simple and elegant reason is the fact that the loop is a loop, i.e. the tubule returns to its beginning: After passing through the medulla, the ultra-filtrate has lost most of its water, and turned into a concentrated salt solution itself, ready to transfer all its salt into the surrounding medulla. Subsequently, it replenishes the high salt of the medulla.

Correspondingly, the tubule is changing its function and accordingly its name by becoming the distal convoluted tubule. Its wall cells can quite easily fulfil their main task, namely to 'grab' small portions of the salt of the concentrated ultra-filtrate and deposit it across their bodies into the medulla.

In this 'circular' technology the high salt concentration in the medulla condenses the ultra-filtrate in order to become the source of that very salt.

There are other biological examples of this kind of technology, called 'counter-flow multiplication', which sounds like pulling oneself up by one's own bootstraps. For example, arctic animals use this technique to return much of their body heat back into the body before it reaches the skin where it would be lost to its very cold environment. Another example is the function of the *rete mirabilis* of fish, which forces gas into their swim bladder, no matter how high the surrounding hydrostatic pressure is. The student is greatly encouraged to look up these stunning masterpieces of biological engineering.

The vasa recta ('vr' in figure 9-01). Of course, the distal convoluted tubule cannot continue to deposit salt into the medulla forever. Once it has reached a sufficiently high level, the surplus salt must return into the body.

Yet another specialization of the walls of the blood vessels (endothelium) accomplishes this task. After allowing the blood to contribute some 10% of its volume to the ultra-filtrate, and guiding the rest out of the glomerulus and back into the main circulation, their job should effectively be done. In principle they could now leave the kidney altogether.

Yet, with stunning engineering elegance, they open up their vessel walls with numerous tiny pores, and traverse the medulla in straight ('recta') lines. This specialized form of blood vessels is called the *vasa recta.* They establish the right salt concentration in the following way.

Inevitably, some of the salt-rich solution in the medulla diffuses through the pores back into the passing blood, which transports it back into the body. As the salt concentration of the medulla increases due to the work of the distal convoluted tubule, more and more of the salt returns to the blood stream, until the medulla has reached a steady state level between salt deposition and retrieval.

Secondary Task: Blood Pressure

The only driving force for the entire ultra filtration by the nephrons is the blood pressure. There is no muscle-driven peristalsis to help. Therefore, it is 'in the interest' of the kidney to control the blood pressure.

First Level of Quality Control

Naturally, the kidney can influence the all-important blood pressure by adjusting the salt and water concentrations, but this method would be very sluggish because it takes some time to change the osmotic pressure of the blood in the entire body. For rapid responses, there is a much more effective means, namely a specific hormone called angiotensin I. The kidney secretes if by specialized cells called juxta-glomerular cells ('jg' in figure 9-01).

They are strategically located between the afferent blood vessels into the glomerulus and the distal convoluted tubule. As they are located on opposite ends of the nephrons, these two parts of the nephrons that would not naturally meet. However, they bend towards each other until they form a contact area. The resulting 'mini-organ' is able to monitor the pressure at the nephrons entrance on one hand, and the fully processes ultra-filtrate on the other. Thus, the jg-cells carry out a most important quality control between the input and the output of each nephron.

But the engineering perfection of the kidney does not end there! The body does not allow this hormone to change the blood pressure all by itself, because that could trigger a dangerous chain reaction.

Assume that a small bleeding injury or infection of a handful of nephrons would locally drop the blood pressure and therefore trigger the release of angiotensin I by the neighbouring nephrons. If that alone was enough to raise the blood pressure in the entire body, it could merely pry open the internal wound, which would lower the local blood pressure even more, and so forth until it eventually causes a catastrophic hemorrhage.

In order to assure that the drop in blood pressure sensed by the kidney is not merely 'local', the distantly located lungs – they, too, are critically dependent on blood pressure - have to 'agree'. If they also sense a drop, then it is likely to be global. In response, the lung enzymes cleave the circulating angiotensin I into angiotensin II that will then actually raise the blood pressure in the entire body, including in the kidney.

Each of these nephron components is, of course, another mosaic composed of highly specialized cells and extra-cellular matrices with selective filtration properties as their tiles that carry out the described sub-tasks.

Second Level of Quality Control

Naturally, the body must ultimately discharge the residual volume of the ultra-filtrate together with all the unwanted, toxic by-products of our metabolism that inevitably end up in our blood. The 120 l of blood that pass through the human kidney every day generate about 12 l of ultra-filtrate. We can obviously not afford to discard all that, unless we were willing to spend most of the day drinking water in order to replace it.

Therefore, the kidney must retrieve all but a small amount of the water, at least in all land living organisms and especially in all desert living creatures. The situation is even more demanding in birds, which need to keep their body weight at a minimum. They cannot afford to waste energy by lifting water into the air, only to discard it afterwards (see 'cloaca').

As discussed above, it is the task of the loops of Henle to reclaim most of the water in the ultra-filtrate. However, they have no control of the uptake or the discharge, which the changing biological situations require.

Consequently, a number of land-living animals (mammals and birds) have developed an additional fine-tuning control of the amount of water they need to reclaim. Not surprisingly, this final adjustment of the discarded water volume is no longer determined by the kidney, but by the central nervous system. This last step in the function of the nephron is yet another impressive feat of biological engineering.

The processed ultra-filtrates of all nephrons flow together into a number of common collecting ducts. However, instead of discharging their contents altogether, the collecting ducts traverse the salt-rich medulla one more final time ('cd' in figure 9-01). Their wall cells are specialized cells in a new way. Their water permeability is exogenously controlled by the level of the short-lived so-called anti-diuretic hormone (ADH, 'vasopressin'), which is released by the posterior pituitary gland in the brain. It controls how much water returns to the medullary space by this last passage of the ducts. In this way, the organism can produce hypo- or hyper-osmotic urine depending on its biological situation.

The Outside World of the Kidney Mosaic and Its Immensely Complex 'Frame' Topology

All these and further functions are not carried out by the kidney at large, but by the walls of the countless tubules that form the interface between the nephrons, i.e. the 'tiles' and the outside world. In other words, *the frames of the kidney's living mosaic tiles carried it out.*

The kidney's frame interfaces with 3 different outside world compartments ('white' and 'light gray' areas in figure 9-01).

The body cavity. Its frame/interface is the capsule of the kidney built by connective tissue. The frame's main task is mechanical suspension and protection.

The blood volume. Its frames/interfaces are the walls of the blood vessels (endothelium). The frame's main task is the supply of all kidney tissues with oxygen and nutrients, the extraction of the blood plasma into the ultra-filtrate and adjustment of the vital salt concentrations in the medulla.

The ultra-filtrate. Its frames/interfaces are the walls of the collecting tubules. The frame's main task is to reclaim the valuable components and expel the unwanted filtered substances.

The latter two compartments penetrate deeply into the body of the kidney and interweave as a precisely patterned fabric of various kinds of tubules.

Frames as Living Mosaics with Merely Different Sub-Tasks

The frames of living mosaics such as the kidney are obviously much, much more than mechanical, functional, and behavioural constraints of mosaics. Repeatedly, the above description of kidney function had to mention separate domains and groups of cells that had special functions. They were part of the interface between the kidney and its outside world, and changed abruptly into other domains. Therefore, the frames of living mosaics contain many distinct domains as tiles, i.e. they are immensely complex and sophisticated living mosaics in their own right.

More specifically, the walls of the bewildering array of tubules, which separate the 'inside' of the kidney from the 'outside' world, are the actual 'frame' of the kidney. Paradoxically, the 'inside' of the tubules is part of the 'outside' world, while the 'outside' of the tubules faces the 'inside' of the kidney and is part of it.

Accordingly, the frame of the kidney not only is an immensely sophisticated living mosaic by itself consisting of sub-mosaics as tiles, and sub-sub-mosaics as sub-sub-tiles,…. It is also a patterned array of millions of capillaries and tubules, which are in mutual register while densely interwoven with each other. Its three-dimensional topology is almost incomprehensible.

Therefore, the frames of the kidney are not simply some inert border fence at the outermost edge of the mosaic. Frames can be found everywhere on the inside of the organ, too. In fact, most of the parenchyma of the kidney consists of frames, or more accurately of their inside faces.

Although we had introduced nephrons as the tiles of the kidney, they have a dual character. On one hand, they are part of the 'inner' parenchyma of the whole organ. On the other hand, they form the interface with the outside world. As a result, nephrons are actually divided tiles: Right through the middle of their walls runs the borderline between nephrons functioning as the tiles of the kidney mosaic and nephrons functioning at the same time as tiles of the frame mosaic.

In this case, it would be most difficult to maintain that frames are fundamentally different from the internal tiles of a mosaic. In fact, both are just living mosaics surrounding one another. It may still be practical to describe one as the 'frame', but in reality, the 'frame' tiles are simply *different* from the 'inside' tiles. To be sure is not much of a distinction, because there are frequently differences between different 'inside' tiles, anyway.

In addition, we saw how frames mediate various kinds of interaction between different mosaics. In the example of the kidney, the blood and the ultra-filtrate are themselves fluid mosaics with which the kidney frame necessarily interacts. Furthermore, we saw that the kidney frame contains the juxta-glomerular cells, which represent tiles that mediate the kidney's interaction with another, remote mosaic, namely the lungs. In addition, the kidney secretes erythropoietin through its frame, and thus can signal to yet another remote mosaic, namely the bone marrow.

Finally, the urine produced by this immensely complex organ, is not simply waste. As every dog lover knows quite well, the urine of their pet plays a very sophisticated role in the communication between dogs and the marking of their territory.

In summary, mosaics with specialized tiles and functions mediate the interactions between different mosaics. We may conveniently call them 'frames', but what really changes as we approach the edge, i.e. the 'frame' of a living mosaic, is the pattern of the sub-tasks of the mosaic. The specializations near the edge of the patterns of the sub-tasks of living mosaics are merely one of their many variations. Thus, the functional and logical patterns of the sub-tasks offer a reasonable way to unify the analysis of living mosaics.

If, indeed, the main difference between different living mosaics is the pattern of their tasks and sub-tasks, it seems mandatory to classify these tasks. The next chapter will attempt to do so.

CHAPTER TEN

A TASK-BASED TAXONOMY OF LIVING MOSAICS

THE LINNAEAN TAXONOMY ORDERS LIFE FORMS BY THEIR COMMON ANATOMICAL PROPERTIES; LIVING MOSAICS NEED TO BE ORDERED ALSO BY THEIR COMMON FUNCTIONAL PROPERTIES

The Linnaean Characterization of Living Mosaics by Their Anatomy

Naturally, people had always known that there are countless, vastly different looking organisms on Earth. However, it was up to the genius of Carl Linnaeus (1707 – 1778) to explore systematically what anatomical features the different looking ones had *in common*. He established that each of them shares at least one anatomical feature with one of the others. So, he could sort them into groups that shared one or more anatomical features. The result was the well-known Linnaean taxonomy of organisms that sorts the different kinds of organisms into a hierarchy of properties designated as kingdoms, phyla, classes, orders, families, genus and species.

These categories are complete and mutually exclusive. There is always a unique string of categories for every organism, while no organism can ever belong to two different strings of categories.

Most importantly, the categories are 'nested' like Russian dolls. For example, each kingdom contains several phyla, and each phylum contains several classes, and so forth.

In practice, it is sufficient to classify most organisms simply by the names of genus and species, because common knowledge supplies the rest. For example, everybody knows that tigers are animals, vertebrates, mammals, carnivores, and huge cats. Therefore, we never need to list the complete string of 'Animalia', 'Chordata', Mammalia', 'Carnivora', and 'Felidae' as their Kingdom, Phylum, Class, Order and Family. It suffices to call them 'tigers' or zoologically more accurate as 'Panthera tigris'.

It testifies to the genius of Linnaeus that he did all this one century before Darwin. Today we know that his anatomical criteria and their hierarchy presciently reflected the evolution of the species.

Obviously, Linnaeus' work is one of the main pillars of biology. It is solidly scientific, because all his anatomical criteria are objective. Yet, from the perspective of this book, which emphasizes the actions and interactions of semi-autonomous units, it disregards an important aspect: The organisms as living mosaics not only have common anatomical, i.e. structural features, but they have common functional features as well. Therefore, we will try to formulate a supplemental kind of taxonomy that evaluates the common actions of living mosaics and their modules.

The Usefulness of a Task-Based Taxonomy

Of course, it makes little practical sense applying a taxonomy of tasks to characterize every individual action of living mosaics. However, numerous species are known to carry out habitual actions that are genetically fixed and thus are species-specific.

There are numerous examples among arthropods, such as ants, wasps and spiders. To name a few, there is the characteristic fungus-cultivating actions of leaf-cutter ants, the stunning migratory behaviour of Monarch butterflies, the behaviour of various solitary wasps to capture caterpillars or roaches, paralyze them, lay eggs into them, and bury them as living food for their larvae, and the vastly different, ingenious ways of spiders to build traps for their prey. These and many other cases of the habitual behaviour of living mosaics indicate genetically fixed tasks that qualify for a task-based taxonomy.

In some cases, the addition of a taxonomic task-analysis may be superfluous, because the Linnaean anatomy-analysis is sufficient to characterize the species unambiguously. However, if organisms are not distinguished anatomically, but have different genetically fixed behavioural patterns, it may be necessary - and revealing - to apply a task-based taxonomy.

A particularly striking example is the fungus *candida albicans.* The individual cells are anatomically and genetically identical but aggregate into different shaped colonies (in our terminology, we would describe the colonies as living meta-mosaics) that exhibit different pathogenic, even lethal properties in humans [29]. The example showed for the first time that anatomically identical pathogens with the same Linnaean classification, nevertheless can exhibit different pathogenicities depending on their behaviour. In other words, cells can differentiate without changing their anatomical markers, and yet act as if they were different organisms.

Bacteria express a related phenomenon. Without changing their individual anatomy, i.e. without qualifying for a different Linnaean classification, a number of bacterial species, including myxobacteria, form very differently shaped and even moving 'meta-mosaics' that may be considered as multi-cellular organisms [30].

In these and other cases, a taxonomic analysis of their tasks seems quite justifiable. However, once we have accepted the usefulness of a task-based taxonomy for the above cases, we may have to accept it for a much larger number of examples, namely for all differentiated cells.

After all, the Linnaean classification of an organism applies to all of its cells. For instance, all cells of *mus musculus* are classified as *'mouse' cells*, in spite of their vastly different habitual tasks in the fully developed organism. The tissue culturists certainly apply such a taxonomic convention to characterize cells by the organism of origin, e.g. when they refer to *3T3 cells* as *mouse* fibroblasts, to *BHK21 cells as hamster cells*, or to *HeLa cells* as *human tumour cells*. The various differentiated cells may not mate, but a great many of them proliferate, nevertheless. Further complicating for an anatomy-based taxonomy is the fact that their anatomical features are quite variable, especially in tissue culture. Although in practice differentiated animal cells are further characterized as *epithelial, fibroblastic, lymphoid*, etc., the description leaves them under-determined in the sense of the Linnaean taxonomy. A supplemental task-based taxonomy that identifies their habitual migration behaviour, cell-cell contact behaviour, phagocytic behaviour, etc. may help create a more unified picture of them, than histology alone is able to provide.

Besides the above cases, where anatomically identical individuals of the same species have different, yet heritable behaviours, there is the opposite situation, too. It is well-known that the individuals of different species may express common, yet heritable behaviours of hunting, defending, mating, offspring rearing, food gathering, migration, navigation, etc. For example, insects and plants of very different species have often surprising behaviours in common. I am thinking of the light pulsing behaviours of fire flies, the initiation of galls by solitary wasps, the web-construction of spiders, but also the rather mysterious nutation movements of many plants, and the searching movements of vines. Focusing on behavioural commonalities may offer novel insights into the 're-surfacing of ancient traits' or 'cross-fertilization' during evolution. Obviously, such studies would go far beyond the scope of this book, and will not be discussed any further.

The Legitimate Use of Subjective Criteria

Naturally, we will use the Linnaean taxonomy of organisms as a template for taxonomy of tasks. Therefore, we would need to identify – in addition to the Linnaean criteria of anatomy - a number of *functional* categories that (a) are mutually exclusive, (b) are nested like Russian dolls, and (c) allow us to classify every task.

Unfortunately, such functional criteria are not as objective as anatomical ones, because the interpretation and identification of actions and functions are time and context dependent. They are often overlapping, ambiguous and always depend on the present level of the categorizer's technical understanding. In addition, they invoke considerations of 'purpose', which have long been banned from science. In short, they are clearly subjective criteria.

Still, subjective judgements are not necessarily worthless in science. Sometimes they are even inevitable. Consider their important roles in fields like medicine, forensics, law, history, or economy. True, the workers in these fields base their decisions also on objective data such as blood composition, magnetic resonance images, crime scene data, case collections, excavations, or economic statistics. Yet, in spite of all this objectivity, many of their momentous, sometimes life-and-death decisions draw ultimately on the experience, the consensus of experts, or simply on 'sound' human reasoning. As evident by disagreements among the experts and revisions in spite of the objectivity of the data, they are quite subjective, albeit indispensible instruments of decision-making.

Five Basic Elements of Tasks

What features of living mosaics should be the basis of their functional classification? The mentioned equivalence between tasks and living mosaics allows us to classify the functional properties of living mosaics by classifying their tasks. Therefore, we will draw on the everyday experience with the different kinds of tasks for our method of taxonomy. And, of course, we will call the categories by their English names because today's universal scientific language is no longer Latin as it was in Linnaeus' time.

As in the case of the Linnaean taxonomy, the suggested categories to classify tasks should be nested and mutually exclusive. However, right from the onset, we have to admit that they cannot be complete, because the number of different tasks is growing worldwide every day with new ideas and actions. At least, we can demand that the categories should be able to incorporate every new idea and activity.

The method of classifying tasks as proposed in this book tries to fulfil these requirements by simply formalizing the way we talk about tasks in common language. After all, a task is always 'something that is to be done', and no matter what the new ideas or activities are, we should always be able to talk about them in this format.

How do we usually describe tasks? Assume, we want Tom Sawyer to whitewash Aunt Polly's fence. We would begin by specifying the **performer** of the task by stating that 'Tom Sawyer's task is...'. Next we would identify the required **action** by a verb '...to whitewash ...'. Subsequently, we would add the **target** of the task, '..Aunt Polly's fence...' followed by the **means** '...with paint and brush...' and the **manner** '...on a Saturday morning'.

Occasionally, the performer only is 'implicit', as in 'The task is to fulfil the campaign promise of reforming the school system.' In other cases the performer is 'anonymous' as in 'The task is to volunteer for welcoming the attendees....'. If we include such implied names, the task is always able to specify its performer, even when the performer is not named explicitly.

Hence, borrowing from the format in which common language would describe a task, we may divide 'tasks into 5 categories

1. **Performer** ('Who or what carries out the task?')
2. **Action** ('What is done to perform the task?')
3. **Target** ('At which object is the action directed?')
4. **Means** ('What procedures, instruments, or strategies are applied?')
5. **Manner** ('What are the circumstances, under which they are applied?')

To be sure, these categories are not mutually exclusive. In other words, it is quite normal that a task may have e.g. a target *and* a means of action. On the contrary, every task is expected to contribute to all five categories. In other words, they form the blanks of a questionnaire. The taxonomy of a given task has to begin by filling in all of the blanks.

Also, let me repeat an earlier caveat, namely that the classification of the tasks of living mosaics does not imply that some superior being 'ordered' the execution of the task. In this kind of taxonomy, we merely classify what the mosaics do, not why they do it.

Outline of a Taxonomy of Tasks

The five elements belong, linguistically speaking, to the classes of nouns, verbs, adjectives, and adverbs. Therefore, we may consider borrowing their classification from the field of Linguistics, which has already worked out extensive systems of classification of all elements of common language. An example may be Beth Levine's taxonomy of verbs [12], which is widely recognized as a standard work.

In addition, we can tap into an even more recent resource, which resulted from today's growing need to render dictionaries and thesauruses machine-readable. It has prompted detailed and comprehensive taxonomy-like digitized schemes for all elements of language, and we may adapt either of these for our purposes.

However, neither Linguistics nor Computer Sciences focus primarily on biology, whereas this book does. Therefore, the following subdivisions of the categories of the elements, although guided by the work of linguists, differ from the linguistic sub-divisions in order to accommodate the practice of identifying tasks in biology.

The basic arrangements of the various categories are shown in figure 28-01a. It depicts how a 'task' is analyzed by a tree consisting of 5 levels of branching, one level for each of the 5 elements of definition 28-01. In addition, each of the five elements will use its own specific tree of mutually exclusive categories for the analysis. The various trees will be described and illustrated in Appendix E.

THE 'SEEDS' OF LIVING MOSAICS

CHAPTER ELEVEN

THE 'SEEDS' OF MOSAICS

THE SUCCESSFULLY FITTED TILES OF MOSAICS PROJECT THEIR SPECIFIC INNER LOGIC; NOTHING ENCAPSULATES THIS LOGIC BETTER THAN THEIR 'SEEDS'

*NOTE: One of the key concepts of this book is the 'seed' of a mosaic, i.e. a mosaic with empty spaces that can be completed to a full mosaic without any holes. More interesting are the seeds that we will call **key-seeds**, which only can be completed in one way. They contain a subset of tiles that determine uniquely the final solution. **Their assembly is a matter of necessity.***

*Explaining their peculiar properties will force the student to work through one or the other stretch of rather dry arguments. Nevertheless, it may be worth the student's efforts, because key-seeds offer a more transparent handle on the set of all solutions and, as will be shown later, they are **indispensable to the methods of replication, expression, and information storage of mosaics.***

The solution of a problem in physics requires deciding
(a) the universal laws of physics, which may determine the solution,
(b) the specific functions or fields, which are appropriate, and
(c) the specific boundary conditions, initial values, and constraints, which may apply.

The first two steps do not concern biology. In the first place, we know of no universal laws of biology. This is understandable. As mentioned before, biology is about the engineering details of the specific technologies of terrestrial life forms. In contrast to the laws of physics, other forms of life in the galaxy or beyond may not use the same technologies.

In addition, the second step is foreign to biology because, as we mentioned repeatedly, this book it is much more concerned with mosaics of discrete tiles.

In contrast, the third step concerning the boundary conditions and constraints of a problem is also important for biology. It may be even more important for biology than for physics. After all, without laws of biology driving evolution, the present day terrestrial life must be a product of the serendipitous boundary conditions and constraints imposed by life's history and our planet's history, composition, geology and orbital parameters.

If, indeed, we can interpret biological objects as tiles and mosaics that evolved from certain initial boundary conditions and constraints, we have to ask how to formulate the initial conditions of mosaics. We begin with a very pragmatic question.

How Many Initial Tiles Are Required to Specify a Particular Solution?

No matter how simple it sounds, the above question will turn out to be actually very complex. Let us look at the pentomino model to illustrate the difficulty we have to face.

In the case of the pentomino model-frame the shapes of the 12 pentomino tiles are the same for all sibling-solutions. Also the frame poses the same constraint for all of them. Why then are not all solutions the same? And if not, what could possibly distinguish between each of the 64 sibling-solutions and the astronomical number of non-solutions?

Let me repeat. These boundary condition and constraints certainly do not determine the success of the search for solutions, because **they are the same for all the different solutions and for all non-solutions**. Therefore, we have to look elsewhere. Obvious candidates for the determinant of success are the names, locations, and orientations of the first pentomino inside the frame. Similarly, we may try to link success to the first two, or three initial pentominoes to find out which initial condition determines whether a solution results, and which one it is? If there is such a group of initial tiles that determine the outcome, we will call it the 'seed'.

In principle, any group of initial tiles can be the seed of a solution. Unfortunately, their numbers are astronomically large and the vast majority of them will turn out to be the seeds of no solution at all. We need to look more specifically for seed groups that lead eventually to one or more particular solutions.

A possible approach would be to take the opposite route: Begin with a solution and strip it increasingly of its tiles, while considering at every step the remainder as a seed group for this solution. There are

$$N(s) = \binom{12}{s}$$ ways to remove s tiles from the 12 pentominoes of each

solution. As mentioned, we consider each such group as a seed group, which amounts to a total of

$$N = \binom{12}{1} + \binom{12}{2} + \ldots + \binom{12}{11} = 2^{11} = 2048$$ possible seed groups for

every single solution of the model-frame. As shown in Appendix C, there are 64 different solutions for the model-frame. Hence, the model-frame has maximal 64 · 2048 = 131,072 seeds. The actual number will be smaller, because not all these seeds will be different.

Obviously, not all of them will lead back to the solution where they came from. If the number of tiles is small enough, they are likely to be shared by other solutions, and will no longer be able to function as a seed group.

Therefore, we must ask, which of the group of tiles left by the stripping procedure will still be able to specify their solution? Or in more biological terms, what is the germinal configuration, from which a mature mosaic only can grow with necessity? Such seeds will be called 'key-seeds'.

def. 11-01
1. **A 'key-seeds** M̶ **of a solution mosaic M' is a seed, whose completion leads unambiguously to the solution mosaic M.**
2. If the seed configuration contains exactly 'm' tiles, it will be called the **'m-key-seed** ᵐM̶ **of a solution mosaic M'**.
3. The integer 'm' of an m-key-seed ᵐM̶ will be called the 'rank' of the key-seed.

The Large 'Spectrum' of the Initial Seed Tiles of Each Sibling-Solution

Assume we found a 5-key-seed similar to the group in Figure 11-01a. Given the large degree of freedom of assembly of general mosaics, usually there are thousands of ways of placing a next pentomino in the empty spaces. However, it is not true in this case!

The space marked 'x' in panel a only can be filled with a 'P'. After filling in the 'P', we arrive at the configuration of panel b.

Again, the new configuration also has a place (marked 'x') that only can be filled with a particular pentomino, namely the 'X'. If the 'X' was put in any other place, it' would create space fragments around itself that are smaller than 5 and, therefore, cannot be filled with any of the other pentominoes. Hence, we place the 'X' in its only possible place and arrive at panel c.

Continuing in this way, a chain of logical inevitabilities ensues: Only unused tiles may fill the marked spaces. Based on the inevitability of placing unused tiles into the configuration of panel (a), one can consider this configuration as a germinal set of 5 initial tiles that leads inevitably to the solution in panel h. In our terminology, it is a 5-key-seed of the solution in panel h.

However, it is not the only one. There are many more such key-seeds. For instance, after filling a 'P' into the marked space of panel b, the resulting tile configuration is obviously a 6-key-seed. However, since *all* the empty places only can be filled with one of the unused tiles, we could just as well have used as the first tile any other of the 12 − 5 = 7 unused ones.

In other words, every 5-key-seed gives rise to n = 12-5 = 7 different 6-key-seeds. Each of them, in turn, gives rise to n = 12-6 = 6 different 7-key-seeds, but since the order of filling unused tiles into the marked spaces does not matter, we can add the first and second tile in any order.

Thus we obtain a total of $n = (7 \times 6)/2 = \binom{7}{2} = 21$ different 7-key-seeds.

Likewise, we obtain $\binom{7}{3} = 35$ different 8-key-seeds, $\binom{7}{4} = 35$ different 9-key-seeds, $\binom{7}{5} = 21$ different 10-key-seeds, and so forth.

Including the original 5-key-seed, the grand total of key-seeds that is derived from the one 5-key-seed in panel a is

$$N(7) = \binom{7}{1} + \binom{7}{2} + \ldots + \binom{7}{7} = 2^7 = 128.$$

In more general terms, every m-key-seed gives rise to $2^{(12-m)}$ key-seeds. If there is more than one m-key-seed, i.e. a solution has N(k) m-key-seeds, it has

$$N = \sum_{1}^{12} N(\kappa)\, 2^{(12-\kappa)} \text{ key-seeds}.$$

Not all are different, but each of them specifies the solution unambiguously. Still, there are many different pathways from a key-seed to its mature mosaic.

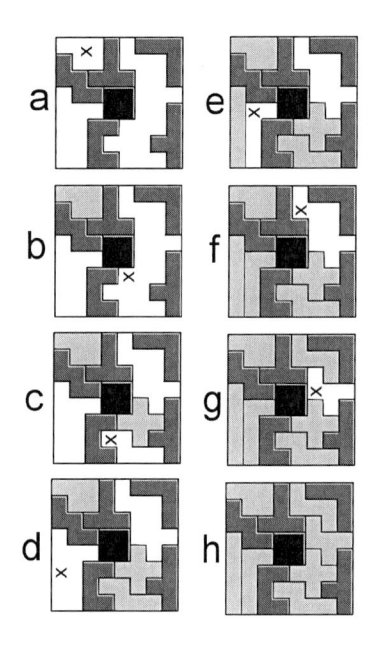

Fig. 11-01. Example of an incomplete mosaic (panel a) whose completion is dictated by necessity. The locations marked with an 'X' can be filled with only one of the missing tiles. The configuration of the 5 tiles in panel (a) is called a '5-key-seed'.

The Multiplicity of the Assembly Pathways (How Many Different Mosaics Can Develop From the Same Seed?)

(Please note, we are talking about simple 'seeds' not about the key-seeds of a solution; By definition, only one solution can be developed from a key-seed!)

As argued in detail in Appendix C, the fitting algorithm to find solutions for the assembly of a mosaic makes us stop and retrace our last steps if we are entering a search path that could not possibly lead to a solution. This strategy of so-called 'truncated searches' speeds up the search for solutions considerably.

Of course, this strategy only applies after we found out that all other remaining options must fail. It would be wonderful if the algorithm could tell us early on that we are approaching such a dead end. Unfortunately, no method, which is based only on the present state of the mosaic, can predict whether we are heading the 'right' or the 'wrong' way.

This uncertainty is due to the ambiguity of the possible outcomes. We demonstrate this by the following example of the successful assembly of a particular solution, based on a 'truncated search'. However, we will do much more than assemble this particular solution. After each step we will stop, and let the assembly program find *all* solutions that begin with the presently assembled tiles as their seeds. In this way, we find out at every stage, how many different solutions are still possible that share the same group of initial tiles.

Although it is a rather tedious one, this procedure also qualifies as a method of finding a key-seed of the solution: As we keep applying it, we cannot help but discover that sometimes the number of possible alternate solutions shrinks to 1. In that case, we have obviously found a key-seed.

The step-by-step assembly of every solution, such as the one shown in panel of figure 11-02 starts with the empty frame. At this stage we know that there will be exactly 64 solutions.

However, after placing in the left upper corner an 'F'-pentomino as the first pentomino in a particular orientation (Figure 11-02 a), we find that only 8 solutions are left among the total of 64 solutions. They are all the solutions that have the 'F' in the exact same position and orientation. In other words, the placement of the first pentomino as the initial condition restricts the number of possible fitting solutions, but it does not determine them uniquely.

152   Chapter Eleven

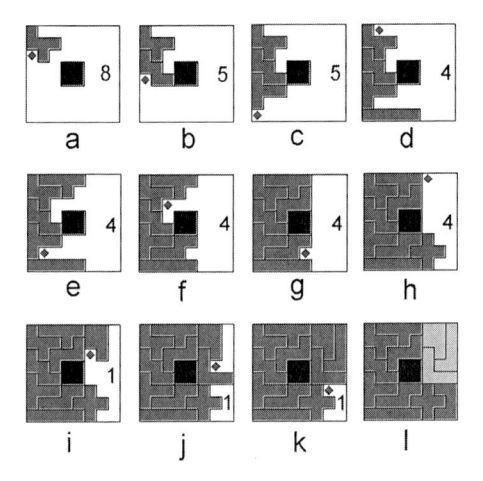

Fig. 11-02. Ambiguity of the continuation of the assembly steps of a model-frame solution (panels a-l; the diamond indicates the location of the next added tile). The numbers indicate at each stage how many different sibling-solutions share the already assembled tiles. Hence, the assembly could continue in as many different pathways, and lead successfully to a solution.

Next, we fit a 'W' as the second pentomino. We can still find 5 different solutions with these two tiles in their exact orientation in the left upper corner (figure 11-02 b). Continuing this way and fitting another 3rd, 4th, 5th, and 6th pentomino, there still remain 5, 4, 4, and 4 possible solutions (panels c – f). Only the fitting of an 'X' and a 'P' as the 7th and 8th pentomino (panel i) finally reduces the ambiguity to one.

The slow reduction of the ambiguity of possible solutions is due to a mirror symmetrical group of tiles (panel l, coloured group). Like all other rectangular shapes, it can exist in 4 different orientations, each corresponding to a different solution. Only the placement of one of the group's tiles (the 'P' in panel i) finally determines the orientation of the group and, thus, decides which of the four the final solution will be.

'Key-Seeds' and their 'Complements' for Every Sibling-Solution

If one wants to be sure to find all solutions, the method of 'growing the solution from a corner' is the systematic method of choice. On the other hand, in this chapter about key-seeds we want to identify the particular tiles that 'nail down' the final solutions early on during the assembly process. As we have seen in figure 11-02, assembling a nucleus of tiles around a single starting tile is too slow a method to restrict the possible outcomes, because it adds tiles to the expanding periphery of the nucleus regardless of their possible power to exclude alternate solutions.

A better method to find seed tiles should spread the tiles strategically all around. Only when they are sufficiently distant from each other, the initial tiles become most effective in preventing alternative placements and orientations of the remaining ones and, thus, qualify as tiles of key-seeds.

Understanding mosaics requires understanding the logical relationship between their key-seeds on one hand, and the properties of all the tiles, the frame, and the solution on the other. Hence, we need to study the key-seeds in more detail. They should help us recognize and predict the growth and properties of mosaics, be they pentominoes or biological objects.

How should one space the initial tiles inside the frame? Obviously, there are astronomically large numbers of possibilities to do it randomly. Hence, random placement of tiles would certainly not be an efficient approach.

In the case of the pentomino mosaics there is, indeed, a much better way to find key-seeds. In general, the method will work for other mosaics, provided their solutions are known.

Since all solutions of the model-frame are known, we can *reconstruct* their key-seeds by dismantling each solution 'S', one tile at a time, while leaving the remaining tiles spread strategically apart. At every step, we test whether we can still reverse the process and rebuild this and this solution 'S' alone. Eventually, the procedure should lead us to a set of tiles that still specifies uniquely the solution 'S', whereas any smaller set would not. Then we have found a key-seed of 'S'.

The approach assumes that we can define what 'rebuilding' means independent of any particular assembly algorithm. One can derive this definition from the very concept of fitting itself by the following line of reasoning.

Removing a tile from a mosaic must leave a hole. *Regardless of the particular assembly method, it must be able to decide whether a tile fits a given hole exactly.* As long as we dismantle the solutions while only leaving holes, which only match one specific tile, then the rebuilding of the solution does not depend of the assembly method.

This approach should also work for biological objects, provided they are, indeed, mosaics of identifiable 'tiles'. We also know a huge number of the biological 'solutions' already, because the perfectly finished mosaics of organelles, cells, organisms, and ecologies are all around us. Of course, we need to identify their physical and/or functional 'tiles'. However, here we can draw on the extensive knowledge of the components of biological objects, as biologists over several centuries have identified them.

In short, we need to learn to dismantle finished mosaics by creating holes inside them that only can be closed by specific tiles. In the following, we will develop such a method for the pentomino model. As pointed out above, it will have the additional advantage of producing sets of initial tiles **independent of the fitting protocol**.

How does the procedure appear in practice? Assume, we take a fitted solution and remove arbitrarily one of the tiles. Obviously, only the very same tile that we had removed can close the hole, because all tiles have unique shapes.

For a better comparison with the previous method, let us place the first hole in the upper left corner (figure 11-03a). Subsequently, we go around the perimeter removing all tiles that only touch each other by their corners, or not at all (figure 11-03 a-e). Consequently, if we would start with the tile configuration of panel e, there would be exactly one solution for the remainder of the mosaic possible, as each hole demands a unique pentomino that is still available.

Next, we remove also tiles that are adjacent to the previous holes, provided the resulting larger holes still only match the very same pentominoes whose removal had created them. For example, the hole marked with an asterix sign in panel f only can be closed with an 'I' and an 'F', or an 'L' and a 'P'. However, the second option is not possible, because the 'P' is already used.

We may even remove this very 'P' and create a larger hole (figure 11-03 g, marked **). Now we cannot use the argument any more the 'P' is already used. Yet, we still cannot put an 'L' and a 'P' into the place of the asterix sign (panel g), because the 'L' is indispensable for closing the hole marked with an arrow.

Hence, the set of initial tiles shown in panel h, is a correct 5-key-seed of the solution in panel a.

The 'Seeds' of Mosaics

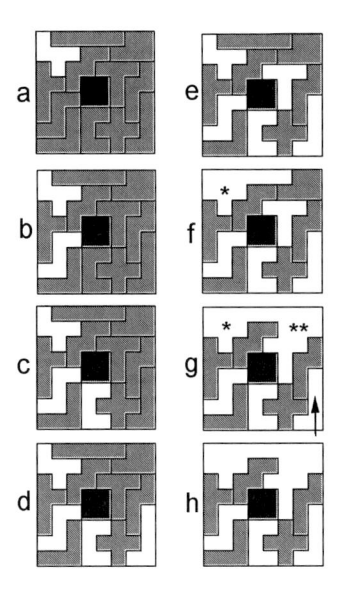

Fig. 11-03. Construction of a 5-key-seed of the solution in panel (a) by removing tiles that can only be replaced by their exact own shape.

The above approach has a major disadvantage. Its success depends too much on the visual inspection of the particular starting mosaic. A more mechanical routine to reduce a mosaic to key-seeds uses the following steps. This method ends up removing about half of the tiles. Other methods may be able to remove even more.

The assembly of a solution starting with the key-seeds is entirely independent of the fitting-protocol, as long as this protocol is able to find the only tile that fits a hole with the exact same shape. Provided, all tiles are different and available, there only is one way to complete the tiles of a key-seed towards a finished mosaic. Invariably, the resulting mosaic is a unique solution (figure 11-04)

156   Chapter Eleven

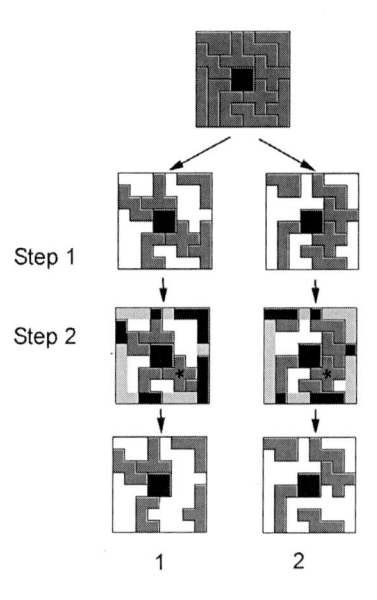

Fig. 11-04. Algorithm to construct key-seeds from given solutions by removing radial layer by radial layer the tiles that do not touch each other (detailed explanation see text).

Step 1. Search along the outer frame and remove every other tile that touches the frame. Obviously, only the very tiles that were removed can close the holes left behind. Hence, the remaining tiles specify the solution. However, there is ambiguity because there are 2 alternate ways of doing it.

Step 2. Define a first inner, concentric frame by shrinking one tile unit away from the frame.

Step 3. Do Step 1 on the first inner frame, by alternate removal of every tile that the inner frame encircles completely, while touching it (only the 'X' qualifies and is marked with an asterix). In the case of other solutions and frames, there could be more than one qualifying piece. Consequently step 3 could yield multiple versions..

There is nothing new about the way in which a key-seed turns into its corresponding mosaic. It follows directly from our main premise that there is a method to find solutions for the mosaic. Consequently, after placing k tiles, the method is always able to place the $(k + 1)^{st}$ tile, provided there is room for it. This must be true, no matter where the previous k tiles are located. In case they have been placed in the k locations of a k-key-seed, there is always room for the (k + 1)st tile, and the method will find and fill it.

Using the above method we must obtain at least 2 different key-seeds, depending on whether we start with the first tile, or with its next neighbour along the rim of the frame. By virtue of their construction, the described **algorithm yields two key-seeds** that can always be made into complementary key-seeds.

The two partial mosaics are complements of each other in the following sense.

def. 11-02

Given a key-seed $^{n}\mathbf{M}$ of the solution mosaic M that contains N tiles, then the '**complement**' ($^{n}\mathbf{M}$)* consists of the same frame and the N-n tiles of M, that are missing in $^{n}\mathbf{M}$ in the exact same order as in M (for an illustration see figure 21-01a). This consequence is very important for a later application (see chapter 14).

Therefore, we arrive at the following definition and corollary.

def. 11-02

Two key-seeds are complementary to each other, if
(a) Both are key-seeds of the same mosaic.
(b) They have no tiles in common.
(c) Together they fill the mosaic precisely.

equ. 11-01

Hence, the 2 key-seeds $^{n}\mathbf{M}_1$ and $^{m}\mathbf{M}_2$ are complementary for the mosaic solution **M** iff
(a) $^{n}\mathbf{M}_1$ and $^{m}\mathbf{M}_2$ are both key-seeds of M.
(b) $^{n}\mathbf{M}_1 \cap {}^{m}\mathbf{M}_2 = \mathbf{M}$.
(c) $^{n}\mathbf{M}_1 \cup {}^{m}\mathbf{M}_2 = \varnothing$, where $\varnothing$ symbolizes the empty set.

The student may be wondering about the need to add the requirements (a) in definition 11-02 and in equation 11-02. Why emphasize that both partial mosaics are actually key-seeds? The answer is given in figure 11-05a, which demonstrates that the complement of a key-seed is not necessarily a key-seed itself.

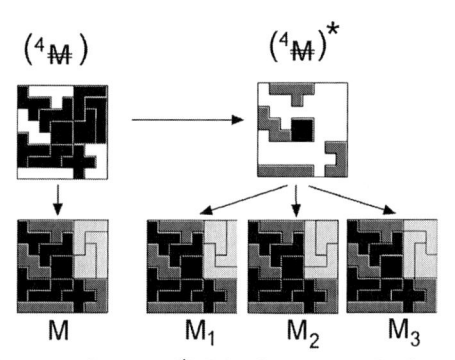

Fig. 11-05a. The complement (^4M)* of a key-seed of a solution M may not be a key-seed, if M contains symmetrical groups. The complement (^4M)* of the above example leads to 4 solutions, which differ by the orientation of the symmetrical group.

(In thinking through the relationships between key-seeds and their complements, the student should also keep in mind, that one cannot construct the complement of a key-seed, unless the completed solution is known beforehand.)

The algorithm removes single tiles in an alternating sequence starting with a certain tile, or with its next neighbour. Therefore, the tiles of the two resulting key-seeds are always mutually exclusive.

As to the number of tiles in either key-seed, there are two possibilities.

1. Both key-seeds contain exactly half of the total tiles of the mosaic. In this case they complement each other to fit and fill the mosaic. (q.e.d.)

2. One or both key-seeds contain less than half of the tiles of the mosaic. Hence, some tiles of the mosaic are missing from both key-seeds. One can always place them back in the same position and orientation as in the mosaic. Adding back a tile to an n-key-seed simply makes it an (n+1)-key-seed of the same mosaic. Hence, after adding back all the missing tiles, the two key-seeds are still mutually exclusive, but now their tiles make up the entire mosaic. (q.e.d.)

Hence, in either case, the resulting two key-seeds are complementary to each other.

If the sizes 'm' and 'n' of the 2 complimentary key-seeds $^mM_1$ and $^nM_2$ add up to the size N = m + n of the mosaic, then no tiles are common to both. Together they fill the frame precisely to constitute the mosaic.

In this case, the 2 complimentary key-seeds contact each other along a specialized inner surface, which will be called the **'inner seam'**. It consists of all the tile-to-tile contacts, where a tile of the first key-seed touches a tile from the second. This case is important for chapter 14 on mosaic replication, and will be illustrated in figure 14-01.

Given a pair of complementary key-seeds, either of them can be completed to the same solution. Naturally, each stage along the way creates a key-seed, albeit of a lower rank, i.e. a 5-key-seed becomes a 4-key-seed, the 4-key-seed a 3-key-seed and so on. The resulting key-seeds can be arranged in a circle of stepwise changing ranks as illustrated in Figure 11-05.

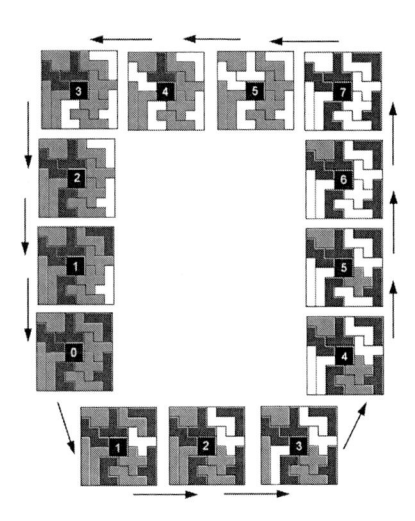

Fig. 11-05b. Circle of complementary key-seeds. White numbers inside the centre indicate the rank of the key-seed (see text).

At this stage, the notion of a mosaic consisting of two complementary key-seeds is no more than an abstract concept. It does not claim to reflect the reality of mosaics. In chapter 14 we will discuss the idea of physically splitting each mosaic into the 2 complementary key-seeds as the first stage of their replication. It will raise a number of important questions about the mechanisms involved. For example, we will need to explain from where the two frames of the two key-seeds may come. After all, the original mosaic, at least so far, only has one frame. I defer this question and other similar ones to chapter 14.

160   Chapter Eleven

What is the smallest number of tiles in a key-seed for a given solution provided all tiles are different? It certainly is not 1 or 2. Such a small number of tiles always lead to multiple solutions or to none at all. In the example of figure 11-06 h, we found already a case of 5. Hence, the question is actually, whether there is a 3-key-seed for the fitting of pentominoes into the model-frame.

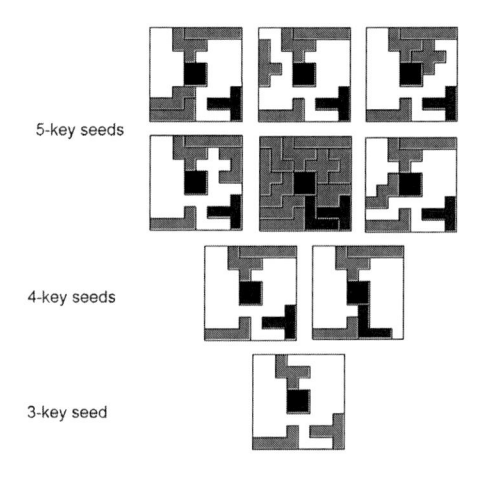

Fig. 11-06. Examples of key-seed with less than 6 tiles. Not every solution has 3-key-seeds, but many exist.

All this begs the questions whether a solution with a 5-key-seed can have more than one, but also whether there are 4-key-seeds, or key-seeds with even fewer tiles.

As shown in figure 11-06, there is more than one 5-key-seed. The figure presents a solution (centre) and 5 of the 5-key-seeds. The student is welcome to verify that the shown germinal 5-sets only have this solution.

The figure also answers the second question. At the bottom, it shows two 4-key-seeds and one 3-key-seed. Hence, smaller sets of key-seeds are possible.

As the number of key initial tiles decreases from 6 to 3, they lead to a number of conclusions about the solutions of the model-frame.
1. While every solution has a 6-key-seed, not every solution has a 4-key or 3-key-seed.
2. With decreasing number of initial tiles, the logic by which a key-seed leads to its solution is changing its character. In the case of 6-key-

seeds the placement of every next tile is unambiguous and automatic. They are truly **seeds**.

In contrast, in the cases of 4-key and 3-key-seeds many positions for the 'next' tile seem legitimate, at least early on, although all of them will lead to failure at a later stage of the assembly. The choice of the 'next' tile is no longer automatic. One has to try several others before one can identify the only viable solution. In other words, the initial tiles have acquired the character of **constraints** of the free assembly process, although they will turn out to be stringent, and only permit one solution.

In summary, initial conditions exist that are able to uniquely specify a solution independent of the fitting protocol. If the solution is known, they can be constructed as key-seeds.

At this stage it is already clear that the possible ambiguities of key-seeds only can be reduced to an acceptable practical level by determining the minimal ranking key-seeds. *These seeds express in the clearest way, how every tile of a solution entail every other, so that the completion of the mosaic follows a path of necessity.*

Of course, the above rationale requires that all the components of the future mosaic are available and ready for incorporation. If some tiles were not yet finished, they would force the partly assembled mosaics to halt their progress and 'wait' until the required components have developed or evolved.

The Number of Different 'Key-Seeds' of Each Solution How Many Starting Groups Develop to the Same Mosaic With Necessity?

How to identify key-seeds.

Earlier, when we discussed the huge failure rate of attempts to find solutions, it seemed little credible that there are sets of tiles, which lead to a mosaic *with necessity.* Yet, the key-seed configurations do just that. And they do it even independent of the assembly techniques.

Perhaps even more surprising was the finding that there are large numbers of such seeds. How many key-seeds does each mosaic have?

Assume you want to find all 4-key-seeds of (say) mosaic #2. In order to understand the procedure, imagine the workings of a slot machine, except in this case we are looking for the ***non-matches*** of the wheels.

The first wheel (Figure 11-07, left column) lists all combinations of length 4 of the 12 pentominoes. On the second wheel each of the combinations is highlighted in mosaic #2. Since mosaic #2, like all solutions, contains *all* pentominoes, it must contain these four. As an example, figure 11-07 uses the combination 'FIPV'.

The third wheel lists all mosaics *except* #2. As we spin the third wheel, it compares the 'FIPV' group of #2 with the positions and orientations of the 'F', 'I', 'P', and 'V' pentominoes in every other sibling mosaic.

If the wheel comes to its end without finding a match, there is no other solution, which has its 'F', 'I', 'P', and 'V' in the same position and orientation as mosaic #2. On the other hand, starting an assembly process with this particular 'FIPV' group must lead to solution #2, but to no other. Hence, **'FIPV' is a 4-key-seed of #2.**

The upper limit of the number of key-seeds.

How many key-seeds does each mosaic have? At this point we are no longer only looking at the solutions of the model-frame, but consider any arbitrary mosaic made of the 12 pentominoes. Let us begin to estimate it with large seed sizes.

Consider a specific solution #Ξ and count the number of its 11-key-seeds. It sounds complicated, but it actually is quite simple: It means nothing more than counting how many different ways there are of taking exactly one tile out of the mosaic 'Ξ'.

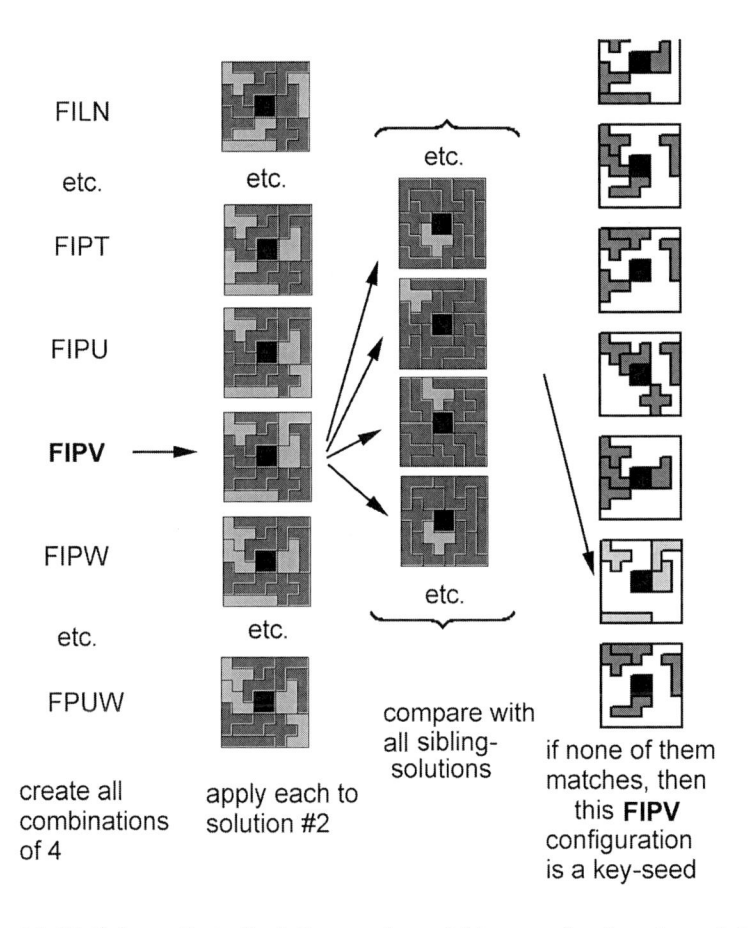

Fig. 11-07. Schematic to find the number of 4-key-seeds of each model frame solutions (explanation see text).

Obviously, there are exactly 12 different ways of removing a tile from 'Ξ'. Each set of remaining tiles represents an 11-key-seed for the following reason. They only allow one way of completing the mosaic by filling the hole. In addition, the resulting mosaic is unambiguously the one called 'Ξ'. The number of different 11-key-seeds N(11) can also be written as N(11) = $\binom{12}{11}$ =12.

The situation is no more complicated if we are to place 10 seed tiles. It is equivalent to taking 2 tiles out of the mosaic. Obviously, there are exactly N(10) = 12·11 different ways, i.e. the number of 10-key-seeds is

$$N(10) = \binom{12}{10} = 66 .$$

Again the remaining mosaics with the 2 holes are generally true key-seeds, because they only can be filled in one way. The only exceptions occur when the 2 pieces happen to be neighbours and present a symmetrical group.

Continuing in this manner we see immediately it follows for all key-seeds of size k (k = 1,2,...,11)

equ. 11-03

$$N(k) \leq \binom{12}{k}$$

Hence, the binomial coefficient presents an upper limit of the number of key-seeds of size k.

Based on the above rationale, we can be sure that the 'equal'-sign is valid as long as k is large enough. However, the situation changes when the number k of seed tiles becomes small.

As mentioned above, in order to construct a k-key-seed of a given solution, we have to take 12-k tiles out of the mosaic. This means that the sizes of the holes left behind in the frame are getting the larger the smaller k is. Eventually, the holes are large enough to permit multiple ways of placing the missing tiles back. Therefore, we expect the actual number of k-key-seeds N(k) becomes considerable smaller than the binomial coefficient $\binom{12}{k}$ in cases where k is small.

In the case of the model-frame, we know the exact number of all k-key-seeds of solution #2 for each size k. Figure 11-08 shows the comparison of this number with the binomial coefficient.

As expected, the number of k-key-seeds is considerably smaller than the binomial coefficient as long as k is smaller than about half of the number of tiles.

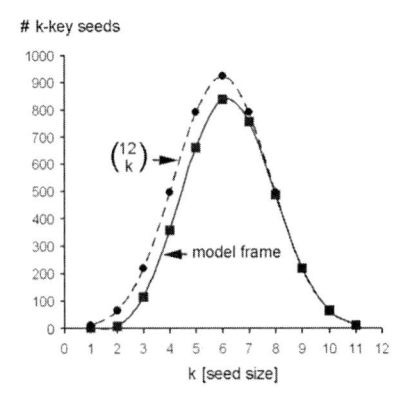

Fig. 11-08. The number of different k-key-seeds of every single model frame solution (solid line) and the binomial coefficient as their theoretical upper limit (dashed line).

We applied these considerations to a number of different frames. In addition to counting the number of k-key-seeds of the model-frame, we tested 3 other rectangular frames with an area of 60 that had different length-to-width ratios (table 11-01). In all cases, we counted the number of various sized key-seeds of a single solution.

The result confirms that the number of key-seeds approaches the binomial coefficient with increasing key size, even if the frames differ from our model-frame. However, the actual deviations from the binomial coefficient depend on the shape of the frame.

The Variability of the Number of 'Key-Seeds'

The number of key-seeds is not the same for all sibling mosaics. For example, plotting the numbers of all 6-key-seeds for siblings yields the somewhat 'noisy' curve shown in figure 11-09. The abscissa in this graph is an arbitrary index number for each solution in the particular order, in which the assembly algorithm had produced them. We could have plotted the numbers of 6-key-seeds also in any other order.

The figure shows the sizable variability of these numbers. Still, the numbers, as predicted by equation 11-03, run up to, but never exceed their upper limit $\binom{12}{6} = 924$.

seed size k	6x10 (2339)	5x12 (1010)	4x15 (368)	model-frame (64)
11	100	100	100	100
10	100	100	100	100
9	100	100	100	100
8	100	100	100	98
7	100	99	99	95
6	100	97	98	90
5	99	90	92	83
4	93	75	76	72
3	64	40	45	55
2	0	0	12	30
1	0	0	0	0

Table 11-01. The percentage of actual key-seeds compared to their upper limit (= binomial coefficient) for different frames (in bracket: number of solutions for the particular frame)

Not only plotting the numbers of 6-key-seeds in this way, but also the number spectrum of 2-, 3-, 4,-, 5-, and 6-key-seeds (figure 11-10) for all sibling mosaics shows two major results.

(a) Quite similar to the curve of the 6-key-seeds, each set of k-key-seeds runs up to but not over its upper limit $\binom{12}{k}$.

(b) Somewhat more surprising, **the curves run essentially in parallel,** which demonstrates that their 'noisy' character is not the result of randomness.

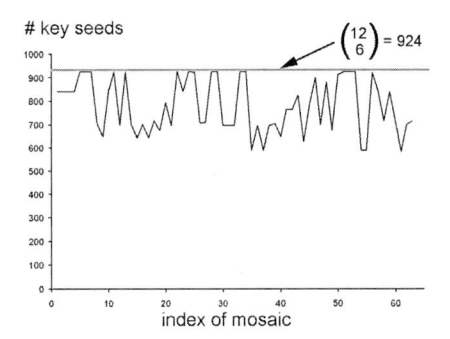

Fig. 11-09. Number of 6-key-seeds of each of the 64 sibling-solutions for the model frame (abscissa: index number of the various siblings; horizontal line marks the binomial coefficient as the upper limit).

Why do the curves run parallel? Earlier we had already mentioned an explanation for this kind of relatedness: Every k-key-seed becomes a (k+1)-key-seed by filling in any one of the missing tiles. After all, if any arbitrary sequence of tiles that are added legitimately to a k-seed leads inevitably to the same solution, then the very addition of one tile leaves the inevitability intact, but also produces a (k+1)-seed. Hence, the (k+1)-seed is a key-seed.

Obviously, one can do this to every k-key-seed, although there is more than one way of doing it: If there are a total of N tiles, then one can add N-k different tiles to a k-key-seed. And each time one obtains a (k+1)-key-seed.

Here we have to caution the student. It does not mean that the number of (k+1)-key-seeds are necessarily larger than the number of k-key-seeds. It is quite possible that the number decreases because the addition of single tiles may generate many of the same tile assemblies. Figure 11-11 shows an example for the pentomino model. Two different 5-key-seeds become the *same* 6-key-seed after adding an 'X' to one and a 'Y' to the other.

Nevertheless, judging by the behaviour of the binomial coefficient $\binom{N}{k}$, the addition of tiles will increase the number N(k) of k-key-seeds if k<k/2 and decrease when k>N/2.

168   Chapter Eleven

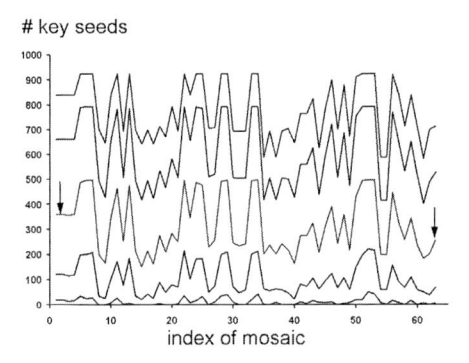

Fig. 11-10. The number of k-key-seeds for different values of 'k' (k = 2, 3, 4, 5, 6) of each of the 64 sibling-solutions of the model frame (abscissa: index number of the various siblings; arrows marks the 4-key-seed curve).

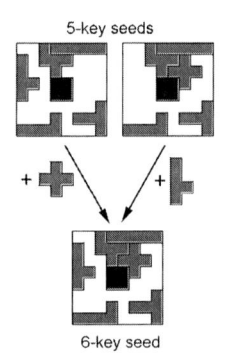

Fig. 11-11. Illustration for the reason that the number of (n+1)-key-seeds can be smaller than the number of n-key-seeds (see text).

Let us now take one of the curves in figure 11-10, e.g. the curve of the 4-key-seeds. By adding to each of the seeds one of the missing tiles, we obtain proportionally more 5-key-seeds if the solution had many 4-key-seeds in the first place. Likewise, we obtain proportionally fewer 5-key-seeds if there were fewer 4-key-seeds in the first place. In other words, we can expect that the 5-key-seeds and 4-key-seeds rise and fall together.

Hence, the changing numbers of key-seeds provide a characteristic expression of the sibling mosaics. They reflect indirectly the specific ways in which the frames and tiles interact in these mosaics.

Seeds with Symmetrical Tile Groups

The finding that some mosaics have fewer key-seeds than others is caused in part by the existence of symmetrical groups inside the mosaic. Consider the seed group 'IYUW' as illustrated in figure 11-12. It can never be a 4-key-seed, because its final mosaic contains a rectangular group composed of an 'L', a 'P', and a 'V' that can exist in 4 different orientations (labelled in light shades in figure 11-12).

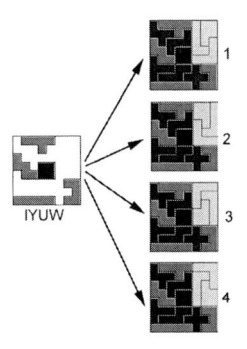

Fig. 11-12. The need to specify at least one tile of a symmetrical group in every key-seed. Otherwise the mosaic fragment (e.g. IYUW) would lead to multiple solutions, and not represent a key-seed.

Any attempt to complete the particular seed tiles must lead to ambiguity as they contain no information as to which of the 4 different solutions (labelled 1 to 4 in figure 11-12) should be assembled. Therefore, the symmetrical group eliminates the depicted 'IYUW'-combination as a potential 4-key-seed.

It is obvious from the example that every symmetrical group of tiles will do the same: If the components of a mosaic are free to rearrange their mutual interactions, each configuration will reduce the number of seeds, which are able to regenerate the mosaic with necessity.

There are a great many symmetrical configurations. As depicted in figure 11-13, symmetrical groups may contain as many as half the number of the tiles (panels c-f), and even contain smaller symmetrical groups inside themselves (panel e) or between each other (panel f).

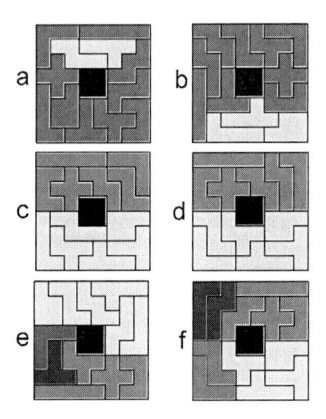

Fig. 11-13. Examples of typical symmetrical groups that occur among the model frame solutions.

In these and similar cases, *at least one of the member tiles of a symmetrical group has to be included in every effective seed group.* Otherwise, the seed group must lead to the ambiguity of as many different solutions, as there are symmetries of their symmetrical group.

There is one kind of exception, though. If the symmetrical group contains half the tiles, but the solution contains no other symmetrical group (panels c, d), then it is impossible for a seed group to avoid 'nailing down' the solution. Even its very first tile has no alternative but to decide where and in which orientation each half-group is.

Seeds in the Presence of Multiple Copies of Tiles

The construction of the seeds of mosaics rested on the argument, that only one specific tile is able to close a particular hole in a mosaic. Obviously, this only is true if all tiles are supposed to have different shapes. If we drop this condition and allow multiple copies of some of the tiles, there is no longer a chain of inevitability for the assembly process that connects a seed to its solution, unless all multiple copies are included in the seed. Hence, *the number of seeds of such solutions is necessarily larger than the seeds for single copy solutions.*

The presence of multiple copies of tiles would not necessarily affect the function of the assembly algorithms. They would still work, even if some of the tiles had the same shapes. All, which the algorithms need, are different **names** for the tiles such as 'F', 'P', or 'X'. Once the names are different, unique search-codes can be formulated such as 'T1 X0 U0 I0 F1 N6 L2 W2 Y4 Z2 V2 P7' and allow the algorithms to test whether the tiles in this code will fit, **regardless of their shape**.

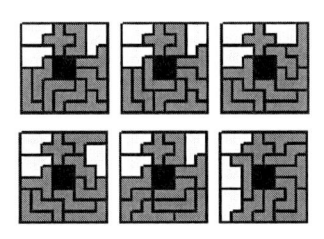

Fig. 11-14. Assembly of model-frame solutions with multiple copies of a specific tile (e.g. 'P' –pentomino; marked bright). For the modification of the assembly algorithm, see text.

For instance, one can assemble pentomino solutions with (say) 2 extra copies of the 'P', while they are missing an 'F' or 'I'. Six examples are shown in figure 11-14. In this case, the names of the 'F' and 'I' appeared quite normally in the search-codes. Yet, all the shape data of the 'F' and 'I' were replaced with appropriately rotated shapes of the 'P'. Hence, whenever, the algorithm tested the fitting of an 'F' or 'I' in a search-code, it actually was 'tricked' into testing whether one of the orientations of the 'P' would fit in their place. If it did, the algorithm had no choice but to place another copy of the 'P' there, and move on to the next test. The actual shape of the tile did not matter.

Examples of Key-Seeds in Biology

The principle idea of key-seeds is the splitting of a mosaic into two complementary halves, which the mosaic's normal assembly mechanism is able to complete to full mosaics. Living mosaics apply this principle in multiple variations, predominantly for replication (see chapter 14). *After completing each complementary half, the initial mosaic has obviously duplicated itself.*

The prime example is DNA replication, where the double helix splits into the two complementary strands and subsequently re-creates two double helices by assembling two matching new strands. Meiosis presents another example, as it creates complementary (haploid) sets of alleles that the process of fertilization will later on grow to the size of the original (diploid) sets.

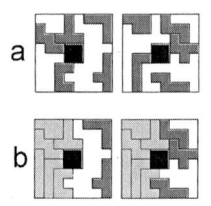

Fig. 11-15. Total and partial complementarity of key-seeds. (a) two totally complementary key-seeds. They share no tile. (b) partially complementary key-seeds: They share the P, W, I, L, T, and U pentominoes.

The complementary key-seeds of the pentomino example did not share any tiles; their complementarity was **total**. In reality, however, the two key-seeds may share several tiles, and their complementarity may only be **partial**. They are complementary only with respect to certain groups of tiles.

In other words, the two key-seeds are quite similar, although each has some tiles that the other is missing. An example may be the division of stem cells. Both sister cells will mature into fully functional cells, complete with membranes, nucleus, mitochondria and cytoskeleton. Yet, they are not the same. One sister returns to the pathway of a stem cell, while the other initiates differentiation.

Pentomino key-seeds may also present partial complements. Figure 11-15 shows an example.

The Variation of Living Mosaics

Chapter Twelve

Re-Direction of the Assembly

One Way to Turn a Fitted Mosaic Into Another Is to Change One or More Steps During Their Assembly; Key-Seeds Need Not Be Explicitly Involved; the Crucial Question Is, Which Tiles May Be Changed Without Risking Failure of the Assembly?

Chapter 11 on mosaic seeds showed how a relatively small number of initial tiles may decide the success of the assembly process. If it starts with such a seed, the path to the eventual solution mosaic is determined by necessity.

Yet, perturbations can change that. The natural target for this kind of change is the assembly process for the following reasons.

In chapter 4 on self-stabilization we pointed out, that completed mosaics may be quite rigid objects, because they leave no room for change. Barring a traumatic event to create room for new and/or different tiles, weaker perturbations only can be effective during the assembly process, i.e. during development, where the incomplete mosaic still has 'holes' that permit deviations.

Naturally, in the vast majority of cases, the perturbation will cause the assembly process to fail. After all, 'blinded' assembly processes should practically all lead back to the 'ocean of non-solutions'. However, in very rare cases, they may redirect the assembly towards a new and very different solution.

Indeed, mutants, variants, or phenocopies occur spontaneously without 'warning'. Similarly, such quantum leaps to new configurations during evolution may explain the absence of fossil intermediates. Stephen J.

Gould's concept of 'punctuated equilibria' describes this behaviour of mosaics.

Let me use the pentomino model-frame in order to illustrate the abrupt, 'punctuated' appearance of new solutions after a small change occurred during the assembly process (figure 12-01). *Deliberately, we are not starting with a key-seed. The next chapter will discuss the case involving seeds.*

Instead, let the process start with the five initial tiles in the model-frame (label a). Next, the assembly process places a 'Y' in the particular orientation, which is shown in label b. Subsequently, the normal assembly continues (label c), and finally arrives at its solution (label d).

Assume two different 'mutations' occur at the second stage (label b): The assembly process chooses an 'N' (label e), or a 'V' (label j) instead of the 'Y'.

In the first case, one may add 4 more tiles (label f). The open space permits the completion of a new solution by adding the remaining 'P' and 'Y'. However, the right hand frame could accommodate either of them (white colour in labels g and h). Because of the original 'mutation' of replacing the 'Y' (label b) with an 'N', we end up with two different 'viable' variants. If this solution was the representation of a genome, one may interpret the two solutions (labels g and h) as 'transpositions'.

The other 'mutation' (label j) that resulted from the replacement of the 'Y' with a 'V' suffers a quite different fate. Immediately afterwards, a 'P can be placed on either side adjacent to the starting field of tiles (dark colour in labels k and l). The completion of the configuration labelled k yields yet another 'viable' variant (label m). In contrast, there is no solution for the configuration with the label l. It cannot be completed. One may consider it as a 'lethal' mutant.

All these changes are necessarily discontinuous. They would appear spontaneous and abrupt, if they were occurring in reality. Of course, our present pentomino model does not yet include a mechanism that one might call a 'pentomino-genetics', and that could explain how the various described choices of tiles were made. Therefore, using terms like 'mutants', 'variants', or 'phenocopies' is a stretch. Nevertheless, the inclusion of such a genetic-like mechanism to the model would not alter the abruptness of the changes. They are inherent in the nature of mosaics.

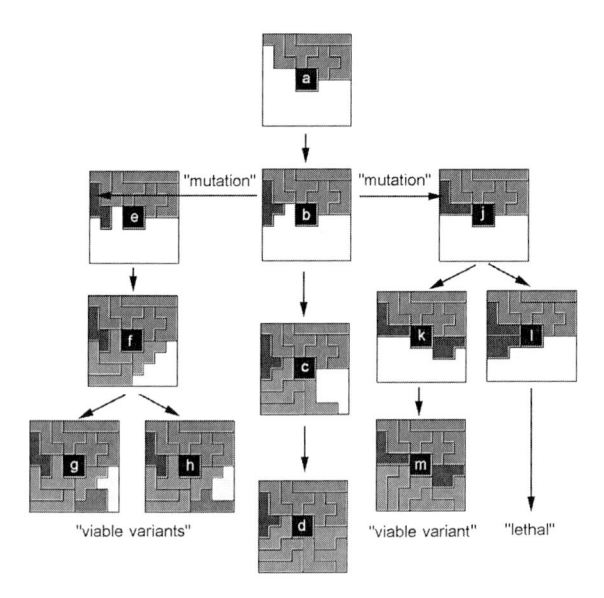

Fig. 12-01. Pentomino model of the spontaneous appearance during the assembly process of 'mutations' that lead to 'viable' or 'lethal variants' (see text).

Chapter Thirteen

'Mutation' of the Seeds

Altering Key-Seeds Explicitly Is Another Way of Turning a Fitted Mosaic into another; Again the Question Is, Which Ones May Be Changed Without Risking Failure of the Assembly?

The Basic Steps of Changing Mosaics

The previous chapter illustrated how variants of a mosaic can form as the result of the assembly process. Obviously, their existence depends critically on the fact there is *room for change* inside the frame of the mosaic during the assembly process.

In contrast, there is no such room in finished mosaics. Any mechanism that can modify an existing mosaic must first make room for new tiles. In other words, every change must begin with a *partial disassembly of the mosaic*.

Step 1: Reduction to the Key-Seed Stage

Destructing a mosaic ('degeneration') in order to prepare it for change could mean that either a few or even many tiles are removed before the rebuilding process ('re-generation') begins. In most cases, the removal of only a couple of tiles offers little opportunity for creating a significantly altered mosaic. Most likely, the rebuilding process will have no choice but to restore the removed tiles exactly as before.

Significant changes are much more likely to occur, if one removes a sizable fraction of the tiles. However, instead of leaving their number to chance, we will standardize the process by removing as many as are needed to return the mosaic to its key-seed stage.

Since key-seeds are equivalent to the finished mosaic, this step leaves us with the same mosaic, except for two important differences.

a. There is now room for change, and

b. By the definition of key-seeds, any change of tiles ('variation') must either lead to failure, to a different solution, or in rare cases back to the same solution.

(The last possibility applies, if there are multiple seeds for the same solution. It is entirely conceivable to change a seed in such a way it leads back to the same solution, nevertheless.)

Step 2: Variation at the Key-Seed Stage

Having reduced the initial mosaic to one of its key-seeds, one can now simply start rebuilding it. If nothing is changed, the rebuilding process will regenerate a mosaic that is identical to the initial one, however, at least it receives new tiles. In other words, a procedure like this may represent either the 'repair' or the 'turn-over' of a mosaic.

Alternatively, the key-seed may actually suffer a change. After all, it has gained ample room within its frame to do this. Still, there are several possible outcomes.

1. If the altered key-seed turns into just another one of the many key-seeds for the same initial mosaic, nothing new will result, except again repair or turn-over.

2. If the altered key-seed has turned into the key-seed of another mosaic, the rebuilding process will no longer reproduce the initial mosaic, but will turn into another one. The result is generally a variant, but if there is a substantial difference, one would consider it a mutant.

3. After the change of one or several tiles, the remaining tiles form no longer a key-seed, but are merely a more or less random small group. Rebuilding a mosaic from it, could result in unpredictable outcomes, because necessity no longer dictates the assembly mechanism. In this case, an element of *variability has entered the mechanism of mosaic change*. The process has become ambiguous or possibly unstable and, therefore, vulnerable to external influences that are not part of the normal assembly process.

4. It is even conceivable that the change of the seed yielded a group of tiles that cannot lead to any mosaic: *In this case the seed change has turned lethal.*

Mosaic Variation as the Result of Key-Seed Variation through Iteration

The described process of altering key-seeds could operate on a mosaic not only once, but also repeatedly. In this case, it is iterative, because the result of the process initiates the same process again.

In this way, the iteration of key-seed changes can generate a series of varied mosaics. The iteration only targets the few tiles of key-seeds, turns them into the tiles of other key-seeds, and finally completes them (Figure 11-01).

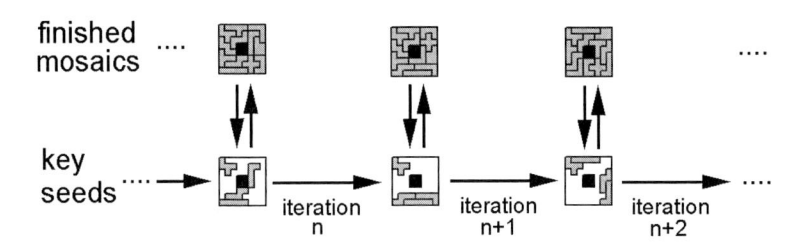

Fig. 13-01. Principle of a mechanism to create a series of variations of a mosaic through iteration. The iteration only targets the key-seeds of each mosaic, which is subsequently completed to its fitted form.

Since, by definition, the path from key-seed to solution is automatic, we can simplify the further discussion by ignoring the rebuilding steps and focusing entirely on the key-seeds. In later chapters, whenever an iteration step with key-seeds is completed, we skip the details of its assembly, and make the tacit assumption that some unnamed mechanism will complete the seed to its corresponding solution mosaic (figure 13-01).

THE REPLICATION OF LIVING MOSAICS

Chapter Fourteen

A Role of Seed- and Scaffold-Mosaics in Replication

Efficient Replication of Living Mosaics Usually Employs Their Key-Seeds, but Also Scaffolds

Replication not only is a hallmark if living things, but next to intelligence it is life's most successful weapon against the destructive powers of inanimate nature. All living mosaics are capable of replicating their modules and eventually the entire mosaic. The replication of their modules provides for the vital repairs and part replacements. And the replication of entire mosaics may pass on their unique tile configurations for millions of years.

However, not only was replication a necessity for the immortalization of phyla, species, and their building plans. It allowed the vast expansion of the very idea of modularity by turning the copies of the living mosaics into the tiles of novel meta-mosaics. Just consider how the replication of single cells became the basis for the replication of entire organisms! Thus, replication led to the growth of larger mosaics from the replica of many small ones. In addition, their post-replication modifications led to differentiation, variation, and the eventual evolution of new phyla, species, and building plans.

Replicating mosaics requires 3 different tasks.
1. Making new building materials such as frames and tiles,
2. Assembling the new tiles inside the new frame in precisely the location and orientation of the original, and
3. Maturing the mosaic to the stage where it can replicate again.

Task 1 of making new frames and tiles usually requires building them from scratch before the actual mosaic replication begins. As to task 3, we will discuss the processes of growth and development in chapter 19.

Hence, this section focuses on the second task, namely the arrangement of the new tiles in corresponding locations and orientations as in the original.

The Control of the Placement of Tiles

There are a number of very different approaches to guarantee the proper tile arrangements. They differ in the ways in which
1. the necessary information is stored and transmitted,
2. the information is implemented, and
3. faulty placements are eliminated.

Delivery of the Information of the Tile Placement

Obviously, the replication procedure must 'know' where to place the tiles in a replicated mosaic. The carriers of this information may be **material** (physical templates), or **symbolic** ('texts' of instructions).

The template may be the parental mosaic itself **('auto-templating'**; e.g. DNA-replication). Alternatively, the template may an auxiliary, intermediary structure **('allo-templating'**; e.g. messenger-RNA synthesis from DNA templates, or protein synthesis from messenger-RNA templates).

Implementation of the Information of the Tile Placement

Most frequently, the new tiles are added **one-at-a-time** to the replicating mosaic. Well-known examples are DNA-synthesis, protein-synthesis, and the assembly of cytoskeletal polymers. Examples from human technology are the assembly-line construction of cars and many other commercial products.

Alternatively, the replication may be **instantaneous**. Human technology uses this method quite frequently. Examples are the printing presses, coin dies, or stencils to replicate currency, books, or composite shapes in a single act. I cannot think of an example for the same method in biology proper. However, populations of locusts, algae, or protozoa may synchronize the replication of their individuals to such a degree, that the population seems to replicate instantaneously.

Quality Control of Tile Placement

It is conceivable that the mechanism of tile placement is error-prone. In extreme cases, the placement may even be left to chance. The more error-prone the procedure, the less effective it is, because it wastes enormous amounts of time and raw materials on failed products.

However, even in the ideal case of implementation with a high degree of precision, errors can and do occur in the real world. Therefore, biology proper as well as human technology always follow up such productions with subsequent **'quality controls'** that use the placement information to eliminate products with incorrectly placed tiles.

Unwanted Replications

So far we have tacitly assumed that the described replications were 'intended' by the mosaic. This is not always the case, because there are mosaics, which cannot replicate by themselves, and therefore need to borrow the replication machinery from a host system. We may call them *'parasitic replication methods'*. As in the many examples of phage- and virus-replications, they are dangerous to the host systems, and in fact may destroy them. In these cases they 'fool' the host to treat them as part of a complementary key-seeds that is to be used as template for partial replication.

In response, living mosaics have developed numerous counter-measures to unwanted replications, ranging from restriction enzymes and innate immune responses, to fully developed immune systems.

Key-Seeds and the Auto-Templated Replication of Mosaics

Auto-templating methods may be the safest and most direct methods, because they require the least number of peripheral machines or host organisms to provide the necessary functions. A supply mechanism for the tiles together with a protective scaffold for the self-templated assembly may be all that the peripheral support needs. It draws upon the assembly and solution-finding mechanisms that gave rise to the original mosaic in the first place.

This method of mosaic replication rests on the truism that one cannot replicate anything that does not exist already. On one hand, it is implicit in the meaning of 'replication'. On the other hand, it leads to infinite regress back in time, because it poses the question, where the pre-existing mosaic template came from, which, in turn raises the question where *its* precursor came from, and so forth. In addition, it implies that all throughout the past there were (and still are) mechanisms capable of making the tiles and assembling them into the pre-existing mosaics.

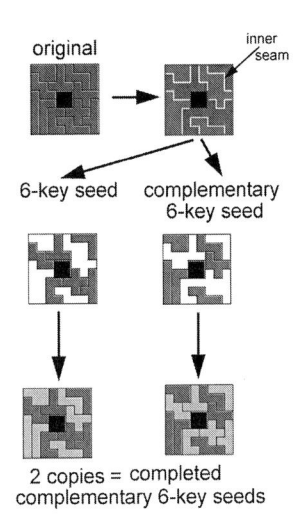

Fig. 14-01. Principle of the 'canonical' mosaic replication. Each mosaic can be interpreted as the union between 2 complementary key-seeds. They interface along an inner seam (marked white). After splitting the mosaic along the inner seam, each of the resulting key-seeds is completed by the normal assembly mechanism of the mosaic. The process results in 2 identical copies of the original. The 'old' and 'new' tiles are marked in different shades for a better visualization and illustrate the semi-conservative nature of the replication (i.e. each replica contains an 'old' and a 'new' set of tiles).

Disregarding the problem for the time being, the most important implication of auto-templated methods is their relationship to the key-seeds of mosaics. After all, what better templates are there than the key-seeds, which assemble with necessity the mosaic from which they came? Furthermore, each mosaic can be split into 2 complementary key-seeds (see (equation 11-02; figure 11-04). Hence, the easiest auto-templated method to duplicate a mosaic is this: **Split the parental mosaic into its 2 complementary key-seeds, and let each of them restore its specific sibling-solution.**

This relationship can be formalized as follows.

Each pre-existing mosaic M is the union of two complementary n-key-seeds $^nM_1$ and $^nM_2$, where n ≈ half the number of tiles (equation 11-02). It replicates by splitting into these seeds along their 'inner seam' (see chapter 11), and subsequently completes both of them.

$$^nM_1 \text{ (completion)} \rightarrow M$$
$$M = {}^nM_1 \cap {}^nM_2 \Big\langle$$
$$^nM_2 \text{ (completion)} \rightarrow M$$

We discuss this theorem by considering three mutually exclusive cases.

Case 1. The mosaic contains an even number of tiles; all tiles have different shapes.

Let N= 2n be the number of tiles. Then we construct two n-key-seeds $^nM_1$ and $^nM_2$ as described in (equ. 11-02).

As mentioned in chapter 8c, by virtue of their construction $^nM_1$ and $^nM_2$ are complementary key-seeds. As a result, one-half of the original tiles make up one key-seed, and the other half makes up the other. No tiles are common to both. Together they fill the frame precisely to constitute the mosaic.

The 2 complimentary key-seeds contact each other along a specialized inner surface, which will be called the **'inner seam'**. It consists of all the tile-to-tile contacts, where a tile of the first key-seed touches a tile of the second. This is illustrated in figure 14-01 in the case of a pentomino mosaic.

Depending on the particular pair of complementary seeds, each mosaic is capable of containing several possible inner seams. Yet, only one of them will become reality. As we will show below, this seam is the one that was used in every previous replication of the mosaic, and is the one that will be reformed upon the next and everyone thereafter.

As pointed out above, auto-templated methods imply that the assembly and solution-finding mechanisms of the mosaic pre-exist and, therefore, these pre-existing mechanisms will complete the seed mosaics.

Furthermore, the process will also reproduce the identical inner seam in both copies. It uses the same mechanisms by which it formed in the first place, namely marking the interface between the tiles of each key-seed and every of their new tiles (figure 14-02).

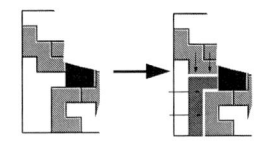

Fig. 14-02. Recreation of the inner seam by forming a seam (white edge) wherever a newly added tile (dark gray) interfaces with an 'old' tile (light gray) of the 6-key-seed.

Of course, the described process requires a sufficiently large supply of tiles.

Case 2. The mosaic contains an odd number of tiles; all tiles have different shapes.

Assume that the extra tile is t_0. Copies of t_0 must be contained in the tile-storage buffer and may be considered as part of the frames. One copy of t_0 can be added to all frames accordingly. Subsequently, the replication of the mosaic involves the replication of an even number of tiles. It can proceed as in case 1.

Case 3. The mosaic contains any number of tiles; some or all tiles have the same shape.

No two tiles of a mosaic can occupy the same location inside the frame. Otherwise, they would sit on top of each other. Hence, even if they have the same shape, at least their location is different.

In other words, the assumption that all tiles have unique shapes only has didactic value, as it facilitates the explanation of the way to construct key-seeds. However, it is not a necessary condition for the construction of key-seeds. It is enough if their location is unique. Figure 14-03 illustrates the cases, where 6-key-seeds only can be completed in one way, although several tiles have duplicate shapes.

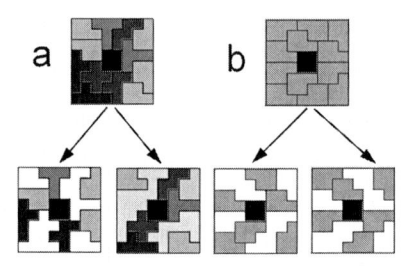

Fig. 14-03. Successful construction of 2 complementary 6-key-seeds in spite of multiple copies of certain tiles in the pentomino mosaic. By completing them, the mosaics can be replicated. Identical tiles are marked with the same colours. (a) a mosaic that contains 5 P, 2 T, 2 Y, F, X, and Y pentominoes. (b) a mosaic that contains 12 P pentominoes. The principle of the construction is illustrated in figure 11-04.

In summary, the above argument of case (1) is valid in all cases of mosaics.

Of course, the student will have noticed that the best-known example of the replication of a linear mosaic via its 2 complementary key-seeds is the replication of DNA. Here the mosaic is the double helical molecule. Its frame consists of the two sugar backbones with their 3'- and 5'- ends. The tiles are the 4 nucleotides. Correspondingly, the two n-key-seeds are the two single strands. (The number n is obviously a very large number for most genomes). The inner seam consists of the hydrogen bonds between the base pairs.

For replication, the mosaic splits along the inner seam into its 2 complementary key-seeds, and either of them will be completed by placing the only fitting (nucleotide-) tiles at the 'empty' spaces located opposite to each (nucleotide-) tile of either key-seed.

The Scaffolding of the Complementary Key-Seeds

As mentioned in chapter 4, the exact fitting of the tiles in a mosaic provides large numbers of contacts between the tiles, which in turn contribute significantly to its mechanical stability. The above discussion has introduced several additional types of contacts, such as tile-to-frame contacts, specialized tile-to-tile contacts (inner surface), and tile-to-inner seam contacts (figure 14-04). Each of these types of contacts may have an enhancing or diminishing effect on the mechanical and functional stability of the mosaic. Yet, the very existence of the mosaic implies that the overall stability is sufficient.

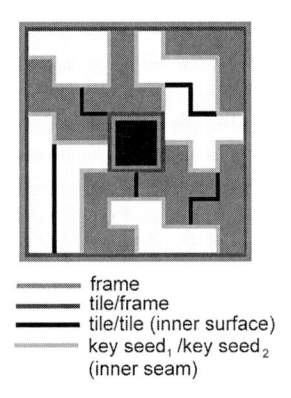

Fig. 14-04. Illustration of the various interfaces between the tiles and frame of a pentomino mosaic.

The splitting along the inner seam reduces the number of inner contacts between tiles and may cause some mechanical instability. Therefore, some additional structural support of the complementary key-seeds may be required.

Of course, whatever stabilizes the key-seeds temporarily must be removed in the subsequent phase, when the holes of the seed are being filled. In other words, the described method of replication may have to support the key-seeds with some kind of scaffolding in order to prevent their disassembly during the separation and any subsequent movements.

Copy Numbers of Replicating Mosaics

As an iterative process, every replication process occurs within a biological context, and needs to implement how fast and how often it occurs. The time interval τ_R between successive replications is the sum of the time τ_{Td} it takes to duplicate and assemble the new tiles, and of the 'inter-phase' time interval τ_{Ip} between the completion of a replication and the onset of the next.

190 Chapter Fourteen

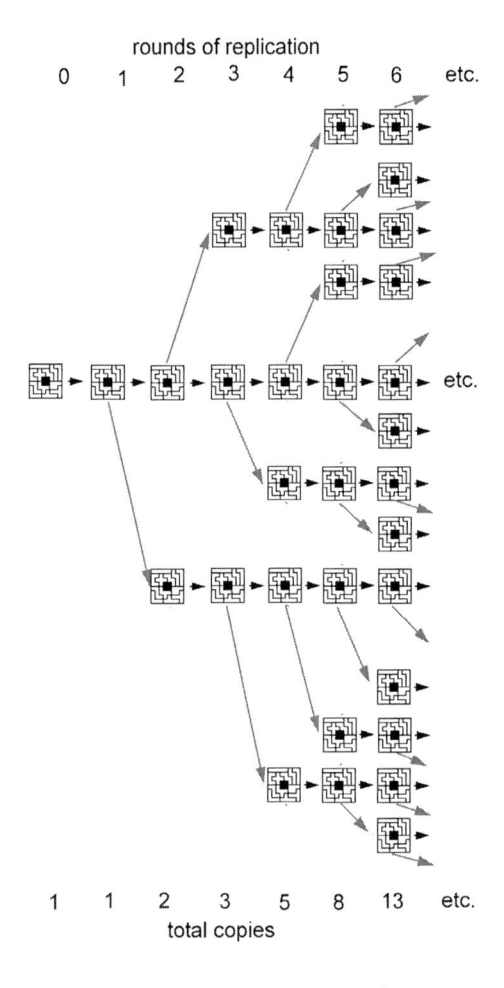

Fig. 14-05. At every round of replication, each mosaic and each of its copies produces 1 new copy (light gray arrows), but also stays around for another replication at the next time point (black arrows). The resulting copy numbers are the Fibonacci numbers.

equ. 14-02

$\tau_R = \tau_{Td} + \tau_{Ip}$

Assume that each replication yields just 1 copy at every round of replication, and that each copy, including the original, stays around to replicate from there on at every round as well. In this case, as is well-known, the total copy numbers including the originals will be given by the famous Fibonacci numbers 1, 1, 2, 3, 5, 8, 13, 21,... (Leonardo of Pisa, called Fibonacci (1170-1250)). Figure 14-05 explains the reason. The total numbers F_k after round k, i.e. the iterative definition of the Fibonacci numbers is given by

equ. 14-03
$F_k = F_{k-1} + F_{k-2}$;   with
$F_0 = F_1 = 1$.

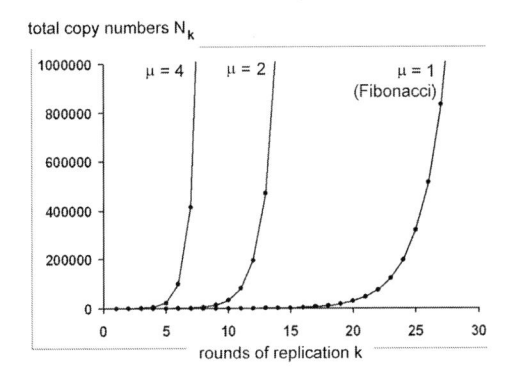

Fig. 14-06. At every round of replication, each mosaic and each of its copies produces μ new copies, but also stays around for another replication at the next time point. The resulting copy numbers for μ = 1, 2, and 4 are calculated according to equation 14-06 and plotted as functions of the number of rounds of replication.

Of course, the Fibonacci numbers F_k are integers. Therefore, equation 14-03 and the following similar equations indicate **iterations**.

Assume that at each round of replication produces $\mu \geq 2$ copies. Let N_k be the total number of copies at the kth round, then

equ. 14-04
$N_k = N_{k-2} + \mu \cdot N_{k-1}$;   with
$N_0 = 1$; $N_1 = \mu$;

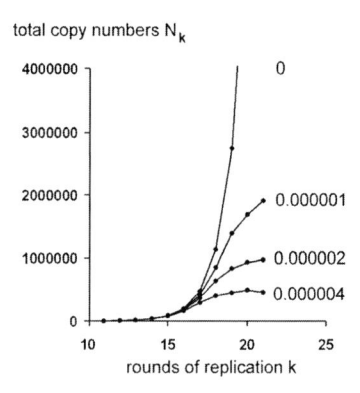

Fig. 14-07. Illustration of the effect of the degradation factor δ on the number of mosaic copies according to equ. [14]-08 for μ = 2 and δ = 0, 0.000001, 0.000002, 0.000004. It shows that the copy numbers flatten off towards a saturation level after their copy numbers had reached a large enough value.

For μ = 2, the numbers N_k are 1, 2, 5, 12, 29, 70, 169, 408, 985, 2378, 5741, 13860, 33461, 80782, 195025, … For μ = 4, the numbers N_k are 1, 4, 17, 72, 305, 1292, 5473, 23184, 98209, 416020, 1762289, 7465176, 31622993, …, and so forth.

As shown in figure 14-06, the latter numbers rise much faster than the Fibonacci series. Yet, even if the copy numbers of mosaics were 'only' the Fibonacci numbers, they would rise explosively to unsustainably large values. In reality, however, this cannot happen, because the replications would rapidly run out of tiles and other raw materials for the copies. Besides, there are always mechanisms to degrade the finished copies, which eventually lead to a certain saturation level of the numbers.

There is also a suitable iteration describing the effect of a concurrent degradation of mosaic copies. As is common for comparable situations (e.g. Volterra equations) we assume that the multiplication coefficient μ decreases with the number of existing copies.

equ. 14-05

$\mu = \mu_0 - \delta \cdot N_{k-1}$, as long as $\mu \geq 0$

Combining it with equation 14-04 yields the more realistic description

equ. 14-06
$N_k = N_{k-2} + [\mu_0 - \delta \cdot N_{k-1}] \cdot N_{k-1}$, (as long as $[\mu_0 - \delta \cdot N_{k-1}] \geq 0$)
$N_0 = 1; N_1 = \mu_0;$

Fig. 14-07 illustrates the case of $\mu = 2$ for various values of δ. It shows that the copy numbers flatten off towards a saturation level after their rise reaches a large enough value.

CHAPTER FIFTEEN

THE STRATEGIES OF MOSAIC REPLICATION

BESIDES BEING ALWAYS 3-DIMENSIONAL OBJECTS, LIVING MOSAICS MAY HAVE DIFFERENT 'EFFECTIVE' DIMENSIONS; THE TOPOLOGY OF THEIR 'EFFECTIVE' SPACE DICTATES THEIR REPLICATION

The 'Dimension' of a Replicating Mosaic

The previous chapter proposed the basic idea of replicating a mosaic by splitting it into two complementary key-seeds and completing each 'half' separately. In practice, the procedure requires a number of additional basic mechanisms such as the formation of an inner seam, the production of a new frame and new tiles, the splitting along the inner seam, locating the holes in the complementary key-seeds, and so forth.

Naturally, it also requires certain movements of all duplicated components away from their originals. Therefore, the details of replication will depend critically on the spatial arrangements of tiles, frames and key-seeds.

Of course, all living mosaics are 3-dimensional objects, be they DNA molecules, metaphase spindles, or animal embryos. Yet, their tiles may be arranged in 1, 2, or 3 independent directions.

For example, regardless of its 3-dimensional double helix and the 2-dimensional bi-nucleotides rungs, a DNA molecule is essentially a *linear* chain of paired nucleotides, which makes DNA a 1-dimentional arrangement of its tiles, i.e. a 1-dimensional mosaic.

Similarly, despite the 3-dimensionality of chromosomes and the mitotic spindle, during mitosis the chromosomes form a 2-dimensional ring in the metaphase plane, and hence we may consider them as a 2-dimensional mosaic. In contrast, the anatomy of a human is a truly 3-dimensional mosaic of its organs, which are 3-dimensional objects themselves.

There are conceivable tile arrangements that are much more complex. For example, a linear mosaic could be folded into a number of knots, or one could even imagine that the tiles formed a Moebius-band or a Klein-bottle.

Such exotic cases are presumably purely academic. In most cases of living mosaics, common sense will suffice to identify and discuss the various classes of their tile arrangements.

Still, it may not always be straightforward. Sometimes the assessment of the tile arrangement may depend on their functions. For example, in view of its essentially cylindrical symmetry, the intestinal tract appears as a 1-dimentional mosaic that extends along its cylinder axis (after unfolding the many coils of this axis). In this view, we focus on the progression of the digestion process along the intestines.

In contrast, if nutrient uptake is the focus, one may consider the intestinal tract as a 2-dimensional mosaic with a cylindrical surface.

In reality, it helps that the number of tiles in living mosaics is always finite. Hence, while trusting the finite numbers of components and the application of common sense in biology, we may define **the 'dimension' of a living mosaic as the number of independent directions which the mosaic uses in order to arrange its tiles without overlap**.

With that in mind, perhaps the most difficult task of the described method of replication is the splitting of the mosaic into the 2 complementary key-seeds. What keeps the tile arrangements stable during the seed separation, and what prevents the duplicated tiles from becoming tangled up?

The latter question is not academic. For example, after several decades of studying biology, I am still wondering about the 'engineering' reasons for the different phases of meiosis.

I suspect that the complex technology of meiosis is needed to prevent entanglement of chromatids. Take the case of human meiosis. A total of 96 chromatids have to be sorted into four different sets of 24 each. Yet, in spite of thermal turmoil and the rather free movements of the 96 long, thin 'snakes' coiling around, it must never happen that any of the four sets of 24 chromosomes accidentally contains two homologues.

Yet, strangely enough, the technology seems to fail in a sense. During pachytene and diplotene, homologues become entangled, until the so-called crossing-over seems to forcefully cut them apart during prophase I and metaphase I.

Nevertheless, upon closer inspection, a seeming failure of nature is rarely one. Crossing-over does not merely correct an unwanted accident of meiosis, but mediates gene recombination, and is essential for the all-important diversity of genomes. This surprising turn of events is possible, because the entanglement of chromatids involves only alleles, but does not tie random genes together!

The Topology of Key-Seed Separation ('Mosaic-dimension + 1' Rule)

According to figure 11-04, the construction of 2 complementary key-seeds from a mosaic may seem simple enough. Just discard every other tile!

Unfortunately, auto-templating does not permit to use this recipe, because one cannot discard any of the tiles. On the contrary, the tiles and their locations missing from one key-seed are precisely the tiles and their locations of the other. Therefore, the first key-seed must be literally 'lifted out' of the mosaic, while leaving behind the other intact.

This requirement has topological implications. The mechanical separation of the 2 complementary key-seeds cannot proceed along the inside of the mosaic. The 'lifting-out 'mechanism requires an additional independent direction at right angles to the dimension of the mosaic (**'mosaic-dimension + 1' rule**).

1-dimensional templates. If the mosaic is 1-dimensional, the requirement to separate the complementary key-seeds at right angles to the direction of their tile arrangement demands a second, independent spatial direction to accomplish that. Obviously, there is no problem finding a suitable direction, because there are 2 remaining dimensions, and either can be used for the separation procedure (figure 15-01a).

2-dimensional templates. If the mosaic is 2-dimensional, there only is one spatial dimension left for the separation procedure (figure 15-01b). Yet, this remaining dimension is sufficient to be used for the separation of the complementary key-seeds.

3-dimensional templates. However, if the mosaic itself is already 3-dimensional, there is no 4th spatial dimension left, along which the complementary key-seeds may separate (figure 15-01c).

Hence, the replication procedure of 3-dimensional mosaics must resort to an entirely different technique. There are numerous alternatives, but they all boil down to selecting a different location to construct from scratch a copy of the original mosaic.

Of course, the method must supply a second set of frames and tiles from which to assemble the copy.

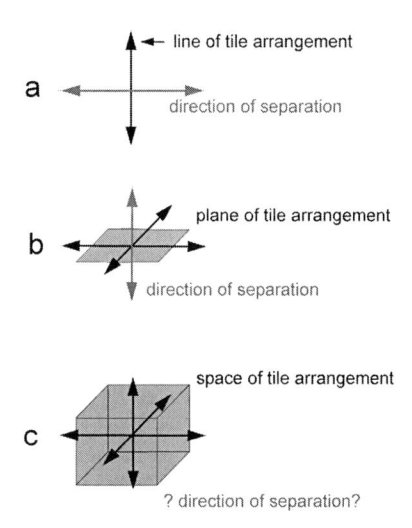

Fig. 15-01. Spatial requirements for the separation of the 2 complementary key-seeds ('mosaic-dimension+1'- rule). The direction of separation must be at right angles to the directions of the tile arrangement. (a) Linear tile arrangement of a 1-dimensional mosaic. The separation requires a second, independent direction in order to separate the complementary key-seeds. (b) Planar tile arrangement of a 2-dimensional mosaic. The separation requires a 3-dimensional space in order to separate the complementary key-seeds. (c) Volume-filling tile arrangement of a 3-dimensional mosaic. The separation would require an impossible 4th dimension in order to separate mechanically the complementary key-seeds. Hence, no simple mechanical separation is possible.

Illustrations of Key-Seed Separation and Completion

In order to illustrate some of these requirements we turn to the pentomino model system.

1-Dimensional Mosaics

The most spectacular example of the auto-templating of a 1-dimensional mosaic is the double helix of DNA. Its principal topology can be simplified by only using one such pair, (say) Adenosine – Thymidine, instead of 2 complementary base pairs. Figure 15-02 shows the simplified topology of a 1-dimentional mosaic with 2 pentominoes as its complementary tiles and a frame element, which form 2 different, yet complementary elements 'A' and 'B'.

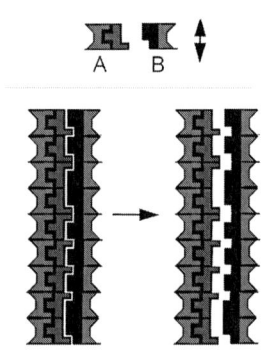

Fig. 15-02. Example of the key-seed separation of a 1-dimensional mosaics shows the non-trivial case, where the 2 matching complementary tiles a and B are different and can have 2 different orientations with respect to the 2 backbones (= light gray), therefore, the self-templating reproduction can produce a vast variety of different linear mosaics.

The prototype is even more simplified, as each strand only contains one kind of tile, which exists in either its 'up'- or its 'down'-orientation. Thus, it generates an information encoding sequence such as 'up-up-down-up-down-down-up-... etc'.

The 2 key-seeds are the 2 single strands. Their frames are provided by the 2 linear arrays of the frame elements (light gray). After the splitting up of the key-seeds, the 'holes' are the missing complementary elements along each single strand.

2-Dimensional Mosaics

Figure 15-03 illustrates the auto-templating of a 2-dimensional mosaic by using the model frame. It produces two identical copies of the original mosaic in two stages: (1). the separation of the two complementary 6-key-seeds, and (2). the filling in of their missing tiles. The tiles of the 2 complementary key-seeds are labelled in different shades. After completion of the key-seeds (panel c), the newly added tiles are labelled in light shades in order to indicate the semi-conservative nature of the replication method.

Consistent with figure 15-01b the key-seeds separate in a vertical direction relative to the plane of the mosaic.

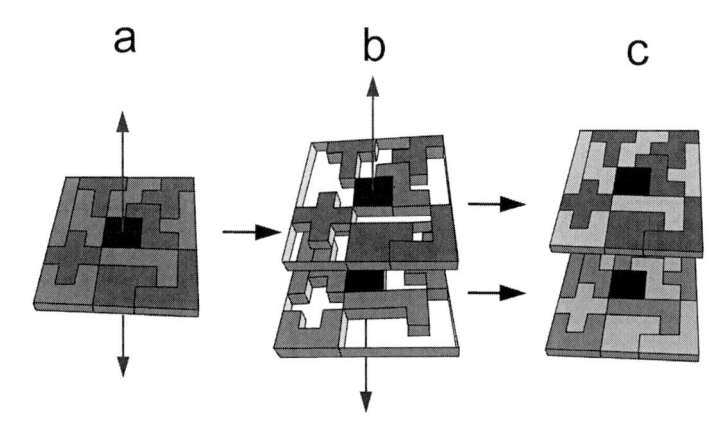

Fig. 15-03. Principle of the synthesis of a 2-dimensional pentomino mosaic (a) The directions of the separation are at right angles to the direction of the tile arrangement. (b) Splitting along an inner seam as shown in figure 14-04 and separation of the two complementary planar key-seeds at right angles to the plane of the mosaic (up or down arrows). (c) Completion of each key-seed with 'new' tiles (light shades) creates 2 copies of the original. The tiles of the original mosaic (dark colours) are still intact, which shows the semi-conservative nature of the replication.

3-Dimensional Mosaics

As pointed out before, the impossibility to untangle the tiles of a 3-dimensional mosaic by moving half of them into a fourth dimension, leaves no other choice but to resort to a *de novo* assembly of a new mosaic from a set of precursors or completed tiles using information instead of mechanics and forces.

Figure 15-04 illustrates the principle of the replication of such a mosaic. It requires two novel elements.

(1) The formation of an 'embryo', i.e. a package of precursors and encoded instructions that can be developed into a copy of the original mosaic.

(2) The processes of growth and development that turn the embryo into an 'adult' mosaic by reading and implementing the information of the embryo.

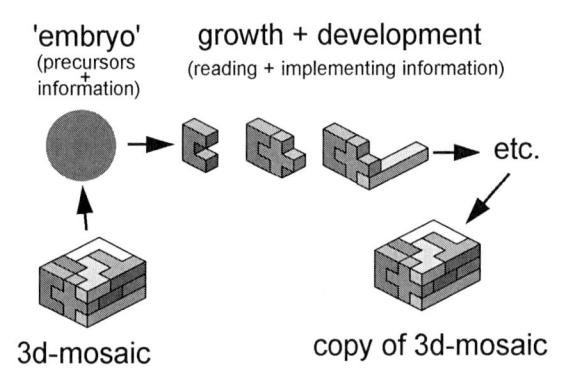

Fig. 15-04. Embryogenesis, i.e. the rebuilding a 3-dimensional mosaic 'from scratch' in view of the impossibility to untangle it mechanically. Since there is no 4th spatial dimension to separate the tiles tangle-free, an 'embryo' forms from precursors of the tiles and frame and an information-rich mosaic that contains all the information to grow and develop into a new copy of the original mosaic.

In these cases, the most important guidance is provided by the information how to place and grow the relevant components. This information is necessarily encoded. A separate set of material carriers contains the codes, but its data and instructions are essentially non-material, i.e. symbolic. Nevertheless, as letters, codes or words they are always discontinuous objects assembled as strings or as sheets that form unique, non-continuous patterns.

In other words, a separate information-rich symbolic mosaic contains the information needed for the de novo assembly of the replica mosaic.

The Expression of Living Mosaics

CHAPTER SIXTEEN

PARTIAL REPLICATION OF FITTED MOSAICS ('CODING' AND 'EXPRESSION')

REPLICATING ONLY SELECTED TILES REQUIRES TO MARK THEM AND TO LINK THE MOSAIC TO 'CODE-READERS' AND 'CODE-PROCESSORS'; EVENTUALLY, PARTIAL REPLICATION OF A MOSAIC MAY BECOME ITS 'EXPRESSION' AND 'FUNCTION'

Partial replication of a mosaic simply means the replication of only a subset of tiles and/or frames. For easier formulations, let me call the mosaic that contains the replicated part, the 'mother'-mosaic.

Marking Tiles for Partial Replication (Entry Point of 'Information')

On the face of it, the concept of partial replication represents no dramatic departure from the idea of replicating a complete mosaic. For example, chapter 8 on fractal mosaics discussed how each tile could be an entire mosaic. Hence, the duplication of such a composite tile was actually the replication of a complete mosaic and at the same time the partial replication of a mother-mosaic.

However, as evolution turned partial replication increasingly into programmed and targeted actions, biology added entirely new interpretations to partial replication. They probably had insignificant beginnings, but eventually led to new and most pivotal biological functions.

For example, the partial replication of genome mosaics presents the core of the all-important mechanism of gene 'expression'. The partial replication of cellular mosaics enables the integration, but also the independent functions of cellular instruments such as centrioles (basal bodies), cilia, mitochondria, and chloroplasts. The partial replication of whole organisms or populations allows the independent actions of stem cells, the neurocrest migration, the regeneration of limbs, the growth of shoots and tubers, the swarming of bee colonies, and the expansion of certain species in an ecosystem, and many more.

As to their beginnings, one may speculate that partial replications began with the accidental appearance of certain 'faulty' tiles, which caused the full replication mechanisms to start or stop prematurely on occasion. If living mosaics survived in spite of producing such incomplete copies of themselves, they may have evolved specialized 'detectors' that allowed the full replication mechanism to skip over the flawed tile-configurations (-groups, -sequences, etc.). In time, some of the incomplete copies may have turned out to be 'beneficial' for mosaics that happened to make their production heritable. Inevitably, however, such changes set the stage for the promotion of the altered tiles from being flawed to becoming 'markers', and their 'detectors' from providing 'treatments' to becoming 'readers' for the 'encoding', 'decoding', and the 'meaning' of the new breed of tile-markers.

'Encoding', 'decoding', and 'meaning' are fundamentals of 'information processing'. As we know from our own technology, once information processing has appeared on the scene, it will not remain in a supporting role, but soon take centre stage. Let us look at a seemingly harmless example of our daily lives: clocks.

Example of the Natural Entry of 'Information' (Clocks)

A clock is a portable device to model the daily Earth rotation. It allows us to synchronize all our actions with the Earth rotation and, thus, with each other.

With the exception of sundials, their basic idea is always the same. Every clock counts the number of repetitions of a reasonably rapid oscillator, be it a dripping water faucet, a pendulum, a vibrating quartz crystal, or the inversion of a molecular structure. If the clock design is elegant, in doing so it apportions at the same time the energy needed to keep the oscillator vibrating.

It was inevitable that information processing emerged from the clock mechanisms. After all, the concept of 'counting something' is inherently a concept of information processing, although initially, it only was a very unimpressive part of the elaborate and magnificent designs of the mechanical clocks of the past.

Eventually, however, the information processing functions of clocks did not remain inevitable technicalities. On the contrary, they moved to the centre of our civilization and exploration of the universe in the form of the Giga-Hertz timers of microprocessors that alternate between data and instruction modes of computers, the synchronization of the internet, and the GPS navigation of cars, ships, and airplanes, and the experimentation with relativistic space-time and the building-tiles of matter.

The Problematic Role of Tile-Marks

As yet, we do not know the actual origin of partial replication in all its forms, but it seems likely that its profound linkage to information crept in as the need of marking specific tiles.

Although clearly needed, tile markings are problematic for several reasons. For once, they could present mechanical obstacles for the normal functions of the mother mosaic.

Furthermore, their possible misplacement could trigger perturbing, even fatal consequences for the mother mosaic. For example, on the molecular level, the misplacement of start- and stop-codons must lead to dysfunctional proteins. On the cellular level, the marking of the wrong subset of cells for mitosis and migration may trigger the devastating phenomenon of metastasis.

However, even if the markings were placed correctly and yielded perfect copies, the copies might actually pose a danger, if they were unmarked, Once released, they could compete with their original templates. As the dramatic effects of RNA interference (RNAi) show, such competition can indeed happen.

Therefore, placing and 'reading' the marks cannot be a simple task. However, once evolved, it offered a most significant leap forward. The selected, marked, and 'read' tiles became the carriers of 'meaning'. If the mother mosaic had not contained codes before, now it did. The evolutionary path was open to add encoded meaning to all their tile groups.

The Need to 'Colour' the Partial Replicas Differently

In chapter 8, we had encountered a similar situation. There we had to distinguish between tiles with identical shapes, and used 'colour' as a non-essential modifier of the tiles. In the present case, we will use the same approach, and say that the **partial replicas must have different 'colours' than the original tile sets**.

Of course, the term 'colour' only has metaphorical meaning. In the example of DNA mosaics, 'colouring' the partial replicas means to replace DNA sequences with RNA sequences. Nevertheless, the partial copies are quite similar to their originals as they still rely on the principle of matching hydrogen-bonded, complementary bases like their original DNA sequences.

Inevitably, though, the RNA copies as 'coloured' and partially replicated versions of DNA were now able to encode something. Such coding-enabled copies even received additional marks in the form of caps and poly-Adenosine tails.

'Transcription', Translation' and 'Expression' of Pure Symbolic Mosaics

The role of information in mother-mosaics grows more dominant as more tile sets express codes rather than physical functions. Eventually the involvement of information in partial replication may escalate to the point where the tiles of the mother-mosaic have no other functions than as codes. Correspondingly, the mother-mosaic, now a 'coded mosaic' must evolve into a pure symbolic mosaic.

At this stage, the partial replication of such code-bearing combinations has morphed from a mosaic copy into the copy of a piece of information. It has become a **'message'** in the terminology of molecular biology. Borrowing more from that terminology, one may call the entire process as the **'transcription'** of the information-rich mosaic.

The partially replicated tiles will trigger and/or implement functions specific for their code. The implementation of the specific functions of the different partial replicas of the information-rich mosaic can be called their **'translation'**.

The combined actions of transcription and translation are correspondingly called **'expression'** of the information-rich mosaic.

The Partial Frames of Partial Replication

A special kind of partial frame must link the tiles for partial replication. Its primary function is to guarantee that replication only is possible for the particular tile groups that are primed and surrounded by such frames.

Since the complementary key-seeds of the mother-mosaic are vital for its replication, it seems natural to employ them also for the partial replication of 1- and 2-dimensional mosaics. Hence, the natural places to install the partial frames are these two seeds. Either of them may direct partial replication. Therefore, both of them must contain pre-installed partial frames.

Obviously, the partial frames must not interfere with the mother-mosaic's assembly algorithms and mechanisms. Moreover, they must be unspecific, i.e. the method must use the same partial frame materials for different groups of partially replicated tiles. Hence, they must represent modifications of a property that *all* tiles have in common.

In the case of information-rich mosaics, reserving certain codes for the partial frames would fulfil all these requirements. For example, in the case of the DNA mosaic such reserved codes are the start- and stop codons.

In cases of 3-dimensional mosaics, partial replication means the growth of 'partial embryos', such as organs, tissues, limb buds, etc. Hence, their partial frames are their natural limitations, such as capsules, basement membranes, body cavities, etc.

The Similarities between Partial and Total Replication

In spite of their necessary differences, the method of partial replication is likely to be a variation of the method to replicate whole mosaics. After all, by increasing the number of partially replicated tiles, their partial replication must gradually become the replication of the whole mosaic.

In the case of the mother-mosaic, the principle of replication uses the fitting of specific tiles into the holes of complementary key-seeds. The same principle should apply to the partial replication.

This is illustrated in figure 16-01 with the pentomino models. It argues that there are 5 phases of the partial replication of 1- and 2-dimensional mosaics:

Phase I: The initial phase is the normal state of a mother-mosaic. For clarity, the inner seams and other tile-contacts are omitted in the illustration.

Phase II: As in the normal replication, the mother-mosaic splits into the 2 complementary key-seeds.

Phase III: The tiles of the key-seeds enclosed or surrounding by the partial frame are primed (light gray). In these cases the partial frames are modified contact areas either between tiles, or between tiles and frame

Phase IV: The matching complementary tiles are fitted to the holes similar to the case of normal replication. However, the matching tiles are modified (changed shade of gray).

Phase V: The partial replicas are released and the 2 complementary key-seeds re-unite to restore the mother-mosaic.

Partial Replication of Fitted Mosaics ('Coding' and 'Expression')   207

In view of the small number of only 12 pentomino-tiles, the pentomino model has only very few and small tile-sets that can be replicated partially. Correspondingly, only very few partial frames can be implemented in the models. Nevertheless, I hope that they can illustrate the principle of the method.

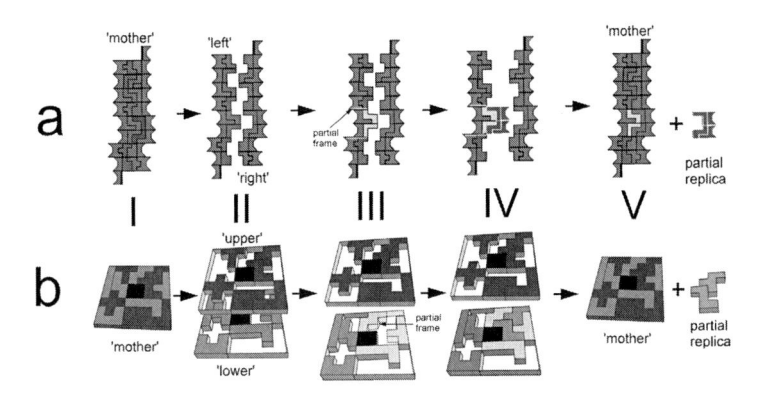

Fig. 16-01. Partial replication of a set of 2 tiles of the pentomino models. (a) 1-dimensional mosaic. (b) 2-dimensional mosaic. The 5 major phases are described in detail in the text.

Of course, the above considerations and illustrations do not explain or predict any technical details how living mosaics implement partial replication. They only are supposed to point out the necessity of *what* they need to do in principle. They may also suggest which variations on the principle theme one may expect.

It is clear that the specific markers and partial frames will involve complex mechanisms. In the case of DNA they take the form of transcription factors, primers, polymerases, etc. In the case of leafs, a gall wasp may prime a few epithelial cells, and cause them to selectively undergo cell division and differentiations in order to build the specific galls for wasp's larvae, and so forth.

The Cycles of Living Mosaics

Chapter Seventeen

The Imperfect Cycles of Living Mosaics

All Living Mosaics Express Cyclical Properties; Yet, the Periods and Amplitudes Are Never Precisely the Same

The Ubiquity of Cycles

Cycles are a very basic class of processes for all forms of life. In fact, life without cycles is quite inconceivable. Cycles need not have fixed periods or fixed amplitudes. For example, the heartbeat may change its period and amplitude every minute depending on adrenaline, sleep/wake state, and environment. Yet, nobody would doubt that they are cyclical.

Hence, we define quite generally *biological cycles as a biological phenomenon that expresses some recurrent features in time*. If we understand cycles in this broader sense, it is hard to find phenomena among bio-molecular mechanisms, cells, organisms, or ecologies that are not cyclical.

Cycles need not involve exclusively the same individual. Consider, for instance, the reproduction of organisms or their turnover. They always involve new individuals. Yet the process of reproduction itself repeats in very much cyclical ways.

To be sure, biology has no monopoly on cycles. Naturally, there are solar cycles and galactic cycles, recurrent ice ages and quasi-periodic reversals of the Earth magnetic field, and many, many more. Arguably, these 'inanimate' cycles may act as the trigger and driving mechanisms for the biological ones.

Consistent with the tenets of this book, all cycling biological objects are in particular living mosaics. Hence, one may define naively the cycle of a mosaic as a repeating series of sibling mosaics, whose members are ordered in such a way that the first and the last mosaic are identical, or at least share a majority of properties.

However, this definition would be far too restrictive if it requires identical repetitions. As we will argue below, nothing in biology is identical. Furthermore, during their cyclical runs the living mosaics may change in many ways, including changes of their frame. As long as the last member of a series of mosaics can function as the 'almost' return to its beginning, we should be able to call the series a cycle. Therefore, we must soften the definition accordingly.

As their component tiles are discontinuous, we cannot expect the changes between sequential mosaics to be smooth and gradual. We must be prepared to observe leaps and bounds of tiles and frames during the cycling of a mosaic.

The 'Almost'-Periodicity of Biological Cycles

No two biological objects, whether they are organisms of the same species, or cells of the same tissue, etc., are ever completely identical. The same is true for all other manifestations of biology. There are never exact identities, only various degrees of similarity. Even the cycles of mosaics are idealizations, because in reality the first and the last mosaic of a cycle is never exactly the same.

Let me repeat the reason for the absence of exact identity. No biological structure, function, or event is ever an exact copy of a previous one; it is always rebuilt from scratch. Reconstructions are subject to variability; the more complex the reconstruction, the more vulnerable to perturbations, thus the more variable the result. To be sure, usually the variations do not render the members of a series of mosaics dramatically different, but they are never identical.

Hence, we may have to accept a less rigorous definition of mosaic cycles and only demand that *the first and last of the series of mosaics are 'almost'-identical* (figure 17-01).

Not only structural mosaics are never identical, but also biological actions fare no better. Legions of interacting organisms and cells do not just follow accurately some universal laws of nature. On the contrary, they may apply individual strategies, or make surprising choices at the crossroads of function pathways as they avoid some options while seeking actively others.

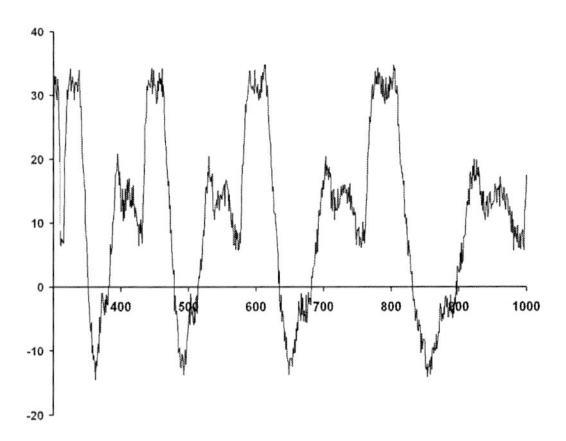

Fig. 17-01. Example of an 'almost'-periodic cycle of a hypothetical quantity with increasing period length and non-repetitive amplitude details.

Obviously, this kind of irreproducibility may be quite meaningful. For example, the basic irreproducibility of heartbeats is the very reason that cardiologists measure and examine the EKG of a patient repeatedly. Other examples of such 'almost'-periodic biological events are found in the addition of the next monomer to a polymer, the circadian rhythms of numerous organisms, the ridges of fingerprints, the swarming of bees, the changing of colours of foliage, the recycling of material within ecologies, the adaptation of ecologies to recurrent geological catastrophes, etc.

It is not just a phenomenon concerning cycles. The 'almost'-identity of biological structures and actions has always challenged biologist to find mechanism to explain them by the randomizing effects of thermal noise, quantum uncertainty, free will, or other such mechanisms that are inherently irreproducible.

As to the 'almost'-identity of cycling mosaics, we may try to set a threshold of similarity, 'sim_{th}', and declare the first and last mosaic of a cycle as 'almost'-identical, if their similarity 'sim' is less than identical, yet remains above this threshold ((i.e. $sim_{th} \leq sim < 1$). Alternatively, we may try to drop the idea of an 'almost'-identity entirely, and soften the concept of 'identity' by tolerating sufficiently large accuracy limits.

Either way, declaring a living mosaic as the repetition of another in a series, will require judgement by the scientist. Arguably, science has always required critical judgement. The handling of mosaics may just require a bit more of it.

CHAPTER EIGHTEEN

ITERATION-DRIVEN MOSAIC CYCLES

THE MOST NATURAL CYCLES OF LIVING MOSAICS ARE THE RESULT OF ITERATIONS

Periodic forces acting upon a mosaic may very well cause its actions to become periodic. However, the reverse is not true. If a mosaic cycles, then its underlying forces may not be cyclical at all.

Take the mosaic of the solar system. The 'almost'-periodic movements of all the planets (= tiles) is not caused by (say) a cycling gravitational field of the sun. On the contrary, the constancy of the sun's gravitational field maintains the planets on their almost perfect cyclical paths.

Yet, all cyclical actions of mosaics have the following properties in common.
1. *A cycling mosaic forms a cycle of its sibling-solutions.*
2. *Whatever forces Φ are capable of driving the sibling solutions along their cycles, the very same forces drive them back to their starting state, whereupon everything begins again.* This only is possible, if each sibling S_n of the cycle is a legitimate target (= input) for those forces. They must be able to act upon each new sibling S_n to create the subsequent sibling S_{n+1}, and so forth. In other words, **iteration can describe the forces Φ that drive the mosaic siblings S_n around in cycles.**

In addition, one may conclude thence, the underlying mechanisms, regardless of their specifics, must at first create room for change in the predecessor, every time they create a new mosaic within the cycling series. In addition, they must guarantee that the successor is a solution. In other words**,** it must involve the **seeds of the mosaics.**

We will even go a step further by trying to derive mosaic cycles not just from the seeds of the mosaics, but specifically from **their key-seeds**. It is an attractive approach because it covers naturally a number of fundamental properties of living systems.

In the first place, it describes them, of course, as mosaics. It also provides a transparent way of deriving each mosaic as a product of its predecessor. By focusing on the key-seeds of the cycling mosaics, it offers a simplified analysis, because it restricts itself to a minimal number of components of each mosaic.

The 'Natural' Iteration

There are countless ways of formulating different iteration mechanisms, by which a key-seed may arise from a predecessor. Depending on the specific biological process in question, each step may be the abstract formulation of a specific gene action, cellular interaction, cellular transformation, organism action etc.

Obviously, the simplest iteration mechanism is a 'non-specific-action' mechanism, i.e. the successor of every seed is determined merely by its close similarity to its predecessor: We call this mechanism the **'natural iteration'**: *The 'next' key-seed is the most similar, yet not identical one.*

The algorithm of every such **'natural iteration series'** is determined by
1. the starting seed,
2. the mechanism by which the seeds are constructed, because it determines which is the 'nearest' with a sufficiently large overlap, and
3. the maximal size of the overlap, which is required for a 'next' seed to qualify as a next closest member of the series.

Figure 18-01 shows 75 stages of a 'natural iteration series' of 4-key-seeds, starting with an arbitrary seed. The overlap, maximal 20, is indicated by dark dots and its count is written into the centre of each seed. The 'F' and 'X' pentomino tiles of the corresponding finished mosaics are highlighted in order to help the visualization of the changing iteration results.

Please note, none of the following examples of iterated key-seeds will illustrate the process of cycling, as yet. We will first analyze the ways in which key-seeds can change, before we identify conditions under which the series will cycle (or 'almost'-cycle).

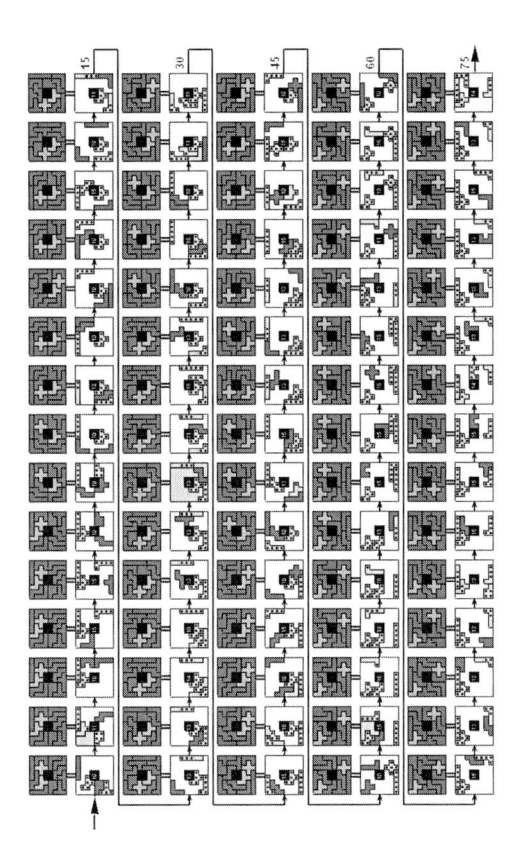

Fig. 18-01. 75 stages of a natural iteration series of 4-key-seeds, starting with an arbitrary seed. In this case the mechanism starts with a 4-key-seed (arrow), completes it into the finished mosaic and selects the next, most similar 4-key-seed, i.e. the key-seed with the largest overlap. The completed mosaics are connected to their 4-key-seed with a double line; 2 tiles are highlighted for an easier detection of the differences, The overlap is indicated by dark dots whose number is written into the centre of each seed. The 'F' and 'X' pentomino tiles of the corresponding finished mosaics are highlighted in order to help the visualization of the changing iteration results. Note: the illustration shows only the progression of the mechanisms; it does not yet show cycling.

The Basic Key-Seed Changes (Replacement, Transposition, Insertion, Deletion)

All members of a series of key-seeds can either have the same rank or else change the rank at some or all iterations.

Series of Key-Seeds with Constant Ranks

Assume, the iteration produces several key-seeds of the same rank m. There are only two ways, by which one m-key-seed can turn into another of the same rank (figure 18-02).

Panel a shows an example of a 'natural' iteration series of 4-key-seeds. Their overlaps are large, ranging from 11 to 18 out of a possible maximum of 4·5 = 20. As before, the corresponding finished mosaics are placed above each seed. Four of the sequential stages are enlarged (panel b), because they illustrate the two of the fundamental possibilities of seed change, namely **replacements and transpositions**.

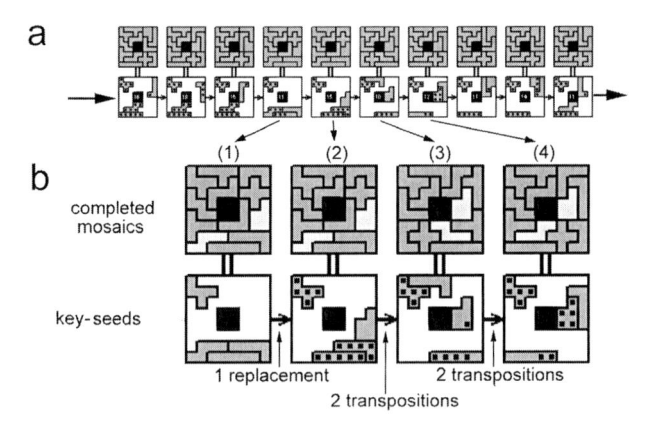

Fig. 18-02. Illustration of the 2 fundamentally different changes of key-seeds, which keeps the rank of the seeds constant, namely 'replacement' and 'transposition'. The labelling convention is the same as in figure 18-01.

In the case of **replacement** (see figure 18-02, marking (1) →(2)) one of the tiles of the seed, namely the 'N', was removed and replaced with a 'P'. The two tiles are highlighted in the images of the corresponding finished mosaics. Obviously, the corresponding finished mosaics are identical, in spite of the differences between the seeds. In short, **replacements leave the solutions intact.**

In contrast, **transpositions** (see figure 18-02 (2) →(3) and (3) →(4)) move one or several tiles of the seed to new locations. In the case of (2) →(3) the 'I', 'L', and 'P' tiles were moved to new locations. In the case (3) →(4), the 'I'- and 'L'- tiles were moved again. Nevertheless, the overlaps remained reasonably large, namely 10 and 12, respectively.

Such altered tile configurations may no longer qualify as key-seeds. However, if they do, then their corresponding finished mosaics must be entirely new ones: Obviously, no solution can have the moved tiles in two different locations at once. Hence, the resulting sibling-solutions can be considered as **mutants or variants** of the preceding mosaic, depending on the extent of their differences.

If the new configuration does not qualify for a key-seed of some other finished mosaic, the iteration ends with a failed assembly. In short, **transpositions enforce new solutions, or else they terminate the iteration lethally**.

Series of Key-Seeds with Variable Ranks

In the example shown in figure 18-02, the key-seeds have the same rank of 4. Of course, an iteration mechanism also may change the rank of the key-seeds. Such methods could remove a tile and make it a 3-key, or add one and make it a 5-key-seed. Correspondingly, we may call these seed alterations **deletions** and **insertions**.

Naturally, the series of key-seed alterations may include all the different types in some random order. Figure 18-03 shows an iteration series with several transpositions (marked 't') and several replacements (marked 'r'), but also with 2 cases of insertions. In the first case it increases the seed size from 3 to 4, and in the second case from 4 to 5. In both cases, the finished mosaics remain unchanged. Moving the same tiles to certain other locations would have destroyed the seed character, and therefore would have terminated the series.

The Key-Seed Pool as the Substratum of the Iteration

In order to simplify the following discussion of mosaic cycles, but especially to discuss the crucial influence of the 'seed-regeneration', it is necessary to introduce the so-called 'key-seed pool'. It is nothing more than the set of key-seeds on which the iteration algorithm draws when it selects its next key-seed.

There are numerous ways to structure this pool.

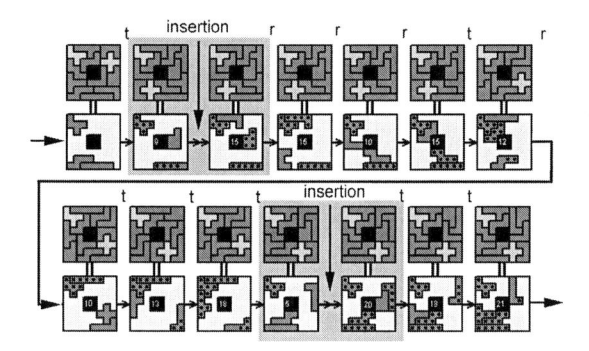

Fig. 18-03. Illustration of insertions ('insertion'), replacements ('r'), and transpositions ('t') into the consecutive seeds of an iteration (deletions are not shown). The labelling convention is the same as in figure 18-01.

1. It could contain all key-seeds for the type of mosaic under consideration.
2. It only could contain the key-seeds of a fixed rank.
3. It could omit specific key-seeds that the iteration mechanism is supposed to avoid.
4. It could eliminate certain key-seeds for only a certain 'regeneration period' in order to guarantee that the iteration would not be able to use them during this period.

Figure 18-04a adds the key-seed pool symbolically in the form of a circle to an iterated series of key-seeds. The arrows indicate that **the iteration procedure returns the previous key-seed to the pool** and retrieves a new one for the next iteration step.

Global Overview of Iteration Results: Plots of Seed and Solution Indices

Unfortunately, depictions of iterations like the ones on figures 18-01 and [18-03 are too busy and hence, ultimately confusing. In order to get a quick and global overview of the iteration results, it is better to represent each key-seed and each solution with a single point.

Naturally, this kind of simplification looses information. However, one can always add back some data via colour codes and labels to indicate e.g. whether a particular point was the result of insertion, deletion, replacement or transposition.

What kind of points should we choose that are able to represent a key-seed or a finished mosaic? It all depends on the information we want to display. In the following applications, we are most interested in the dynamic behaviour of the iterated key-seeds and mosaics. In other words, we want to find out whether certain rules of iteration leave them **unchanged**, while others cause them to **oscillate** and **cycle**, or yet others may render their sequences essentially **chaotic**.

For questions of this nature, it is sufficient to only know the pattern in which an individual key-seed or mosaic re-occurs. It could reappear either randomly, or all the time, or only after regular periods, or never. The necessary distinction can be accomplished by assigning to **each key-seed and mosaic its own unique index number.**

We have introduced earlier such specific indices. They simply indicated the order in which the assembly algorithms created the various mosaics, or the order in which the seed-finding algorithm (figure 11-07) has computed all seeds of a certain rank.

To be sure, the actual index numbers assigned to the key-seeds and solutions are quite arbitrary. Yet, as long as no two seeds or solutions ever have the same index, we can create graphs of the progressions of iterated siblings, and use them to identify patterns and periods in the graphs. We will note these indices as **key-seed #** or **solution #.** For example, figure 18-04c shows the result of the same natural iteration as the one in figure 18-01 using the key-seed #s and solution #s in order to represent the corresponding mosaics. Figure 18-04b explains the relationship between the sibling solutions and their key-seeds on one hand, and the key-seed #s and solution #s on the other.

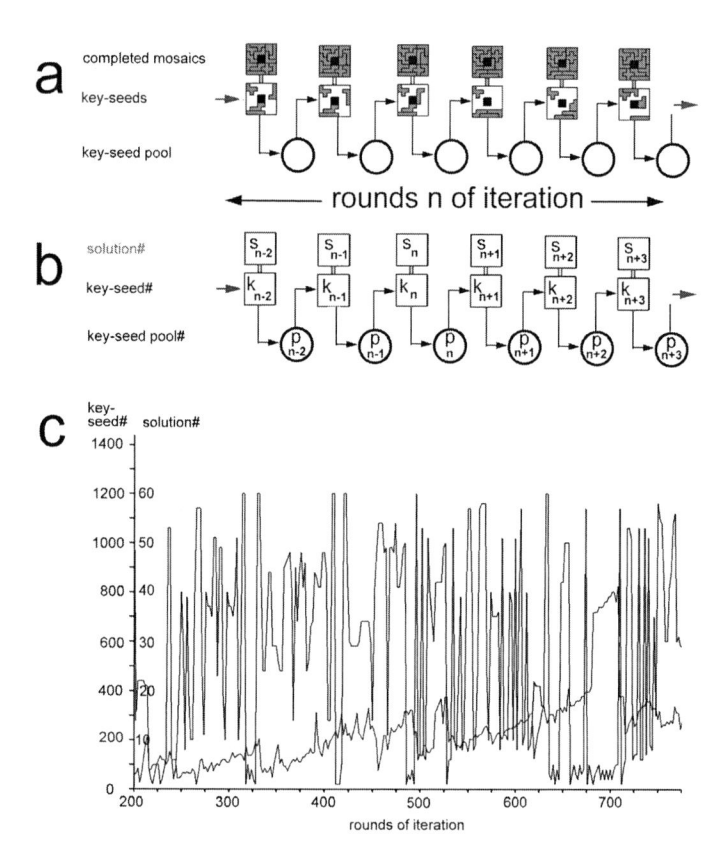

Fig. 18-04. Numerical representation of mosaic iterations based on sequences of key-seeds. (a) Inclusion of a symbolic representation of the key-seed pool from which the iteration retrieves the sequential key-seeds at each round n of iteration. The contents of the pool may change from round to round of iteration. (b) Replacing the actual drawings of the completed solutions and the key-seeds with the solution numbers ('solution #') and the key-seed numbers ('key-seed #'). (c) Actual plot of the result of mosaic iteration in terms of the solution #s (upper curve) and the key-seed #s (lower curve). The graph shows the same iterated sequence as in [18]-01 between rounds 200 and 775, represented not by individual drawings of the mosaics and their key-seeds, but by their unique indices, key-seed # (upper curve) and solution # .

As mentioned above, the actual values of the two graphs are arbitrary. However, once assigned to a key-seed or solution, they are unique. No other key-seed or solution will ever have the same ones. Hence, they are equivalent to any other one-on-one mapping between numbers and key-seeds or solutions that the student prefers to use. Therefore, the relationships between the assigned values are significant.

In this sense, it is obvious that in figure 18-01 the key-seed #s ratchet upwards from the earlier generated key-seeds (with low key-seed #s) towards the later generated ones (with high key-seed #s). Besides the ratchet, there is no periodicity in the essentially irregular sequence of the key-seeds.

The corresponding solution #s are clearly non-periodic, but appear chaotic within the entire range of possible solution #s between 1 and 64. Extending the graph to higher rounds of iteration up to round = 2800 (not shown) confirms the mentioned observations.

Naturally, we assume that the progression of the rounds of iteration occur in time. That does not mean that they correspond to equidistant time intervals. Yet, even if different rounds last different amounts of time in the reality of the biological situation, the abscissa in the above graph is ultimately a measure of evolutionary and/or developmental time. In the following discussions, we will therefore use the terms **'rounds of iteration' and 'time points' interchangeably**.

Regeneration Periods (srp) of Key-Seeds

So far we have assumed that the iteration procedure is able to pick any key-seed as the next one in the series. However, in the reality of biological applications, this is hardly ever the case. Key-seeds represent molecular precursors, zygotes, germinal tissues, or similar such initial structures. After having played their specific role as key-seeds in an iterative series of mosaics, they may require time for their regeneration, repair, reproduction, maturation, etc. before the iteration mechanism can use them again. In order to accommodate this biological reality, we ascribe to each key-seed a *'seed regeneration period, srp',* during which time it is not available, but after which time it (or a new copy of it) may re-enter the iteration procedure.

Technically speaking, the re-entry into the pool of available key-seeds is done in the following way: After every round of iteration, the remaining regeneration time of the key-seed decreases by one unit. When the allotted regeneration time is exhausted, the 'ban' on this particular key-seed will be lifted, and the key-seed may be used again if the rules of the continuing iteration call for it.

Please note, **after the duration of its srp, the key-seed is not guaranteed to enter the iterative series right away**. The iteration mechanism is merely able to find it among the other stored candidates and to call it up, if and when it needs a new key-seed that fulfils the required criteria of a valid 'next' key-seed.

In addition, the regenerated key-seed does not have to be the same one, which was put on hold during the srp. It could very well be a new copy. In this case, the srp means the time interval required to nucleate and grow a new mature copy of the particular key-seed. Formally, the results would be the same.

In this way, every key-seed returns to the pool of available key-seeds after 'srp' rounds of iteration. The illustrations assume the same value srp for every key-seed, although in some biological applications, different key-seeds may have quite different regeneration periods.

However, keep in mind that the regenerated key-seeds return to the pool at different times, even if all have the same value srp. In other words, although the duration of the 'count-down' is the same for all, their individual 'count-down' for the regeneration begins at different times, depending on whenever the seeds entered the iteration last.

For a practical illustration, consider the cycle of everybody's socks. Assume, you have 30 different pairs of socks, which you would like to wear in order. Every day, yesterday's pair goes through a cycle of washing, drying, and ends up in a drawer labelled 'fresh socks'. This drawer plays the role of the 'srp' for socks.

Now assume, you pick from the drawer a new pair of socks every morning. In an ideal world, the socks in the drawer are kept in their exact 'last in – last out' order. However, in reality, they form a heap, and you rarely find the ones that are supposed to be next. In fact, you have trouble finding the matching second sock. Hence, depending on the total number of socks and the size of the drawer, your daily order of socks is more or less scrambled up, although a certain 'almost' periodicity may still be noticeable.

Similarly, from round to round the srp changes its contents, and with it the availability of key-seeds. The selected key-seeds, in turn, determine the next sibling solution # in the series. The results are rather complex series of mosaics. As shown below, they can be constant, strictly periodic or even 'almost'-periodic. Some become aperiodic. Others funnel into almost-periodicities after a time of aperiodic behaviour. In short, as illustrated in the following and in chapter 19, the resulting iterated sequences display a great many behaviours similar to that of living mosaics.

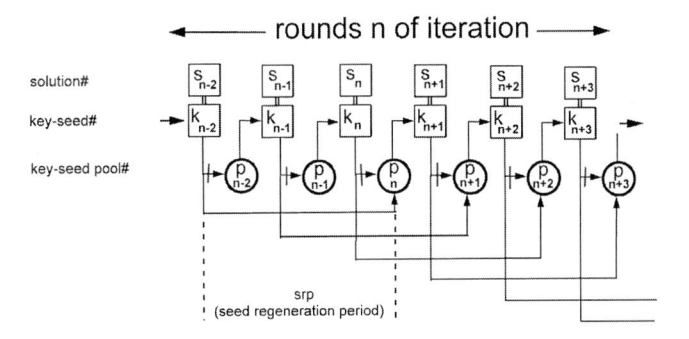

Fig. 18-05. Effect of the 'seed regeneration period (srp)' on the key-seed iteration of mosaics. In contrast to the mechanism depicted in figure 18-04b, after each iteration the previous key-seed is not returned to the pool immediately (see crossed out arrows), but only after 'srp' rounds of iteration. Therefore, in the meantime the subsequent iterations cannot select this very key-seed, until it is returned to the pool. (In the illustration all srp have the same value of 2. This is not a necessary condition.)

Cycles of Key-Seeds

The main effect of the seed regeneration periods srp is the appearance of cycling among the iteration results. The periods of the any cycles depend on the regeneration periods of the key-seeds involved. The following shows some examples based on simulations of fitting pentominoes into the model-frame.

(1). (srp = 1) → Constant Iteration Series

After a key-seed has been used in a natural iteration, its successor is the next, most similar key-seed. Assuming that it recovers right away (srp = 1) its most suitable successor is, of course, the key-seed itself. Hence, in this case the key-seeds never change during the iteration. Figure 18-06 shows this case, where key-seed #s and, corresponding, the solution #s are constant.

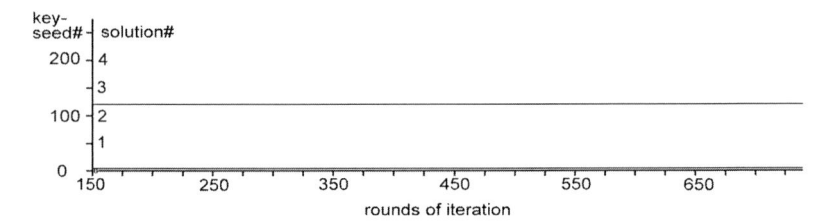

Fig. 18-06. Example of a constant iteration series as the consequence of srp = 1. (Simulated fitting of pentominoes into the model-frame).

(2). (1<srp<5) → Rapidly Oscillating Iteration Series

If the regeneration period is larger than 1, but still small (e.g. 1<srp<5), the iteration procedure has not moved very far away from an earlier seed, when its regeneration period is over. Hence, it is still a good candidate for the next, most similar key-seed, and the iteration may accept it again as a successor. Consequently, the iteration result oscillates (figure 18-07; srp = 5).

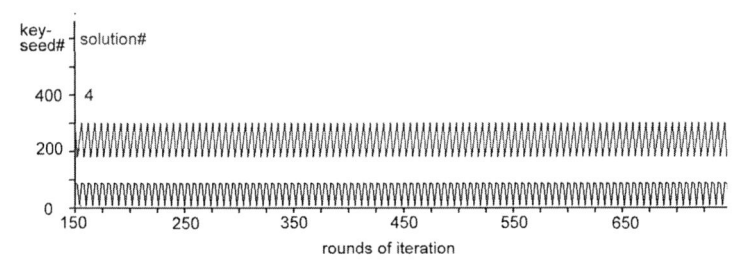

Fig. 18-07. Example of a rapidly oscillating iteration series as the consequence of a short srp = 5. (Simulated fitting of pentominoes into the model-frame).

(3). (10<srp<50) → Cycling Iteration Series with Complex Periods

Longer regeneration periods create more complex cycles. Nevertheless, cycles will result, because eventually one of the earlier key-seeds will be recovered and will qualify again as the next key-seed in line. Inevitably, choosing this particular key-seed again restarts the earlier stage of the iteration, and the cycle repeats (figure 18-08; srp = 20)

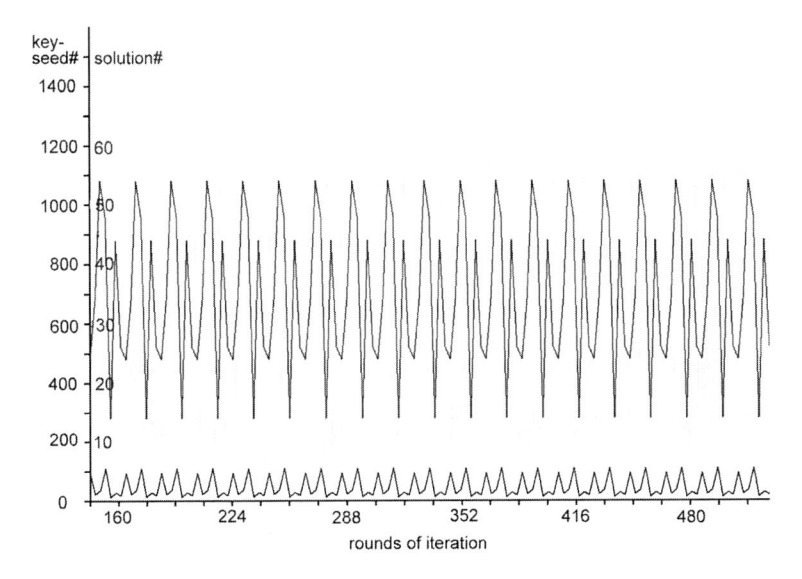

Fig. 18-08. Example of a cyclical iteration series with a complex periodicity as the consequence of srp = 20. (Simulated fitting of pentominoes into the model-frame).

(4). (100<srp)→ Cycling Iteration Series with Long, Very Complex Periods Following an Aperiodic Run

Often, the iteration has to go through rather long irregular, seemingly chaotic runs, before a repetitive configuration of key-seeds occurs. An example is shown in Figure 18-09 (srp = 100), where the iteration runs through an aperiodic stretch of 450 rounds, before a cycling (periodic) configuration is reached.

The iterations create a wide variety of cycling series of mosaics, including aperiodic pulsations that alternated between essentially 2–4 limiting mosaics (figure 18-10).

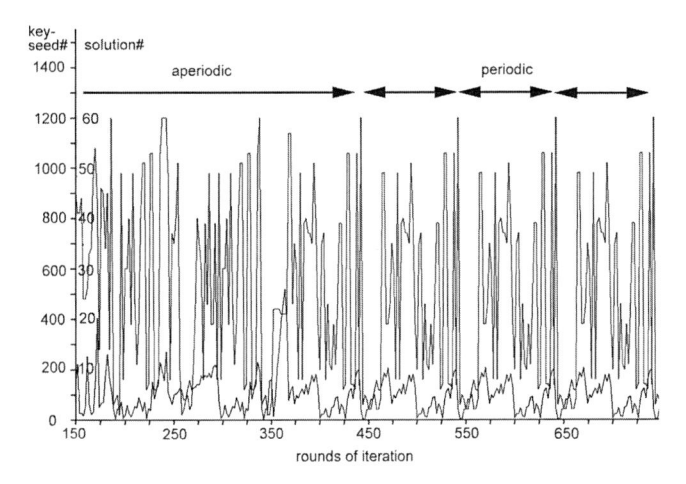

Fig. 18-09. Example of a cyclical iteration series with a long, complex period following an initial aperiodic (chaotic) run. The entire sequence is the consequence of srp = 100. (Simulated fitting of pentominoes into the model-frame).

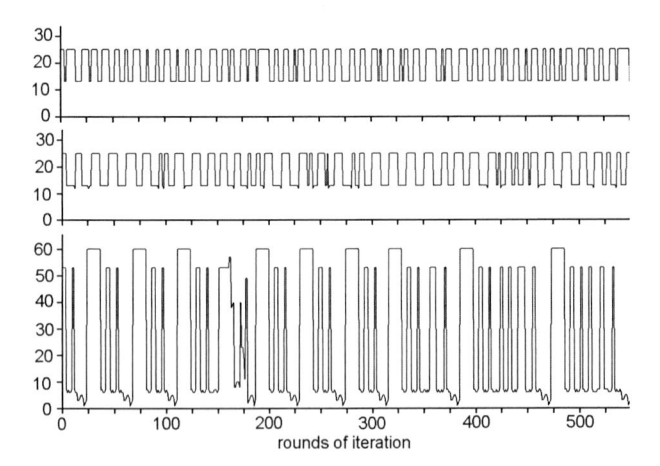

Fig. 18-10. Unusual iteration series. Aperiodic pulsations that alternate between essentially 2–4 limiting mosaics. (Simulated fitting of pentominoes into the model-frame).

The Relationship between Seed Regeneration Period and Cycle Length

Intuitively one may think that the length of the cycle 'T' grows with the length of the regeneration period 'srp'. In a crude way this is correct. However, a more detailed examination reveals a non-linear and much more complex relationship.

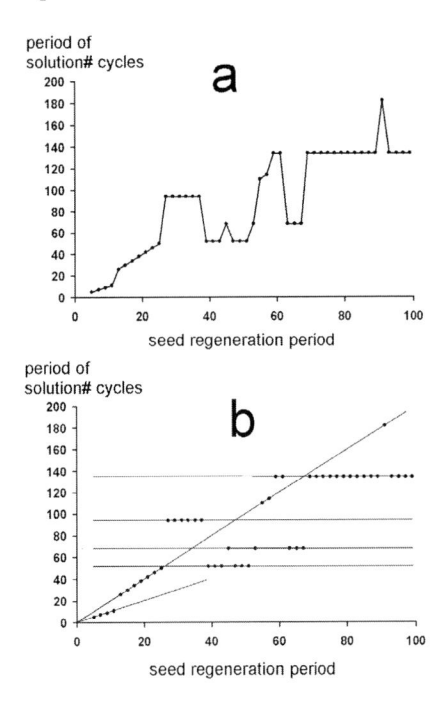

Fig. 18-11. Cycle length as a function of the seed regenerative period, srp. (a) direct plot of the data of a simulation under the condition of nearest key-seed similarity that varied systematically the srp. (b) interpretation of the same plot as the expression of several mechanisms with constant and linearly rising periods (discussion see text).

Translating the result into biological terms, it appears that small changes of the regenerative properties of a key-seed may yield dramatic changes in the all important biological cycles of a system.

Figure 18-11 shows the example of the cycle length 'T' as a function of the regenerative period 'srp' of 4-key-seeds under the condition of a similarity value >10. It appears that the complexity of the result can be reduced to 3 basic mechanisms that determine the length of T.

Consider panel b, which shows the same data points as panel a, although they are connected in a different way! One set of points belongs to the function $T_1 = $ srp, the other to the function $T_2 = 2 \cdot$ srp, and the remaining sets belong to the seemingly preferred levels $T_{3,1} = 52$, $T_{3,2} = 68$, $T_{3,3} = 94$, and $T_{3,4} = 134$.

Chapter Nineteen

The 'Almost'-Periodic Behaviour of Iterated Cycles

The 'Seed Regeneration Period (SRP) that was Introduced in Chapter 18 Offers a Simple, Yet Deterministic Mechanism for the 'Almost'-Repetitions of the Cycles of Living Mosaics; No Need to Resort to Thermal Noise or Other Random Perturbations as the Causes

As pointed out earlier, nothing in biology repeats exactly. There is no mechanistic explanation for this ubiquitous phenomenon, except that the number of variables in living mosaics is so large, that their possible combinations exceed the numbers of actual individuals by far. Furthermore, there is a universal inter-connectedness between everything biological. Once such dense interconnections are established, the 'almost'-periodic behaviour of only a single system may influence every other to various degrees and prevent that it repeats precisely whatever cycle it is aiming to complete. Hence, each of the connected systems, among them those with strong influences on others, become 'almost'-periodic, and so forth.

But, how exactly can the 'environment' around an iterating system influence which key-seed it chooses next? For a possible answer, let us look more closely at the cyclical phases of iterations as well as at their non-repetitive, aperiodic phases.

The Periodic Behaviour of Iterated Cycles

Initially, it may seem logical that all iterations must be strictly periodic, or else stop altogether. After all, every 'next' key-seed of the series comes from a finite supply of possible key-seeds. If every next key-seed is to be different from its predecessor, eventually the number of different key-seeds must be exhausted. After having arrived at such a point the iteration must either come to its end, or select as its next key-seed one that has already occurred. Once a particular key-seed repeats, all its subsequent key-seeds must repeat, as well, because the logic of picking the next is still the same as it was the first time around. However, the effect of the regeneration periods 'srp' changes all that.

The Aperiodic (Chaotic) Behaviour of Iterated Cycles

As we have seen there is a third, more interesting type of behaviour of the iterations, which is due to the existence of the seed regeneration periods 'srp'. Figure 18-09 showed that the iterated cycles often begin with a phase of aperiodic behaviour, until they funnel into a strictly periodic continuation. How is the aperiodic phase linked to the role of the srp?

Regardless of the complexity of the relationship between the seed regeneration period and the cycling period, the beginning of the iteration series is always independent of the regeneration processes. As long as the first key-seed had no reason to regenerate, or else has started but not finished its regeneration, the iteration mechanism cannot even be 'aware' of the fate of this or any other of the earlier key-seeds. *Only after the first key-seed returns for the first time to the pool of available key-seeds, can its regeneration, i.e. its re-appearance have an effect on the continuing iteration.*

Here is the reason for the aperiodic phase. Since no two key-seeds can be the same, the iteration series cannot contain two copies of the same key-seed, as long as the first regenerating key-seed is not finished with its regeneration. Afterwards, the iteration may still not repeat this key-seed, but from now on, it is a valid candidate as the next key-seed. Hence, until it happens to be actually selected, the forgoing phase is necessarily aperiodic. As a further consequence it follows, that *the initial iteration series is seemingly chaotic the longer, the larger the seed regeneration period 'srp' is.*

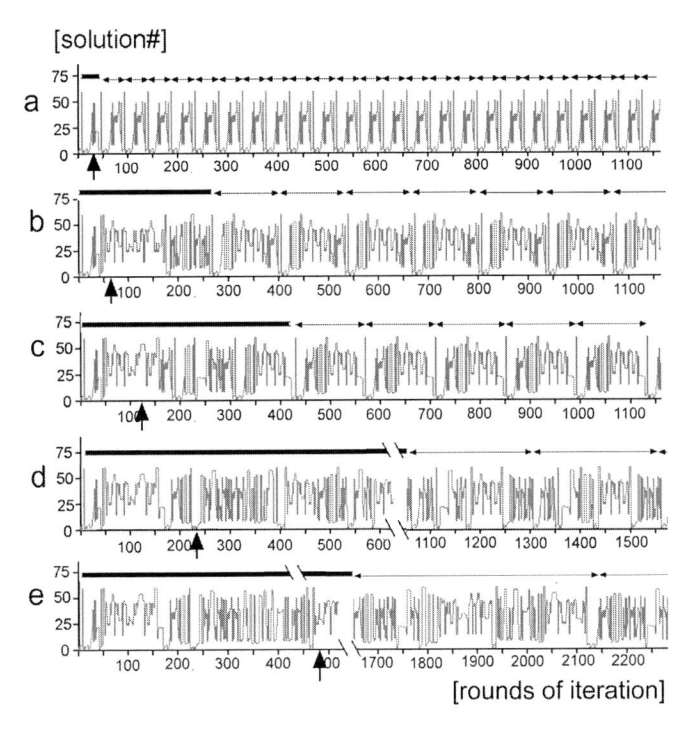

Fig. 19-01. Comparison of iteration sequences that use the same parameters, but have different srp's. (a) srp = 30, (b) srp = 60, (c) srp = 120, (d) srp = 240, (e) srp = 480. Vertical arrows indicate the srp levels. The horizontal double-arrows mark the resulting periods. The horizontal heavy bars mark the aperiodic phases. The larger the srp is, the longer is the aperiodic phase, which is always longer than the srp. Each aperiodic phase shares the aperiodic phase of the previous (smaller) srp. Therefore, the aperiodic phases are highlighted in alternating gray-levels in order to show their common segments. Also the length of the periods grows with increasing srp, until there is no longer an exact repetition of the solution numbers (marked ('almost')). Eventually (panel e) there is only aperiodic iteration left. (Simulated fitting of pentominoes into the model-frame).

Let me emphasize again, that any subsequent periodicity does not necessarily begin immediately after a formerly used key-seed returns to the srp. Returning to the srp only means that the key-seed becomes eligible. The next period can start only after the iteration happens to pick it as its next key-seed. *The aperiodic, seemingly chaotic phase of the iteration lasts at least as long as the srp, although in most cases it lasts much longer.*

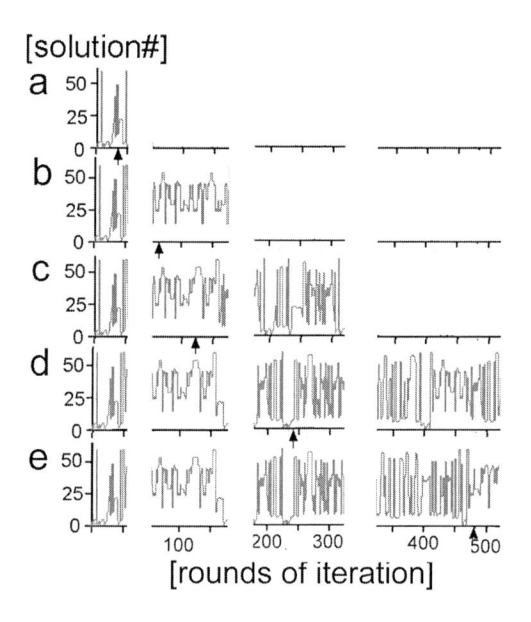

Fig. 19-02. The incremental structure of the aperiodic phases of iteration sequences as a function of the size of the srp. The figure shows the aperiodic phases of the iteration sequences of figure 19-01. Arrows indicate the sizes of the srp's. Whenever 2 iteration sequences have the same general parameters but different srp's, then their aperiodic phases are identical for the duration of the smaller srp.

Figure 19-01 illustrates this conclusion by comparing several iteration sequences that have all the same parameters, except the values of their srp's increase from 30 to 480. As mentioned above, the iteration cannot show any effect of the regeneration processes until it has exceeded at least the smaller of two srp values. Consequently, the iterations are identical for the duration of the smaller srp.

In general, *whenever 2 iteration sequences have the same general parameters, but different srp's, then their aperiodic phases are identical for the duration of the smaller srp* (figure 19-02).

Take the iteration sequence (d) with a $srp_{(d)} = 240$, and compare it with the sequence (c) with a $srp_{(c)} = 120$. Since $srp_{(c)} < srp_{(d)}$ we should find that both sequences are identical for the first $srp_{(c)} = 120$ iterations. The reader can easily verify this conclusion and several others like that.

It is important to note all these conclusions only are valid, if the srp is constant for the entire length of the iteration.

How the 'Environment' May Create Deterministic 'Almost'-Periodicities

Let us return now to our question how the 'environment' consisting of interlinked iterating systems, is able to modify an iterating mosaic in such a way as to render it 'almost'-periodic. Of course, if a mechanism would result from the strictly periodic behaviour of all interlinked iterating systems, it would hardly create 'almost'-periodic behaviour. After all, the 'environment' only contains a finite number of iterators and, as one can show easily by Fourier analysis, the finite sum of strictly periodic functions, is always strictly periodic itself. Therefore, we look towards the aperiodic phases of the interlinked systems as the generators of 'almost'-periodic behaviours.

Furthermore, if a hypothetical mechanism is capable of modifying the key-seeds directly by deletions, insertions, transpositions or replacements, it would have to be mutagenic. Yet, mutations would be far too powerful cannons to shoot at subtle modifications such as 'almost'-periodicities. Imagine what our life would be like if it would require mutations of cardiac cells to explain our variable heartbeats!

Therefore, if we want to understand how strict periods may turn into 'almost'-periods, the srp appears as the likely targets for changing the parameters of iteration. In other words, we propose a mechanism that links the amplitudes of a natural iteration, including its aperiodic phases to its own srp or to those of another. This effectively scrambles up the srp's in a deterministic way and renders the phases of strict periods of cycling mosaics into 'almost'-periodic phases. In other words, ***we interpret the 'almost'-periodic cycling of biological mosaics as the result of cyclical natural iteration sequences whose srp is varied by the key-seed or mosaic indices of their previous own, or another natural iteration.***

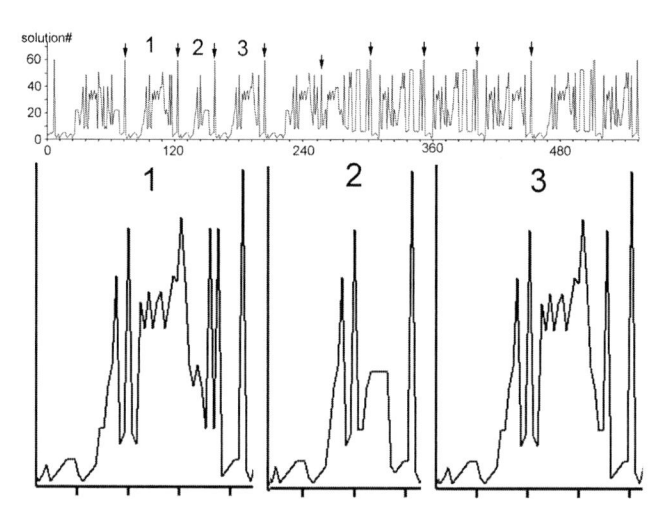

Fig. 19-03. Self-randomized 'natural' iteration using the iterated solution indices and a weight factor f = 0.005 to modify srp₀ = 50. A sequence of varying periods (starts marked by arrow) results. Three of the examples are marked with numbers and enlarged. Abscissa: rounds of iteration. (Simulated fitting of pentominoes into the model-frame).

In a symbolic formulation,

equ. 19-01

Given the k-key-seed indices $^kM_{i,n}\#$, the seed regeneration periods srp_{i0} of the k-key-seed-driven 'natural iterations

$$^kM_{i,n+1}\# = IT_i(^kM_{i,n}\#), (I = 1, 2)\text{ then}$$

1. Their self-randomized periodicities are generated by
   $srp_i = srp_{i0} + f_i \cdot {}^kM_{i,n}\#$, (I = 1, 2).
2. Their mutually-varied periodicities at every stage 'n' of the iteration IT() are generated by
   $srp_1 = srp_{10} + f_1 \cdot {}^kM_{2,n}\#$, .
   $srp_2 = srp_{20} + f_2 \cdot {}^kM_{1,n}\#$, with $f_1, f_2 \ll 1$.

Instead of the k-key-seed indices $^kM_{i,n}\#$, one can also use the solution indices $S_{i,n}\#$ with appropriately modified weights $g_i \ll 1$.

Figure 19-03 shows an example of a 'natural' iteration with a self-randomized srp = 50 + 0.005·$S_n\#$. A sequence of varying periods (starts marked by arrow) results. Depending on the magnitude of the feedback, the periods are between constant and unrecognizably different from each other.

The Imaging of Mosaic Cycles

Using the convention, in which the chapters 18 and the present chapter depicted iterations, one can easily identify periodicities and various other patterns. However, as emphasized earlier, the actual values in the figures were entirely arbitrary, although they were uniquely linked to each mosaic and its key-seeds.

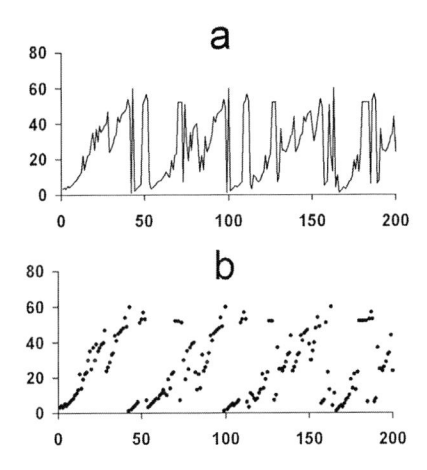

Fig. 19-04. Depictions of the iteration results as described in Chapter [19]. (a) Sequential data points are connected, giving the (misleading) appearance of a continuous curve. (b) Since the results of a series of iterations are only defined on integer values on the abscissa, the disconnected points of the same data yield a more realistic but less readable representation.

However, this is not the only arbitrary feature of the figures shown in these chapters. They show graphs like figure 19-04a as if they were continuous curves.

Of course, the actual results of the iteration only make sense for rounds of iteration, i.e. for the values on the abscissa that are integers. Therefore, they should be depicted more accurately as in figure 19-04b. The connecting line segments between data points in figure 19-04a do not depict any results of the iteration. They merely indicate which point precedes which other

Nevertheless, we will continue to use graphic representations with connected data points like in figure 19-04a, because many people, including the author, cannot relate very well to disconnected clouds of points. The main purpose of these graphs, namely to help simplify the interpretation of the results of iterations, is better served with connected data points.

There are many more representations than those of Figure 19-04. Depending on the requirements of real-life application, it may be necessary to image the data in various other ways.

One-on-One Mapping of Iterations

Unless one is prepared to lose data points, alternate ways of graphic representations should be based on a one-on-one transformation of the given data points into new ones.

It is easy to prove that a one-on-one mapping of a one-on-one mapping is also a one-on-one mapping. In other words, as long as we transform the key-seed indices or solution indices in a one-on-one fashion into some other numbers, everything remains valid, except the graphs will look different. Examples of such legitimate transformations are all permutations of the indices, or monotonous functions such as exponential functions, logarithms, etc.

In practice, one can set up a matrix and mark the number that the transformed index has for every key-seed index or solution index (table 19-01). As long as there only is one mark in every column and in every row, the matrix provides a valid one-on-one transformation.

Effectively, each such matrix describes a permutation of the data points. Unfortunately, the vast majority of permutations simply scrambles up the order of the data points and destroys every pattern that the human eye is able to recognize. If successive points are connected, the resulting graph of permutated data points usually appears as noise. Therefore, in practice, one would prefer matrices like table 19-02, which leaves whole segments of the data intact, and merely changes their order.

0	1	2	3	4	5	6	7	8	9	10	11	12	13	etc
1	X													
2			X											
3														
4		X												
5								X						
6							X							
7									X					
8													X	
9											X			
10				X										
11														
12														•
13					X									
14														
15											X			
16													X	
17				X										
etc.														

Table 19-01. Example of a transformation matrix that defines a one-on-one mapping between the numbers in the rows and the numbers in the columns. This matrix maps 1→1, 2→4, 3→2, 4→10,.. As long as there only is one mark in each column and row it is a valid transformation.

Transformation by Monotonous Functions

Another simple way of transforming the iteration data while preserving the information is the application of monotonous functions i.e. functions that that either fall or rise all the time. In other words, whenever their abscissa values are different, then their ordinate values are different, as well.

238   Chapter Nineteen

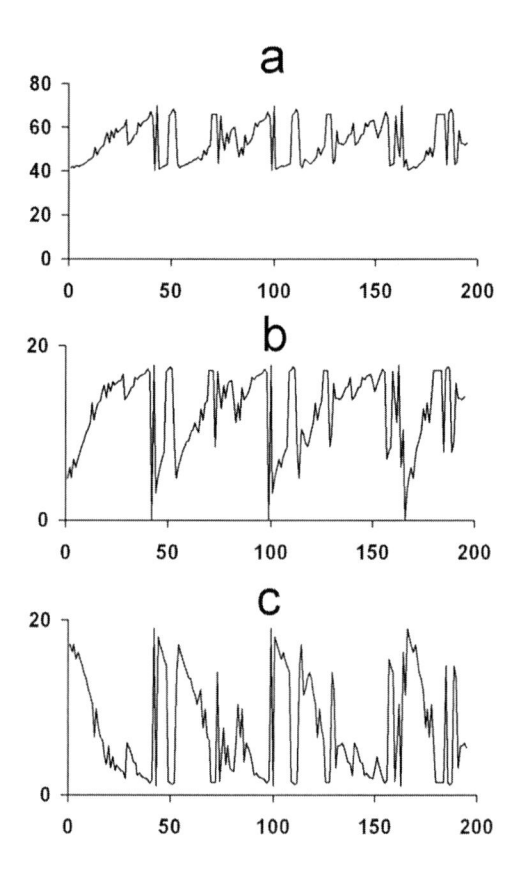

Fig. 19-05. Transformations of the data of figure 19-04a by monotonous functions. (a) linear functions ($\alpha = 0.5$; $\beta = 40$). (b) logarithm ($\alpha = 10$). (c) exponential ($\alpha = 20$; $\beta = -0.05$)

Let w stand for the (always positive) values of key-seed #s or solution #s, and z for their transformed values, then examples of suitable functions are

equ. 19-01
1. $z = \alpha \cdot w \, \beta$,
2. $z = \alpha \cdot \log(w)$,
3. $z = \alpha \cdot \exp(\beta \cdot w)$,

with appropriate constants parameters α, β.

Figure 19-04 shows examples of these functions transforming the iteration series of figure 19-04a.

0	1	2	3	4	5	6	7	8	9	10	11	12	13	etc
1	X													
2		X												
3			X											
4				X										
5					X									
6						X								
7										X				
8											X			
9												X		
10													X	
11														
12														
13														
14							X							
15								X						
16									X					
17														
etc														X

Table 19-02. Example of a transformation matrix that preserves certain stretches of the data but changes their order.

Imaging with Loss of Data

Imaging methods that suppress or lose data for other reasons are not necessarily bad. For example, one may want to only depict the data points that exceed a certain threshold, stay below a threshold, or fall into the window between 2 thresholds. Figure 19-06 shows examples of these 3 cases, using again the iteration series of figure 19-04a.

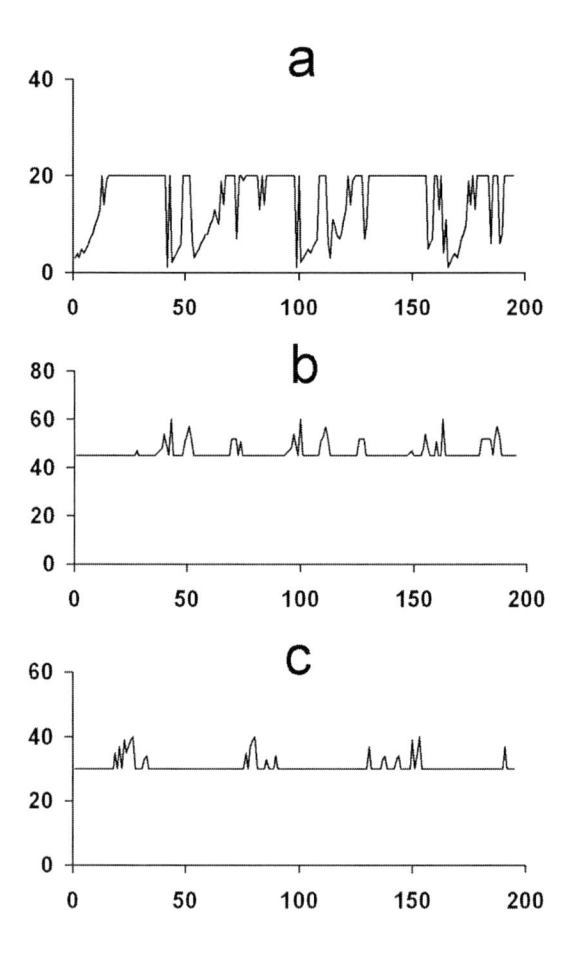

Fig. 19-06. Imaging of the data of figure 19-04a while imposing thresholds th. (a) Values below th = 20. (b) Values exceeding th = 45. (c) Values in the window between th = 30 and th = 40).

NUMERICAL DESCRIPTIONS OF THE FUNCTIONS OF MOSAICS AND TILES

Chapter Twenty

Measuring Distances, Contacts, and Functions of Tiles and Mosaics

Given a Mosaic, How Does One Determine the Distance between Two Tiles, Or the Function of a Tile?

The Distances between Tiles

If you walk from tile to tile inside a mosaic, there only are two outcomes. Either you move within the area (volume) of the same tile, or you step over to another tile.

Crossing a tile border, does not take you to 1/10 or to 1/4 of the next tile, but you arrive in a single step at the entire next tile. It is like crossing the dateline on Earth: After stepping the slightest distance over it, your date is altogether 'yesterday' or 'tomorrow', not 'a little bit tomorrow' or '1/10 of yesterday', no matter how many inches or yards you went over it.

 Therefore, the geometric length of the steps is irrelevant as long as it exceeds zero. It only matters whether the step has reached another tile. Consequently, the sizes of the steps are measured in terms of integers 0, 1, 2,…, provided you stay inside the mosaic and do not step outside the frame.

Step size = 0 means 'no step'; step size = 1 means a step over to an adjacent tile; step size = 2 means a step to the adjacent tile of an adjacent tile, and so forth (see figure 20-01). 'Walking along a mosaic' means to step from tile to tile along a certain path. There are many different path-lengths between two tiles of a mosaic.

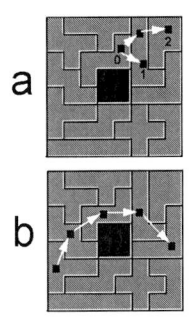

Fig. 20-01 The definition of stepping through a mosaic pattern. (a) Definition of distance between tiles that differs from the usual geometric distance. Adjacent tiles (0 → 1) have a distance of 1; Tiles that are adjacent to an adjacent tile (0 → 2) have distances of 2, and so forth. (b) Every path through the patterns consists of steps from a tile to an adjacent one (black marks connected by white arrows).

Consequently, the 'distance' between tiles of a mosaic can be defined as the shortest path length between them.

The Continuity of Tile Patterns

Besides moving onto or beyond the frame, there are other possibilities for a step not to land on a tile.

It can arrive at a spot where there is a gap between tiles (figure 20-02a). This case may occur because the tiles cannot fit, or because the fitting process is incomplete, as in the case of key-seed mosaics.

Case (1) applies, if the mosaic is a key-seed mosaic (see figure 11-01). It also applies, if the fitting of the mosaic is not perfect, but leaves 'dead' spaces between tiles. Topologists have classified objects with 'holes' and, depending on the number of disconnected holes, they describe them by the degree of their 'connectedness'. In the case of figure 20-02a, the mosaic is still connected, because there is another path that circumvents the gap. In contrast, the mosaic of figure 20-02b, is disconnected, because there is no continuous path between the depicted tiles.

equ. 20-01

In analogy with the mathematical definitions of 'continuity', we may suggest defining the 'continuity of the tile pattern of a mosaic' by requiring that there is at least one path *between any 2 tiles that does not need to leap across a gap between consecutive tiles.*

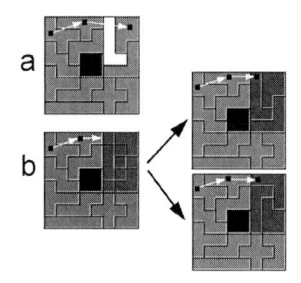

Fig. 20-02 . Possible discontinuities of a pentomino mosaic pattern. (a). A path encounters the gap between tiles (white area). (b). A path encounters an ambiguity, i.e. a spot where 2 different patterns overlap (dark gray area). In one case the path arrives at an 'L-pentomino', in the other at a 'V-pentomino'.

Hence, we define 'continuity' of tile patterns as follows.

The Distance between Mosaics

How do we find the 'closest' next mosaic to a given one? In terms of our metaphor that compares the successful tessellations to rare and isolated islands in a vast ocean of non-solutions, we need to measure the 'proximity' between any two such 'islands' in order to find the nearest.

Our metaphor of 'solutions as islands in a vast ocean of non-solutions' is actually a direct reflection of the assembly protocol. Its running algorithm tries to fill the frame with a large series of tests that turn out to be non-solutions most of the time. Rarely, they are 'interrupted' by a perfectly fitting sibling solution. Sometimes the algorithm finds sibling-solutions in rapid succession, in spite of their rarity; other times large pauses occur between one solution and the next. The ones that come in rapid succession are likely to look quite similar because the protocol did not need to change a solution very much in order to come up with the next. Therefore, we will identify the 'proximity' between solutions with their 'similarity'.

One measure for 'similarity' could be derived directly from the tessellations. Assume we place the two sibling-solutions M_n and M_m on top of each other. Wherever parts of identical tiles overlap, we paint it white (figure 20-03). Obviously, the more similar the two mosaics are, the more white areas will result.

Measuring Distances, Contacts, and Functions of Tiles and Mosaics   245

Of course, we have to standardize the relative orientations of the two solutions when we place them on top of each other. It is no problem, as long as the frames have no symmetries (cf. figure 26-01 c). In that case one can agree on a standard orientation of the frames to be used before counting any overlaps of the tiles.

However, in the case of symmetrical frames, such as our model-frame, rotations and/or reflections of the entire pattern would change the tiles, but not the frame. Consequently, the overlap areas between the tiles inside the frames would change with every such symmetry operation, even though the tiles have not changed. Hence, such frames require a special convention to standardize their orientation.

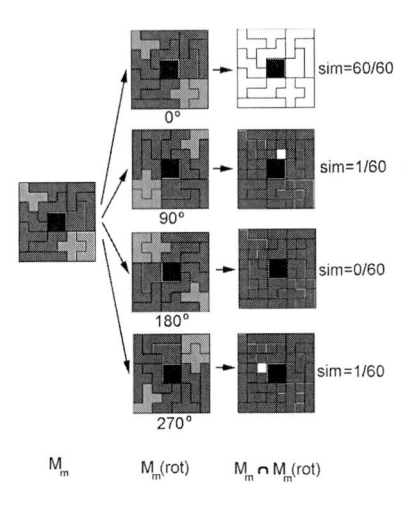

Fig. 20-03. The problem of using the overlap between identical tiles in order to define the similarity between 2 sibling-solutions, if rotation (and reflection, but not shown) of the solution are allowed. The illustration shows how the rotation by 90°, 180°, and 270° destroys the similarity of even a solution with itself.

Figure 20-03 illustrates the problem in the special case where we measure the overlap of a mosaic with itself. Dramatically different overlap areas result if one of the model-frames undergoes the four possible rotations. Obviously, such differences must not occur if we want the measure of 'similarity' to be meaningful: No other mosaic can be more similar than the same mosaic. Therefore, we must place the two sibling-solutions in standardized orientations before measuring their overlaps.

One way, is to measure the overlaps for all possible orientations of the two siblings, and take the maximum value. Unfortunately, it is possible that two different orientations yield the same maximal value, especially if the maximum is a small number. Although this method does not define the best orientation of the two siblings, it still yields an unambiguous value for their similarity.

Alternatively, one can make it a rule to leave all siblings in the particular orientation in which the assembly algorithm had produced them. After all, the algorithm must exclude the assembly of mirror images and rotations of each solution in the first place. Hence, if we place the mosaics on top of each other in the same 'native' orientation as the assembly algorithm did, then none of the alternate rotations or mirror images will be able to enter the measurements.

Of course, the latter convention only works if we know the assembly algorithm that generated the sibling-solutions. In the case of naturally occurring living mosaics this is hardly the case. In these cases, we could use the 'maximal-overlap-method'. The naturally occurring living mosaics rarely have such a wealth of body symmetries that would make it unfeasible.

Hence, we assume that there are practical conventions to place two sibling solutions on top of each other in order to measure the similarity, i.e. the closeness between two standardized mosaics M_n and M_m, by their mutual overlap area. Naming this similarity as

| equ. 20-02 |

$sim(M_n, M_m) = area(M_n \cap M_m)/area(frame)$
(assuming standardized orientations of the mosaics)

This quantity has some desirable properties.

1. It is independent of the individual steps of the assembly protocol, because it is derived from the completed mosaics, where these steps have no longer any direct influence.
2. It is symmetrical, i.e. $sim(M_n, M_m) = sim(M_m, M_n)$.
3. It includes a partial overlap between identical tiles in both mosaics. Hence, the measure allows changes in amounts that are smaller than 1 whole tile, which makes it as gradual as possible.

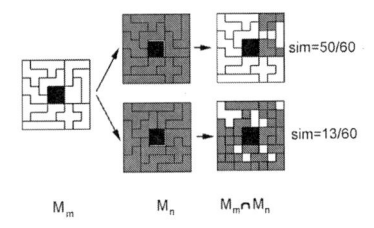

Measuring Distances, Contacts, and Functions of Tiles and Mosaics 247

Fig. 20-04. Example of a similarity determination based on equation 20-07 between a standardized mosaic M_m and 2 different standardized mosaics M_n.

Figure 20-05 shows the overlap patterns between the solution #14 (highlighted at the left upper corner) and 31 consecutive other solutions. Although redundant, it is meant to explain the concept of the overlap in more examples, and demonstrates the diversity of the patterns and sizes of the overlap areas.

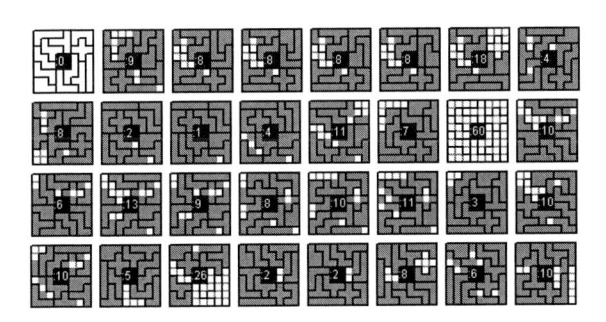

Fig. 20-05. Overlap patterns between the sibling-solution #14 (shown at the left upper corner) and 31 consecutive other sibling-solutions (including solution #14).

Figure 20-06 shows a plot of sim(M_n, M_m) for all mosaics of the model-frame. The plot marks the diagonal as the highest level, because sim(M_n, M_n) = 1 for all n. Furthermore, property (b) makes the values symmetrical about the diagonal.

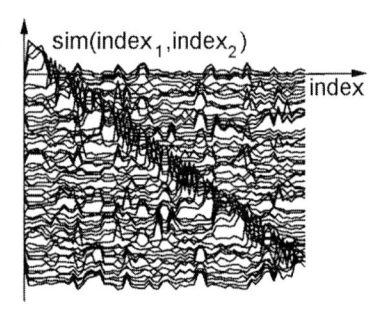

Fig. 20-06. Similarity value sim(M_1,M_2) between each sibling-solution of the model-frame and every other. (M_j are sibling-solutions of the model-frame). Ordinate: value of sim(M_r,M_k) with fixed 'r' (r = 1,...,64) and (k = 1,...,64). Abscissa: the indices 'k' based on the order in which the assembly algorithm created them. The results for different reference solutions #r are staggered equidistantly along the ordinate in the same order as on the abscissa.

Using this quantity, we can now proceed to select a series of mosaics that differ either minimally or maximally from a reference solution.

The Value of Mosaics

How can one determine quantitatively the differences between any two sibling mosaics?

Presumably, the first step would be to determine the 'value' of each mosaic, and then somehow 'subtract' one from the other. Such an approach, in turn, requires that we find an algorithm to compute the 'value' of a solution from the 'value' of each tile and the 'weight' of its location inside the frame. Let us use the pentomino model to suggest a simple way to do this. Afterwards, it will be rather obvious how to generalize the approach to other situations.

However, it is important to note, that our proposed 'mosaic-value' may assign the same value to two different mosaics. The alternative to guarantee that the value of different mosaics is always different, would force us to use extremely large numbers that are probably the unique expressions of prime numbers. However, there is no need for such an elaborate approach. The situation where different objects have the same values occurs quite frequently in the real world, where e.g. two different cars may cost the same, or two diamonds have the same value.

In order to **characterize each tile and its location**, we assign each of the 5 squares of each pentomino p the same number d_P (p = 1,..,12) (figure 20-07 a). For example, we assign the numbers 1, 2, 3,..., 11, 12 to the squares of the pentominoes F, I, L, N, P, ...,Y, Z in order. It means that d_F = 1, d_I = 2, d_L = 3,..., d_Y = 11, d_Z = 12. Then we place these numbers into the tiles of a solution as in figure 20-07 c. In this way, each square of the frame in position [j,k] with j,k = 1,..,8 is given a number $p_{jk} = d_P[j,k]$, depending on the location of each pentomino P within the solution.

In order to **characterize the weight of each location**, we assign each of the squares of the frame a weight q_{jk} with j,k = 1,.., 8 (Fig. 20-07 b).

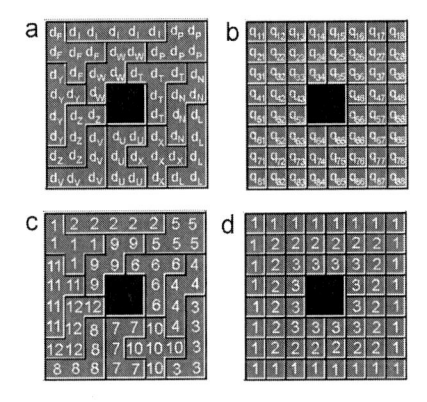

Fig. 20-07. Example of the evaluation matrices for the model-frame. (a) Assignment of an identifier d_i for each element of each tile (b) Assignment of a weight coefficients q_{ik} for each position inside the frame. (c) Numerical example of the identifier matrix of panel (a). (d) Numerical example of the weighting matrix of panel (b). In this case each square is given a weight from 1 to 3.

Figure 20-07d illustrates an example of such a **weighting matrix**. In this case each square is given a weight from 1 to 3.

The assigned coefficients q_{jk} and $d_P[j,k]$ are arbitrary values at this point. Yet, there is an important difference between these assignments and the arbitrary numbers of the key-seed indices and solution indices, which we had introduced earlier. Although necessary for the quantitative handling of mosaics, they had no direct physical or biological meaning.

In contrast, it is possible interpreting the coefficients of the weighting matrices in physical or biological terms. After all, they represent the effect of the presence of a particular part of a tile in a particular location within the frame. This effect on the system may be measurable by appropriate experiments, which makes them similar to many other types of empirical constants. Furthermore, we may wish to express certain symmetry conditions. For instance, the weighting matrix reflects the symmetry of the frame. Using this assignment, the rotation or reflection of the matrix relative to the solution would not alter the result.

Now let us calculate the weight s of the shown solution as

equ. 20-03

$$s(\text{solution \#}) = \sum_{j=1}^{8} \sum_{k=1}^{8} q_{jk}\, p_{jk}.$$

In the above example, w(solution #) = 721. Obviously, if all weights $q_{jk} = 1$, then $s_0 = \sum_{j=1}^{8} \sum_{k=1}^{8} p_{jk} = 390$, independent of the solution #, because it simply adds up 5 times the numbers 1 to 12, no matter where each pentomino is located. Therefore, a natural way of standardizing the weights of the solutions would be to define

equ. 20-04

$$\sigma(\text{solution \#}) = (\sum_{j=1}^{8} \sum_{k=1}^{8} q_{jk}\, [p_{jk}/ \sum_{m=1}^{8} \sum_{n=1}^{8} p_{mn}].$$

All what we said earlier about the imaging of mosaic cycles by mosaic indices, also applies to mosaic values *mutatis mutandis*. Of course, in addition to the described ways, there are many other ways to image the results of iteration series if they are in the form of numbers like the mosaic indices of the key-seed # or the solution #. Yet, the above examples may suffice to reassure the student that the use of the arbitrary mosaic indices does not prevent a biological interpretation of the data.

The Inner Structure of Mosaics

In order to understand the architecture of mosaics and their behaviour, we need to characterize each mosaic's unique inner structure together with its functional consequences. The most basic structural information of

Measuring Distances, Contacts, and Functions of Tiles and Mosaics   251

mosaics is, of course, the shape and location of each of the tiles. We may call it the **anatomy of the mosaic**.

Assume that there are τ tiles and a frame Φ. Even though our model-frame is 2-dimensional, in reference to the 3-dimensionality of biological objects we will call the inner areas 'volumes' and the perimeters 'surfaces. Each tile has a volume of v_τ and a surface of s_τ. In the case of the pentominoes there are $\tau = 12$ tiles. By definition, each tile has the same volume $v_\tau = 5$. It turns out that almost all of them have also the same surface $s_\tau = 12$, except the 'P' which has a surface of only 10.

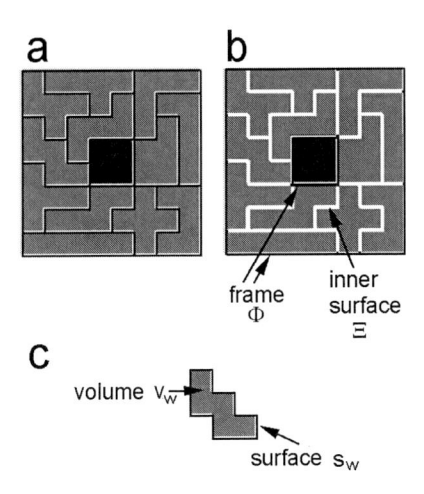

Fig. 20-08. Explanation of nomenclature used to derive the important relation that the total surface of a fitted mosaic is given by equation 20-12.

Similarly, the frame has a total volume of V_Φ and a surface S_Φ. In the case of our model-frame, $V_\Phi = 8 \cdot 8 - 4 = 60$, and $S_\Phi = 4 \cdot 8 + 4 \cdot 2 = 40$.

The definition of the exact fit of a solution requires that the tiles fill the volume of the frame exactly. Hence, for fitted mosaics it follows that

equ. 20-05
$$V_\Phi = \sum_w V_w$$

An important quantity is the total inner 'surface' Ξ of all contacts between the tiles. It determines the energies that hold the mosaic together and, thus, its mechanical stability (see figure 20-08). Let the total surface of all tiles be

equ. 20-06
$$S_T = \sum_w S_w$$

Hence, the total available surface S of all tiles and the frame is $S = S_\Phi + S_T$. The condition that the tiles fit each solution exactly entails that each segment of every tile's surface matches exactly either another tile's equally large segment or an equally large segment of the frame. In other words, the fitting process converts every two segments of the total available surface into a single one. Therefore, the total available surface S_M of every solution must be exactly half the amount S of all its components.

equ. 20-07
$$S_M = (S_\Phi + S_T)/2$$

Now we can calculate the total inner surface Ξ of the fitted mosaic by subtracting the frame surface from S_M.

equ. 20-08
$$\Xi = (S_\Phi + S_T)/2 - S_\Phi = (S_T - S_\Phi)/2$$

In the case of our model-frame where $S_T = 11 \cdot 12 + 10 = 142$ and $S_\Phi = 40$, the total inner surface is the same for all solutions, namely $\Xi_{\text{model-frame}} = (142 - 40)/2 = 51$.

Of course, the total surface S_T of the 12 pentominoes is the same for all pentomino mosaics. Hence, the inner surfaces Ξ of different pentomino mosaics only depend on the surfaces S_Φ of their frames. For example, every 6x10 frame has an inner surface of $\Xi_{6\times10} = (142 - 32)/2 = 55$, every 4x15 frame has an inner surface of $\Xi_{4\times15} = (142 - 38)/2 = 52$, and so forth.

The Contacts of Tiles

All sibling-solutions have the same frame. Likewise, their volumes, surfaces, and inner surfaces are the same. Hence, their individual properties must be sought in the specific contacts between their tiles.

The Contact Matrix

The contact-matrix can identify the patterns in which the tiles may interact with the external and internal environment. It also points to the formation of local clusters or chains of tiles that may create and amplify individual properties of the mosaic, or support transport and signalling across its body.

We had described earlier a way to characterize which particular tiles are touching each other. It applied a scanning procedure to the pattern of the tiles and derived from it a solution code. This code contained implicitly all the information needed to identify each solution.

Although this method is quite useful for the search of solutions of the fitting problem, its dependence on the arbitrary scanning procedure may obscure deeper, underlying relationships between the location and orientation of the tiles of fitted solutions. Therefore, we will introduce a **'contact-matrix Ω'** for each fitted mosaics that does not depend on any arbitrary evaluation method. One obtains this matrix by simply listing at position $\omega_{i,k}$ how often tile #i contacts tile #k or the frame q along a full segment. However, there is no 'self'-contact of the tiles: $\omega_{i,i} = 0$.

equ. 20-09
$$\Omega = \{\omega_{i,k}\, ; i,k = 1,2,\ldots,q;\, i \neq k \}$$

Obviously, whenever #i touches #k, it is inevitable that #k also touches #i. Hence, the contact-matrix is always symmetrical:

equ. 20-10
$$\omega_{i,k} = \omega_{k,i},\, i,k \in \{1,2,\ldots,q\}.$$

Or, expressed in the nomenclature of matrix algebra, the matrix Ω is equal to its transposed matrix Ω^T.
$$\Omega = \Omega^T.$$

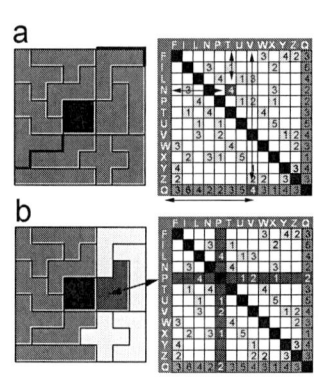

Fig. 20-09. The contact-matrix for the solutions of the model-frame.

(a) Example of a solution and its contact matrix. Each matrix elements $\omega_{i,k}$ counts the number of contacts between the tile 'i' and the tile 'k'. The diagonal elements (marked black) are excluded, as self-contacts of the tikes are ignored.

(b) The tiles with non-vanishing matrix elements in each column 'k' or row 'k' define a local cluster of the tiles and of *only* those that surround the tile '#k'. such clusters will be called 'base-complexes'. Example: The tiles 'V', 'L', 'U', and 'X' form a local cluster with the 'P'.

Note: For reasons of simplicity, we assumed here that all contacts were equivalent, and we only needed to count their numbers. Of course, it is possible that different points along the surface of a tile make different kinds of contacts. If this happens, and two adjacent tiles make different kinds of contacts, we may number the contacts individually for each tile, and expand the contact matrix accordingly. In this case, instead of characterizing the elements $\omega_{i,k}$ only with two indices I and k, we would have to use three or more indices i, k, m, n,…, where m,n,… are the coordinates of the tile surfaces.

Local Clusters

The contact-matrix is able to identify the patterns in which the tiles may interact with the external and internal environment. It also points to the formation of local clusters or chains of tiles that may create and amplify individual properties of the mosaic, or support transport and signalling across its body.

Figure 20-09 illustrates contact-matrices for two solutions of the model-frame. The rows and columns of the matrix elements are labelled with the letters F, I, L,…, Z for each pentomino and Q for the frame. The values of each matrix elements count the number of segments along which the tiles touch each other or the frame. For example, panel (a) highlights in dark gray the 4 contacts between the 'T' and the 'N' pentomino, and the 4 contacts between the 'V' and the frame. Correspondingly, the values of the matrix elements are $\omega_{T,N} = 4$ and $\omega_{V,Q} = 4$ (see the arrows in panel (a)).

Panel (b) demonstrates the symmetry of the contact matrix by highlighting the entire row and column of the 'P'. Obviously, the matrix elements of both are the same. They represent all contacts and the only ones that the 'P' makes with other pentominoes.

The left side of panel (b) highlights the neighbours of the 'P'. They all touch the 'P', and they are the only ones that do. Hence, they form a local cluster of tiles surrounding the 'P'.

The same is true for all other rows and columns. Their matrix elements identify the members of different local clusters surrounding the tile with their particular label. Figure 20-10 shows two other examples of such local clusters.

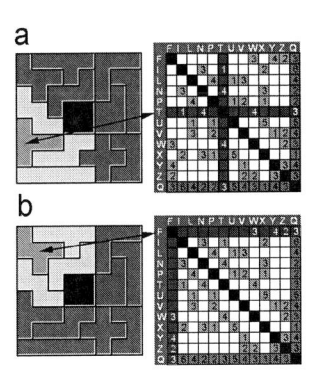

Fig. 20-10. Two further examples of base-complexes represented as rows or columns in the contact matrix.

The clusters that arise from all contacts contained within a column or row represent a special class. They are the complexes that contain **a specific tile together with all tiles and only the tiles that touch it.** We will refer to them as **base-complexes**. By definition, there is one base-complex for each tile.

A contact κ(m,n) between the tiles #m and #n or the frame q exists if and only if $\omega_{m,n} > 0$.

equ. 20-11
κ(m,n) = 1   if ($\omega_{m,n} > 0$, with m, n ∈ { 1,2,…, q});
κ(m,n) = 0   otherwise.

The size Σ_P of the base complex of tile #P, i.e. the number of tiles that belong to the base-complex of tile #P is the sum of all contacts in column or row P (excluding the frame contacts).

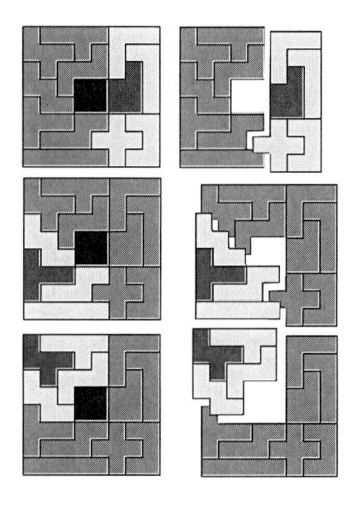

Fig. 20-11. The significance of base-complexes in defining fragments of mosaics that can be considered fitted mosaics in their own rights.

equ. 20-12
$$\Sigma_P = \sum_{n} \kappa(P,n), \quad \text{with} \quad n \in \{ 1,2,…,\tau \}$$

The Base-Complexes Are Different for Different Sibling Mosaics

In the reality of a living mosaic, each tile 'R' represents a fundamental function or task η. Combining every tile with all the other tiles that are in direct contact is important for the analysis, integration and coordination of this function η. Thus, the base-complexes are the functional compounds of all tiles that directly involved in the function η.

As mentioned, the number of base-complexes is equal to the number τ of different tiles. One may add a special base-complex that arises from the Q-column or Q-row. It forms the complexes of tiles that are in direct contact with the frame and, thus, with the inner or outer environment. They are important for the analysis of the interactions of living mosaics and their surroundings.

Another aspect of the importance of base-complexes concerns the duplication of a particular function η in a living mosaic. If development or evolution duplicated a function η, one may search for the duplication of its entire base-complex, as it may represent its roots and integration lines. It is even conceivable that the outer 'surface' of each base-complex may function as a separate frame for the enclosed tiles and form a fracture line in case the living mosaic has reasons to split up.

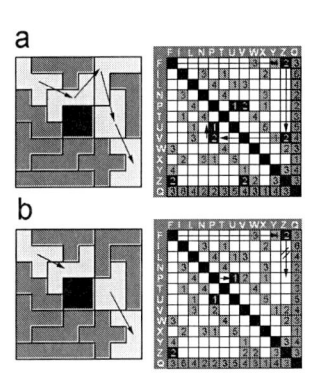

Fig. 20-12. Representation of chains of tiles in the contact matrix.
(a) **Uninterrupted chain of tiles.** It forms a path in the contact matrix that connects each successive matrix elements along a row or a column.
(b) **Interrupted chain of tiles.** There are at least 2 matrix elements in the that cannot be connected along a row or a column (see crossed-out arrow).

Figure 20-11 illustrates three of the base-complexes of the solutions in figures 20-09 and 20-10 around the 'P', 'T', and 'F' pentominoes.

In contrast to clusters whose contacts occupy the same column or row of the contact matrix, **chains of tiles** are distributed over several columns and/or rows. Figure 20-12a shows an example. Obviously, any two adjacent tiles of the chain must touch each other. Hence, *it must be possible to go from one contact (matrix element) to the next by moving along a column or a row.* The arrows in the example of panel (a) follow the chain in the solution display as well as in the contact matrix. If the chain is interrupted, no such path can be found, as shown in panel (b), where there is no contact between the 'Z' and the 'P'.

Therefore, one can establish all possible chains of tiles, i.e. all possible paths of interactions or communication between tiles and/or the environment by an easy algorithm. One merely has to list systematically all possible paths from every contact (matrix element>0) to every other contact (matrix element>0) while following strictly along rows and columns.

Special Matrix Sums

There are properties of certain sums of the matrix elements worth noting. Let us add up the elements $\omega_{i,k}$ in each row k, excluding the frame, and call it σ_k.

equ. 20-13
$$\sigma_k = \sum_\tau \omega_{k,\tau}$$

If we would also add the matrix element $\omega_{Q,k}$ that counts the contacts between this tile #k and the frame, we would obtain the number of *all* contacts that the tile #k is capable of making. That number is, of course, its total surface S_k. Hence,

equ. 20-14
$$\sigma_k + \omega_{k,Q} = S_k, \qquad \text{or}$$

equ. 20-15
$$\sigma_k = S_k - \omega_{k,Q}, \qquad (k = 1,2,\ldots,\tau)$$

Equ. 20-15 is the local version of the global equation 20-08, because the summation over all tiles yields

$$\sum_k \sigma_k = \sum_{k,\tau} \omega_{k,\tau} = \sum_k S_k - \sum_k \omega_{k,Q} .$$

Since $\sum_{k}\sigma_k = \sum_{k,\tau}\omega_{k,\tau}$ is the sum over the possible contacts of all tiles excluding any with the frame, its value is twice the inner surface of each solution, $2\,\Xi$. Furthermore, $\sum_{k}S_k = S_T$ is the total surface of all tiles, and $\sum_{k}\omega_{k,Q} = S_\Phi$ is the total surface of the frame. Hence,

$2\,\Xi = S_T - S_\Phi$, which is equation 20-08.

Physical Properties

The strength of the interactions may vary between tiles. As a result, the local properties of a group of tiles may vary as well as certain over-all properties of the entire mosaic. One can use a slight modification of the contact-matrix to arrive at quantitative expressions for the total effects of the individual tile-tile interactions. In its simplest, linear formulation we may define the **'specific interaction per contact between tiles #i and #k in this order'** with a coefficient p_{ik}.

, the total strength of interaction between tiles #X and #Y is

equ. 20-16

$$E_{X,Y} = \sum_{\tau}\omega_{X,\tau}\,p_{\tau,Y}$$

Please note that the interaction-matrix is not necessarily symmetrical, because the specific interactions may be directional, i.e. in general

equ. 20-17

$p_{ik} \neq p_{ki}$

I left the term 'interaction' deliberately unspecified, as it can mean different physical interactions, including physical strength, conductivity, optical transmission, etc.

The Functions of Tiles

Assume we have a living mosaic and its tiles, how can we determine the functions of each tile? The natural, time-honoured method was to damage or remove individual tiles and observe how the mosaic changed or lost function.

An alternative approach - sometimes the only approach for practical and even ethical reasons - is to map meticulously every contact between the tiles, and try to infer the unknown tile functions from the patterns of their contact partners.

Observing the effect of tile damage on mosaic function is largely an individual approach that precludes general rules. It varies from tile to tile and mosaic to mosaic. The alternative, namely the mapping of all the tile connections is more generally applicable.

The mapping of local contacts between tiles may sometimes not seem sufficient. Take the example of the global release of neurotransmitters or neuropeptides in the brain. Here, neurons interact with other cells, indiscriminately and across large distances without any physical contact. Hence, how can mapping contacts reflect such interactions without contacts?

Global signals flood the entire mosaic and thus contact every possible tile, delivering the same input to all, as if through an actual contact. Therefore, we may simply augment the pattern of the local tile contacts by adding an extra, 'virtual' contact for each global signal.

These global signals are able to deliver *different signals to different tiles with great specificity, even without exclusive physical contacts*. The 'trick' is to attach target addresses to every global signal.

The best-known example that specificity does not have to contradict global flooding of a mosaic is the 'World-Wide-Web' of the internet. Every package of information arrives at billions of computers (tiles) at the same time and, yet, only one specific target computer will admit the package. It bounces off all tiles, except the one with the correct target address. In this way, the source-to-tile contacts are effectively specific. They are no longer physical or functional, but entirely symbolic. We can account for these specific, yet global signals, by adding an extra, albeit 'symbolic' contact area to each tile.

In summary, the pattern of the physical, functional, and symbolic contacts can explore the function of each tile, even in the cases of signals that flood the entire mosaic. The following describes some more advanced methods to establish all tile-to-tile and tile-to-frame contacts.

Serial Contact Matrices

The above text has already described the so-called 'contact matrix' as a way to formulate the entirety of the tile-tile- and tile-frame-contacts.

However, in the case of nested mosaics, their contact matrix would contain far too much repetitive information. It is more economical to describe such mosaics by a series of different contact matrices that depict the nesting procedure. Figure 20-13 shows an example of the first 3 serial contact matrices based on one of the sibling-solutions of the model-frame.

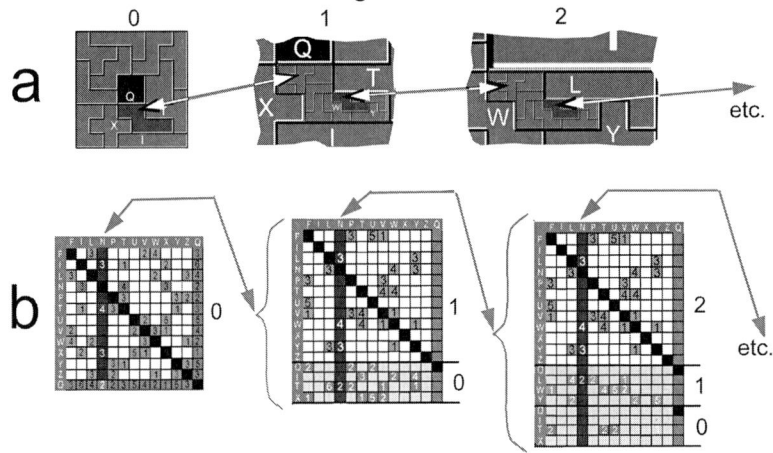

Fig. 20-13. Three stages 0, 1, and 2 of the step-wise development of an expanded contact matrix. The starting mosaic is a solution of the model-frame. The example proceeds by a level-independent tile dissection of the 'N'-pentomino. (a) Tile pattern of the mosaic and successive excerpts of the tile dissection. (b) The contact matrices that belong to the corresponding stages. (More details see text).

Panel a of the figure illustrates how the 'N'-pentomino of a mosaic is selected (stage 0) and recursively replaced by the level-independent tile-dissection of the 'N'-pentomino (stages 1, 2,…). Although the tile-dissection is always the same, the tiles surrounding the 'N' change from stage to stage, as indicated by the white letters in the figures.

Panel b of the figure shows the contact matrix of the 'N'-pentomino at each stage. As we use always the same dissection, the corresponding part is always the same. However, an additional, appended part (in light gray) reflects the new contacts with the surrounding tiles. This part changes from stage to stage.

The vertical column (dark gray) depicts the 'N'-pentomino that is successively replaced. The next contact matrix in the series represents this very column (see arrows).

The example only replaces the 'N', but not the surrounding tiles with a series of dissections. If they were also nested, the size of the appendages of each contact matrix of each tile would be considerable larger.

Eventually, this kind of graphic representation would become quite unwieldy as the stages proceed. It is presented here as a didactic tool. In practice, a multi-dimensional data array may serve as a better formulation.

Further Expansion of the Expanded Contact Matrices

The described expansion of the contact matrix to a series of matrices still only contains as elements the number of contacts between tiles. There is no need to restrict them this way. Of course, one can replace the contact counts with other quantities, if they would be more meaningful for the experimental and theoretical scientist. Obvious replacements for the number (sizes) of the contacts are the strength, amplitude, intensity, or chemical composition. The elements need not be single numbers, but can be n-tuples of parameters that specify the contact. One such added parameter may express stochastic properties by adding the probabilities for the validity of some of the parameters. In addition, there may be codes that indicate the specific contact type. Finally, one may even break the symmetries of the matrix, by adding directional parameters that indicate whether the interaction between of tiles 'A' and 'B' is the same as between tiles 'B' and 'A'.

SPECULATIONS ABOUT LIVING MOSAICS

Chapter Twenty-One

Notes About the Origin of Living Mosaics

The Ultimate Reason for the Composite Character of Living Mosaics Is Most Likely the Universal 'Lumpiness' of the World

Every scientific paper ends with a 'Discussion' section. It allows the authors to escape the strict demands of rigor and precision of the 'Methods'- and 'Result'-sections, and to strike a more conversational and thought provoking tone with the reader. No longer is it necessary that every formulation is 'legally unassailable'. Instead, the authors may admit to unresolved questions, paint with rather broad strokes, and speculate about the meaning and the future of their subject.

The following chapter is written in this spirit of a 'Discussion'-section.

The 'Lumpiness' of the Real World

Distinctiveness is not a unique property of modules. On the contrary, practically every object of the real world, be it alive or not, ends abruptly somewhere and/or sometime. In fact, this abruptness is the very reason that we recognize it an object.

Depending on the level of resolution, even atmospheric gases create seemingly distinct objects such as clouds, hurricanes, tornadoes, hail and snowflakes. At least, if the level of resolution is low enough, each object seems to end abruptly while another one begins. In other words, the whole world may be considered as a mosaic of mosaics, although not all of its components are living modules. This phenomenon has been called the 'lumpiness of the universe' [114].

Why is the world like this? Should we not expect it to be the opposite? After all, quantum mechanics teaches us that the ultimate building blocks of nature at her most basic level are not discrete. On the contrary, depending on their interaction with others they are more or less 'smeared out' over all of space and time. So, why are they forming distinct objects as they aggregate into macroscopic groups?

An important reason for the universal lumpiness seems to be the synergy between two natural phenomena that apply to particles, even if they were sub-atomic and smeared out in space and time.

(Phenomenon a) Whenever there are attractive forces between particles, they increase as they come closer to each other. (Occasionally, they may flip into repulsive forces at very close range, but they do not extend into the distances, where the attractive forces rule.)

(Phenomenon b) Whenever particles follow the pull of the attractive forces and reduce their relative distances, they lose energy. The lost energy must go somewhere. Usually, it escapes by radiation or diffusion. In order to separate the particles again, they need to be given back the same amount of energy. Unless this energy is returned to them later, they are not able to separate again.

The probability that the particles recapture accidentally the exact amount of energy required for their separation is extremely small. Hence, their mutual approach turns into a 'clumping together', which may last quite a long time: Smeared out or not, the particles become more concentrated in a certain portion of space and time, while they are rarified elsewhere.

Interface Sharpness

The result of these mechanisms is a re-distribution of particles. In order to explain lumpiness, we need to explain also that the areas of high particle concentration seem to have sharp borderlines.

The ultimate reason is statistics. The larger the number of interacting particles is, the larger the differences between the portions of space, where they are effectively present, and the portions, from which they are effectively missing.

Let us assume that we have N particles. If left to themselves, they will form a 'cloud' or a 'swarm', which are centred somewhere, and whose density N(x) decreases as we go a distance x away from that centre. For reasons of simplicity, we assume that the particle density follows a normal distribution. In that case, the width of the transition range is proportional to the standards deviation $\sigma \sim \sqrt{N}$. Hence, the relative fuzziness fz of the transition between N particles and zero particles is

$$fz \sim \sigma/N = \sqrt{N}/N = 1/\sqrt{N}.$$

In other words, the larger N is, the smaller is fz, and the sharper the border of the 'cloud' appears.

Considering the typically huge numbers of atoms and molecules, their moieties appear to form quite distinct objects with seemingly discontinuous borders. For example, a rain droplet of 0.02 grams contains of $5 \cdot 10^{20}$ water molecules, whereas the same volume of the surrounding air at room temperature and 100% humidity contains approximately 4000 times fewer molecules. These differences create correspondingly sharp differences of the optical and mechanical properties between water droplets and air, and create the appearance of a clear border between them.

In reality, the surface of the droplet is a chaos of tumbling water molecules surrounded by a shell of dense water vapour, which is surrounded by a shell of lesser dense water vapour, and so on. Yet, if viewed from a distance, the droplet still appears to have a sharp borderline.

Classes of Lump Sizes

We still need to explain, why the lumps come in different sizes. One reason is, of course, statistics. However, there is an important other reason: The different mechanisms of mutual attraction between components create different classes of lump sizes.

Large Scale Lumpiness by Gravitational Attraction

Gravity attracts particles that have mass, and it aggregates them into dust clouds, which aggregate into stars, which cluster into galaxies, which unite into galactic clusters. Each of these objects form because they attract the more masses from their surroundings, the more massive they have already become. Thus, they create discontinuous interfaces with an increasingly depleted space around them. During the aggregation, the gravitational energies transform into heat and radiation. Once formed, these objects cannot re-disperse unless the lost gravitational energy is re-supplied to them in the right form and amounts.

Medium Scale Lumpiness by Electrical Attraction

Similarly, electrical forces are able to cluster particles with opposite charges. However, in contrast to gravitation-created ones, the resulting aggregates are usually small, because the aggregation of particles with opposite charges neutralizes each other's electrical fields and, therefore, leave no net attraction for other, more distant charged particles to follow. Correspondingly, the results are not huge stars or galaxies, but small precipitates, flakes, drops, clumps, colloids and so forth.

Atomic Scale Lumpiness by Quantum Mechanical Uncertainty (Chemical Bonds)

A rather paradoxical reason for lumpiness is the quantum mechanical uncertainty, as it creates chemical bonds. The uncertainty relation states that the average momentum Δp of a particle is the larger, the smaller the confines Δx of its location are. Mathematically formulated, it says that $\Delta p \geq h/2\pi\Delta x$ where h is Planck's constant. In more practical terms, it means that it takes on average the energy $\Delta T_e \sim (\Delta p)^2$ to keep a quantum mechanical particle confined to a small space Δx. And it takes the more energy, the smaller the available space.

Assume, the electron of a hydrogen atom has a mass of m_e and is confined to the atom with the dimensions of Δx. Hence, its momentum is approximately $\Delta p_e = h/2\pi\Delta x$, and its kinetic energy is approximately $\Delta T_e = (\Delta p_e)^2/2m_e$.

Now assume that 2 such hydrogen atoms meet close enough to allow their electron spaces to overlap. As a result, both electrons are no longer confined to the space of their own atom, but occupy the additional space of the other: Each electron has acquired twice the space, namely $2\Delta x$.

Hence, each electron's average momentum becomes half of what it was before, namely $\Delta p_e' = h/2\pi(2\Delta x) = \Delta p_e/2$. Consequently, its approximate kinetic energy shrinks to 1/4 of its previous value, ΔT_e, i.e. $\Delta T_e'=\Delta T_e/4$, and the combined kinetic energy of both electrons is $2 \Delta T_e' = \Delta T_e/2$.

In short, after the electrons of the 2 hydrogen atoms were allowed to share each other's spaces, each lost half of its previous energy. The surplus energy is radiated away in form of a photon. Consequently, the 2 hydrogen atoms H have to stay together until they are able to encounter and absorb a photon of the right size. Only this photon can split the pair back into its 2 atoms by providing the energy needed to confine each electron again to the smaller space of a single atom. As long as this does not happen, they remain bonded by forming the tiny lump of a hydrogen molecule H_2.

Spinning 'Lumps'

The lumpiness of objects does not only create their structural borders, it creates dynamic borders, as well. Before a group of particles coalesced into a denser object, each had some speed relative to their centre of gravity, even if this speed was only miniscule. By the laws of mechanics, their total momentum as well as their total angular momentum must remain the same, no matter how far their aggregation by gravity has progressed.

The angular momentum of the entire aggregate is proportional to the diameter of the group multiplied with the collective angular speed. As the particles aggregate to form clusters and lumps, the diameter of the group will shrink. In order to keep the total angular momentum constant, the collective angular speed must therefore increase.

Every ice skater knows this phenomenon quite well. Pulling one's arms closer and closer makes a pirouette spin faster and faster. The same applies to a group of particles that coalesce towards their centre of gravity. Eventually these aggregates become planets, stars, or galaxies and will rotate quite noticeably, even if their component particles, initially spread over vast volumes of space, had very little net movement at that time.

Living Mosaics

The described primitive lumps are far from being the tiles of a mosaic, let alone of a fitted mosaic. However, they fulfil one of their important criteria, namely to be discrete objects, with no continuous transitions between them.

How were some of the inanimate mosaics able to come alive? This core question of biology has no answer yet. Of course, the present book does not pretend to offer an answer, either. However, I should at least try to formulate in the vocabulary of mosaics some of the steps that may have led lumps to become modules and living mosaics. Still, a speculation about some of the logical steps towards living mosaics *cannot* claim to suggest the mechanisms required to implement them.

From Lumps to Mosaics; the Huge Evolutionary Leap of Replication

Lumps, just like particles can clump together, but in spite of their aggregation they can retain their individual sizes and compositions. Furthermore, lumps need not be solid. Some of them may have large enough holes to be able to trap others inside them, and thus effectively function as the frames of a mosaic. Hence, lumps may aggregate into mosaics.

Such mosaics and their inner structures – far from being living mosaics at this stage! - are likely to be initially quite crude and disordered.

And they will remain this way, unless some of their components discover how to replicate. Afterwards natural selection and evolution can begin to drive their entire population towards individuals with much higher levels of sophistication.

Hence, the next required step towards living mosaics is the discovery of some mosaics how to replicate. By that, I do not mean that growing lumps exceed the limits of their stability and break into two or more fragments, which in turn will grow until *they* break up, and so forth.

Replication is a gigantic and hugely consequential step that requires the discovery, storage, and processing of information. Without it, the mechanisms of evolution and natural selection would have no handle on replicating mosaics.

For example, every drying lakebed creates thousands of almost identical mud cracks. Yet, they are not subject to natural selection. No specific mud crack will ever proliferate preferentially, and thus become the ancestor of 'better' mud cracks.

Nobody knows the path that led from the textbook phenomena of 'equilibrium physics' and 'equilibrium chemistry' to the entirely non-equilibrium, information-driven, extremely low entropy processes of today's biological replication mechanisms. Moreover, the unknown path is hardly a plausible one, because not all its steps could have been at equilibrium. On the contrary, some agents occupying states far from equilibrium must have catapulted parts of it into low entropy. Excellent candidates are the violent conditions of early Earth as demonstrated by the famous Urey-Miller experiments. In addition, we may need unique geological catastrophes of the past that created irreversibly compounds and conditions, which ordinary organic chemistry could never sustain. I hope that we will be lucky enough – and observant enough - to detect someday their fossil records.

From Lumps to Mosaics; the Replication/Evolution Conundrum

Without lots of replica to test variants, evolution would not work. Only replication can offer evolution the handle it needs to change things.

But there is a problem. Why should the 'old' replication mechanism still work for any of the changed and evolved replica? And if the 'old' replication does not work for the mutant mosaics, then there is no evolution, either.

This is a very real problem. As many experimental geneticists know all too well, no matter how difficult and expensive it was for them in the laboratory to generate certain desirable mutants, there is no guarantee that they can replicate. Much to the scientists' disappointment, in a great many cases the desirable mutants are sterile and hence useless for genetic experiments.

The situation is even more precarious. Not only will evolution change the replicated mosaics, it may also change the very mechanism of replication itself. Considering how extremely unlikely it was to come up with a functional mechanism of replication in the first place, it seems suicidal to vary and mutate it. Yet, evolution does and did exactly that.

In order to escape such self-annihilation, mosaics and their replicating mechanisms had to evolve in parallel, or at least in tandem towards quite different results. If we look into nature, we find indeed the living mosaics of different kingdoms such as animals, plants, fungi, protista, and bacteria rely on vastly different methods of replication of their bodies.

However, as we go down the phylogenetic levels, they become more and more similar. Different phyla of animals such as vertebrates, arthropoda, and mollusca use reproductive methods with a number of common features. Already at the level of animal classes such as mammals, reptiles and fish, they are much more similar to each other. Further down the level of order, family and species, there are noticeable and colourful variations, but eventually at the level on animal cells and their genomes the methods of replication are effectively identical.

To be sure, there are differences between (say) the organelles of different cell types. Consider, for example, the large diversity of mitochondrial structures in different eukaryotic cells [31]. Yet, they seem to establish no particular order of evolution, or taxonomy, but share characteristic features at all phylogenetic levels, regardless of the vast differences between the appearances of their macroscopic organisms.

It appears that this is how nature solved the conundrum. At the lowest, most fundamental level of organization (i.e. the level of genomes), the method of replication does not run parallel to the evolution of the living mosaics, but remains essentially the same no matter which organism it is, whose gene or chromosome needs to be duplicated. As the levels of organization rose, small differences evolved among the living mosaics, but they ran parallel to small changes of their replication methods. Eventually, bigger and bigger changes of living mosaics co-evolved with equally large changes of their method of replication.

From Mosaics to Symbols: Segregation and Amplification

The first required step towards handling information is the ability to assign symbolic 'meaning' to certain mosaics. This ability was likely the result of breaking up mosaics into those that 'do something' and those that 'mean something'.

Mosaics, of course, can break up in many ways. Imagine a split into two fragments. We will call one part the 'symbol' and the other the 'reader' mosaic, and abbreviate them as 'Sy-mosaic' and 'Rd-mosaic', respectively.

Assume further that there is some mechanical or electrical repercussion inside the Rd-mosaics whenever it re-unites with a Sy-mosaic. Given the usually quite complex interactions between tiles, this is not a daring assumption.

In exceptional cases, these repercussions may be much too large. This happens, if the Rd-mosaic contains 'amplification' mechanisms, i.e. mechanisms, which use small amounts of energy to open and close gateways ('valves') through which much larger amounts of energy can flow. Conceivably, the presence of such – possibly destabilizing - energy sources was the reason that the initial mosaics split up into Sy- and Rd-mosaics, in the first place.

Knowing how amplifiers in human technology work, it is not too hard to imagine - at least in principle - how some living mosaics were able to assemble crude amplification mechanism, starting with an initial 'spring-loaded release of a lever, later releasing avalanches or chain reactions, but eventually evolving further into more controlled versions. Naturally, such mechanisms would occupy ever more specific domains of tiles, i.e. they would turn them into modules of the mosaic.

By doing so, they achieved more than amplification. They discovered specificity, because they did not react to, and thus amplify just everything. They would only work in response to a limited number of 'inputs', which we will therefore call 'signals', and the amplifiers will be identified as 'modules'.

Obviously, amplification only can grow to higher levels on a foundation of multiple components. Already the handling of the energy delivery requires separate units (tiles) to store it, containers (frames) to confine it, and contacts (valves) to unleash it.

In short, Rd-mosaics were mosaics that contained modules, which amplified specific signals coming from the Sy-mosaics, and influenced the actions and interactions of their other tiles (or conceivably the tiles of other mosaics). By means of these very selective actions, they assigned specific consequences, i.e. 'meaning' to the Sy-mosaics.

Perhaps, under the impact of natural selection the Sy-mosaics developed into ever better defined and more meaningful symbols, while, on the other hand, the Rd-mosaics evolved less error-prone recognition and response modules. In other words, they evolved the ability to process data.

In my earlier publications [102], I called the capacity of 'responding to signals (as opposed to yielding to forces) and to process them' as 'intelligence'. I did not mean 'awareness', 'creativity', or 'mind'. This book understands 'intelligence' in the exact same way.

Furthermore, there is an important aspect of data processing to keep in mind. Intelligence is a static quality, whereas data processing is – as the name indicates – a process, which is anything but static.

Quite a while ago, the emotional meaning of a 'process' was brought home to me, when a courageous flight instructor let me operate the controls of a little 'Beechcraft'. Used to driving cars for many years, I was shocked to realize the obvious, namely that there was no way of pulling over, stopping, and thinking what to do next. One had to fly all the time!

The data processing of living mosaics is like that. They are always processing some signals and data, if only their self-created ones. More accurately, data processing is the result of the continuous flow of information from somewhere to somewhere else, while changing its contents.

The intelligent machines of human technology act the same way. Even during idling, our computers never stop sending data and instructions to the microprocessor, and receiving altered ones coming back from it. They do not even stop when we switch them off. Certain compartments continue to keep track of date and time.

The rationale for the continuity of processing is quite basic. During their booting up phase, data processing systems are their most vulnerable and unreliable. So-called 'hand-shakes' between functional sub-units, access to memory banks, activation and de-activation of interrupt-routines, error-preventing routines, and many more such steps must be timed precisely to prevent that the system freezes up, and to guarantee a smooth operation. While all this happens, the system cannot yet respond to its surrounding.

Imagine the 'Beechcraft' having to request landing permission, to enter holding patterns, land, and then go again through the take-off ritual every time, the pilot wants to think! No. Constantly re-starting an intelligent system is not a good idea.

Detour: Intelligence as a Symbolic Mosaic

Having proposed that the ability of mosaics to recognize and interpret symbols is an act of intelligence, the next logical step towards living mosaics has to be the evolution of intelligent Rd-mosaics. In order to flesh out the argument, we need to go on a little detour. For the sake of readability, we will now drop the terms 'Rd-mosaic' and 'Sy-mosaic', and simply call them mosaics and symbols.

Not all tangible mosaics need intelligence in order to function, but symbolic mosaics certainly do. Some form of intelligence must read and interpret them, in order to be able to operate with the 'meaning' of the symbols. The same is true for tangible mosaics whose actions are linked to a symbolic mosaic, be that a practiced language, specific odours, certain body markings, and other similar symbolic mosaics that by themselves would have no physical impact.

In order to test whether a mosaic is actually able to 'read' an object with a symbolic meaning, one must observe its reaction to it. This, in turn, requires testing its reaction to more than one symbol. Only then can we be sure that any observed effects are due to different 'meanings', and not to different material properties of the carriers, such as their weight, kinetic energy, acidity, reflectivity, loudness, etc.

After fulfilling these and other technical precautions, we may test the mosaic. Assume, that we find different reactions to the different symbols, we may conclude that it

1. perceives the different 'meanings' of the symbols by data processing, because there are no physical forces transmitted,
2. transforms their informational content into some of its own internal symbols, and
3. the internal symbols decide between different response actions.

In short, we ascribe its response to data processing, i.e. to the intelligence of living mosaics.

Naturally, all data processing requires some actual data to process, i.e. symbolic inputs such as letters, numbers, pixels, phonemes, and their patterns as texts, formulas, images, or sounds. In addition, the action of 'processing' them requires their linkage by rules and algorithms that effect the transformation of the input into output data.

In other words, the data processing that we called 'intelligence', appears actually to be a mosaic because it involves 'tiles' in the form of its input and output data, which are linked together by 'fitting' procedures that combine and transform them using its rules and algorithms. Most frequently, the intelligence mosaics are depicted as flow-charts.

In summary, data processing, i.e. *the application of intelligence as we have defined it, is itself a functional, symbolic mosaic*. And it is a living mosaic, because it was created and employed by another living mosaic. In contrast, inanimate objects, such as the moons of Jupiter are not linked to any such symbolic mosaics.

Detour: Intelligence and Individuality of Living Mosaics

Do all living mosaics have the capacity of data processing? The answer is 'yes', for a very simple reason. It rests on our earlier claims (see chapters 1 and 11) that there have never been, nor ever will be identical replicas of living mosaics. Here is our very simple line of arguments.

Premise 1:   All living mosaics must create a 'seed' or a template for their replication.
Premise 2:   the replicas of living mosaics are never identical.
Conclusion 1: Hence, they do not copy themselves either, but assemble the replicated mosaic *de novo* from *de novo* created components, including the key-seeds.
Conclusion 2: The *de novo* assembly of a living mosaic is the process of fitting the tiles by the same rules that yielded the template or parental mosaic in the first place.
Conclusion 3: The *de novo* assembly requires (a) the recognition of pre-placed tiles of the key-seed, (b) the identification of the 'holes' and their similarity with missing tiles, and (c) the knowledge of their proper parameters (e.g. location, orientation and phase), and their directed transport to that location.
Conclusion 4: The functions, which turn the knowledge of these parameters into reality, are inevitably linked to symbolic quantities that carry information. They operate by processing information about space, time, physical parameters, topology, etc.
Final conclusion: Hence, the successful 'almost'-replication of living mosaics by *de novo* assembly is not accomplished by physical forces alone, but must rely on data processing, i.e. intelligence.

As long as living mosaics create key-seeds and use them to re-create themselves (albeit with their characteristic 'almost'-identity) they must apply intelligence. Of course, the above argument says nothing about the origin and subsequent evolution of intelligence, except that it should be inseparable from the origin and evolution of life itself.

In summary: No living mosaic (organism, organ, spoken word, body marking, finger print, hand-written signature, etc.) is the identical copy of a pre-existing template, but is the result of a re-creation, whose mechanisms requires the processing of symbols (markings) and, hence, requires intelligence.

It is likely that the above argument appears too simplistic to many students, who are familiar with the century-old efforts to define intelligence and with the modern search for its neurological substrates. However, my main intention here is to shift the problem of defining and quantifying intelligence mechanisms, to the experimentally much more accessible phenomenon of generating individuality.

At any rate, we cannot formulate the steps that turn complex lumps into living mosaics without including steps that introduce intelligence to them.

Back to the Main Argument: Memory Mosaics

No intelligent data processing is possible without storing symbols, storing intermediate results, storing the end results of the processing, but also storing the series of steps required for data processing (algorithms).

It is not difficult to understand that an array of tiles can store information. The identity and location of each tile are quite suitable to represent specific bits of information.

In fact all human-made forms of memory are mosaics. They consist of written letters on paper, or arrays of optical, electrical or magnetic units that can exist in one of two states. (For the sake of simplicity, we are ignoring here the possibility of mixed states as in quantum computing). Storing information means to place letters on a carrier, or to flip a unit from state '0' to state '1'.

With the exception of the genetic memory, the carriers of biological memory are not understood in detail, but most likely they are based on mosaics, as well. After all, living mosaics contain nothing else but mosaics.

Of course, the methods of encoding, storing and retrieving information differ widely depending on the types of memory mosaic and their contents. However, they all have in common that encoding means to select one of their possible tile configurations, i.e. one of the sibling-solution of the memory-mosaic. Storing the information means to implement this particular sibling-solution. In addition, retrieving the information means to assemble replicas of some or all tiles of this particular sibling.

Storing Information by Generalized Key-Seeds

The key-seeds of living mosaic that we had introduced in chapters 11 and 15 are closely linked to information. They were defined as mosaics, which are missing half their tiles, but which only can be completed in one way, provided one uses their particular assembly-mechanism,. The set of the missing tiles of each key-seed form the complementary key-seed. By completing either of them, one ends up with the same sibling-solution.

We had introduced key-seeds to explain the replication and expression of mosaics. In the present section, we discuss their role in information storage. Here are some of the reasons.

Key-seeds are still mosaics. Like all mosaics, they can store information through the type and placement of their tiles. More specifically, they are economical carriers of memory, because they only need about half as many tiles to characterize a specific sibling solution.

Most importantly, they provide an unambiguous way of 'reading' them. Filling in the missing tiles creates a complement of the key-seed. The complementary key-seed identifies a unique selection of tiles that can be separated from the original solution. Hence, the very uniqueness of the missing tiles represents a specific amount of information.

The creation of key-seeds requires splitting a mosaic into two complementary parts. Once evolution had discovered the complementary break-ups of highly structured lumps, it seemed not too farfetched to assume that nature also discovered how to complete and restore each of the two fragments. However, as pointed out above, that discovery 'taught' evolution how to encode and store information, because only specific tiles in specific locations were able to complete the mosaic. Their unique configuration, as opposed to countless alternative but inadequate choices, carried the particular information.

However, these were not the only evolutionary advantages. As pointed out above, splitting a mosaic into two parts offers the possibility to specialize one fragment as a reader-mosaic, and the other as a symbol-mosaic. As long as there are amplification mechanisms present, it does not really matter which is which. In fact, both can fulfil both functions. In other words, key-seeds can store specific information, while functioning as symbols and readers at the same time.

In view of the information aspect of key-seeds, the methods of replication and expression of mosaics described in chapters 11 and 15 can be interpreted as specialized examples of information storage and retrieval. The present chapter merely generalizes these aspects.

From Information towards Intelligence: Logical Operations

Intelligent mosaics are mosaics that react to and apply rules of logic to the multiple signals received from multiple symbols. All logical operations or rules can be constructed as ordered sequences of certain basic logical operations, such as 'NOT' ('~') , 'IF-THEN', 'IMPLIES' ('⊃') , ' AND' ('∧'), and 'OR' ('∨'). An example could be a string like (~A ⊃ B) ∧(~C ⊃ D) ∨ E…. It shows that even logical rules and operations are linear mosaics.

They operate on symbols A, B, C,…which represent statements (findings, observations, inputs,…) and yield as logical 'conclusions' of other statements or symbols. Books on formal logic explain how one can 'harvest' the result of the conclusion by separating the 'concluded' symbols from the 'concluding' composite formulas by the so-called 'modus ponens'.

However, it turns out that one does not need all the above connectives. Only two basic operations are required to express the other finite logical processes [01]. This is true, because the following equivalences allow us to substitute the connectives '∨' and '⊃' with combinations of the two connectives '~' and '∧'.

equ. 21-01

1. (A ∨ B) is equivalent to ~(~A ∧ ~B). (('For 'A OR B' to be true, it must never happen that both are false.').
2. (A ⊃ B) is equivalent to ~(A ∧ ~B). (for 'IF A, THEN B' to be true, it must never happen that A is true, but B is false.').

From Information towards Intelligence: The 'NOT'- Operation

Let us assume that a particular statement **A** is stored as a mosaic in the form of its key-seed **A**. Consequently, the operation ~(**A**) creates the complementary key-seed **A***. This can be done simply by filling in and subsequently isolating the missing tiles of **A**. Figure 21-01a shows an example, where the complement of a 4-key-seed of the model-frame turns into an 8-key-seed.

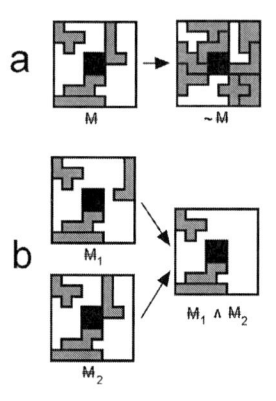

Fig. 21-01. Examples of the effect of the 2 basic logical 'NOT (~)' and the 'AND (∧)' functions on 4-key-seeds of the pentomino model-frame. (a) The NOT-function identifies and unites the missing tiles into a new mosaic, and therefore yields the complementary key-seed. (b) The AND-function eliminates all, but the common tiles of the two key-seeds (see text).

The so replaced key-seeds serve subsequently as inputs for the next link in the chain of basic logical operations.

equ. 21-02

The procedure of **'filling in the missing tiles'** of a key-seed **M** is the same as
1. turning it into its complementary key-seed, i.e. **M → M***, and also
2. turning it into its negation, i.e. **M → ~M**

Since creating the complement of the complement re-creates the original key-seed, filling in the missing tiles of '**M***' yield the original information encoded in '**M**'.

equ. 21-03

The procedure of 'filling in the missing tiles' of a key-seed **M** twice in a row is the same as retrieving the initial information
M → ~(~M) = (M*)* = M

From information towards Intelligence: The 'AND'-Operation (Truth Tables)

The 'NOT'-function is called a unitary function because it only has one argument **A**. In contrast, the 'AND'-, 'IF...THEN'-, and 'OR'-functions, are binary functions, because they operate on two arguments **A** and **B**. They are often described in the form of truth tables, which list the logical result of every possible combination of 'true' or 'false' that the two arguments can assume. Let us take the example of the 'AND'-function

$\boxed{\text{equ. 21-04}}$
The truth-table of the logical operation '**A** ∧ **B**' ('**A** AND **B**') is

A\B	F	T
F	F	F
T	F	T

Table 21-01. Truth table for the AND connective.

It expresses that '**A** ∧ **B**' only are **'true', if both arguments are true**.

Using equations [21-01, one can easily derive the truth tables of 'OR'- and the 'IF...THEN'-functions.

The student's computer uses this definition countless time, when it compares two 16-bit words (=symbolic mosaics), such as

A = 0011100100011101, and
B = 1111001011101011.
Applying the above truth-table yields
A∧B = 0011000000001001.

Admittedly, the zeros and ones do not reflect everybody's intuitive understanding of what 'AND' means, but at closer inspection, it is precisely right. Whenever we think 'A AND B' we mean an experience that only ascertains the features shared by 'A' and 'B'.

The result is a digital word, which only features '1's, where both A and B have '1's. The result is somewhat surprising. Although the truth table expresses exactly the everyday meaning of 'AND', its application yields a mosaic that only contains the tiles that both input mosaics have in common.

(In order to create a word that contains all the '1's of the two inputs, and that seems to reflect the everyday meaning of 'AND' better, one must use the 'OR'-function! The student who wishes more details will find them under the headings of 'Boolean algebra'. The mathematician George Boole (1815-1864) created this branch of mathematics, which more than a century later proved to be central to our digital age.

Similar results apply to other mosaics. For example, the mosaic that represents the conjunction of two 4-key-seeds of the model-frame is a mosaic, which also only contains the common pentomino tiles. (figure 21-01b). Please note: Tiles from both key-seeds that are not identical but only overlap, are missing in the conjunction.

Even though the resultant mosaic is the conjunction of two key-seeds, it is not necessarily a key-seed itself. However, it is certainly compatible with the sibling-solutions of the two key-seeds. After all, it only contains tiles in the exact same location and orientation as their common solution.

We can also express the truth table in terms in terms of key-seeds. Applying the 'AND'-function means that in all cases where the logical operation calls for [A∧B], the key-seed **A** is replaced with the complementary key-seed **B***, except when **A** occurs together with **B**, in which case it is replaced with the key-seed of B, i.e. with **B**.

|equ. 21-05|

The so replaced key-seeds serve subsequently as inputs for the next link in the chain of basic logical operations.

The 'NOT'- and the 'AND'-function require turning the key-seeds **A** and **B** into their compliments. By the definition of key-seeds, the procedure means to fill in the missing tiles, as they represent the complementary mosaic. On the other hand, the procedure can also be interpreted as turning the key-seed **A** into the mosaic that represents the negation ~**A**.

	A	**A***	**A**
B			
B*	**B***	**B***
B	**B***	**B**

Table 21-02. Truth table for the AND connective expressed with key-seeds.

From Information towards Intelligence: The Use of 'Authenticity'-Tags

Whenever we use a statement **A** within a logical operation, we need to know whether it is 'true' or 'false'; whether it is represented by a 'proper' key-seed or by its 'complement'.

How can we know that? Since these alternative qualities flip cyclically from one to the other, there is often no absolute distinction. Most often, we must arbitrarily declare one of the two as 'true' and the other as 'false'.

In reality, living mosaics do not make this assignment arbitrarily. Their survival depends critically on their ability to tell reality from lie and fantasy. Therefore, they usually attach a special marker to the statement, depending on its source and its verifiable consequences. If the statement arises from their own sensory organs, or from the signals of another, 'trusted' living mosaic, they mark it as 'true'. Similarly, if their sensory organs confirm the logical implications of a statement, they mark it as 'true', as well. Humans even assemble groups of people, such as trial juries and expert committees to decide the authenticity of certain statements.

I suggest calling these marks the **'authenticity'-tags** of a statement. *Whenever a statement or key-seed flips into its opposite version, so does its authenticity-tag.*

Unfortunately, the method is far from fail-safe. After all, sensory organs can be fooled. Living mosaics make logical errors, suffer delusions, and apply prejudices that lead to fatal assignments.

Nevertheless, the method is widely used and trusted. In most practical situations, there is no alternative. Therefore, we assume in the following the data processing of the living mosaic in question works with statements that carry authenticity-tags.

From Information towards Intelligence: Analysis- and Action-Mode

Whenever living mosaics use intelligence for a task, they actually have to carry out two very different sub-tasks.

(a). They have to *identify* the various inputs and to *select* between possible responses ('Analysis-mode')

(b) They actually have to *carry out* the selected action ('Action-mode').

Both sub-tasks are acts of data processing. However, the statements 'A', 'B', 'C',... as well as the logical operations 'NOT' ('~') , 'IF... THEN'

('⊃') , 'AND' ('∧'), and 'OR' ('∨') have very different meanings in the two different modes.

For example, in analysis-mode the statement 'A OR B' means that **it is always true, except when both variables are false at the same time** (formally, '~(~A ∧ ~B)'; see equation 21-01), whereas in action-mode it means to **do at least one of them, or both** (either simultaneously or one after the other).

As an example, let us consider a male tiger who encounters an object of about his own size. After classifying its size, his first sub-task is to switch to analysis-mode, identify the object, and select an appropriate course of action: If the object is a rock then continue walking; if it is an adult tigress then prepare for courtship; if it is a deer then hunt it.

His second sub-task is to switch to action-mode and his decision, i.e. to carry out one of the three the series of sub-sub-...-actions that we summarized as 'walking', 'courting', or 'hunting'.

The first sub-task involves mainly symbolic input mosaics, their transduction into internal symbols, and the manipulation of their key-seeds. The second sub-task calls upon sub-sub-...-actions that operate muscles, sensory processing, cardiac and pulmonary actions, and so forth.

Although all this is the response to an object outside the tiger, it is important to remember that most of the data processing involves internal signals and internal processes. Their formats and storage are uniform and controllable by the tiger. In contrast, the external inputs are naturally not under the tiger's control.

It is worth remembering that none of these actions and reactions would ever occur with inanimate objects as automatic consequences of physics or chemistry.

From Information towards Intelligence: Flow Diagrams

Frequently, students harbour the notion that logical analysis and logical behaviour are the result of conscious decisions. Although, conscious decisions may consult mechanical, mindless decision-making networks for confirmation, the basic networks operate quite mindlessly and mechanically. They consist of units, which flip between two states such as '0' and '1', or – if based on fuzzy logic [01] - assume one of several states in between. The state switching of such units is quite fast and as little conscious as the mechanisms that flip a coin between 'head' and 'tail'.

Let me illustrate this with the example of our tiger, who encounters an object. Of course, we will simplify the situation, and use a fictitious tiger in order to depict the principal logical structure behind the tiger's decision to act.

First, we have to introduce the principal workings of 'comparators'. These are units (tiles), which can tell whether a given symbolic mosaic ('input') is equal to another symbolic mosaic ('template'). A comparator matches the two mosaics tile by tile. If even a single tile comparison fails, the comparator flips its state to '0'. Otherwise, its state flips to a '1'. The state changes are entirely mindless and mechanical.

In our example, the input mosaic is the inner representation of the optical image recorded by the tiger. We assume, the tiger has seen rocks, tigresses, and deer before, and holds their inner representation in memory. If he has never seen the encountered object (e.g. a helicopter), or never been warned about it by his parents or teachers, he may ignore it – or attack it - it at his own peril.

Each of the image representations contains a list of properties that the comparator will automatically check one by one. The result is a pattern of outputs that are also 1's and 0's, but which are stored inside the comparator's output unit. Depending on a rigid, pre-programmed set of criteria, the output pattern will be 'evaluated' and causes the comparator to set its own, overall state to '0' or '1'.

Therefore, each comparator is actually a complex mosaic of elementary comparators that determine the identity of single tiles in the image representation. They effectively answer simple questions like 'Is the colour yellow (Y/N)?', 'Is the colour brown (Y/N)?', and so forth.

Usually, all of this is abbreviated by a 'branching'-symbol as in figure 21-02a. Its component, which asks the question 'IS X=Z'?, indicates the logical element 'IF =Z, THEN Y'. The series of '∧'s and '~'s described in equation 21-01 implements all this. As the input undergoes these logical steps, it ends up as either 'Z' (i.e.'Yes') or as ~'Z' (i.e. 'No'). Its second part, which is labelled as a 'Do A' command, indicates the initiation step of the action 'A'.

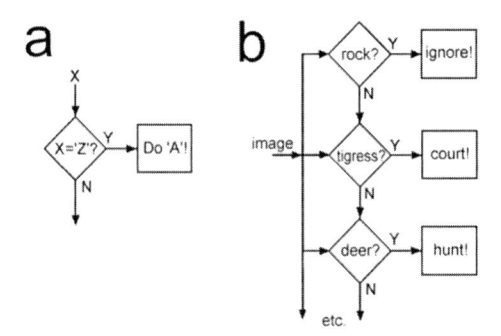

Fig. 21-02. Flow diagram of the' tiger'- illustration used in the text as an example of the analysis-mode and subsequent action-mode of logical decision making. (a). Symbolic representation of a comparator with a subsequent action. (b). The whole flow-diagram of the example (see text).

Having completed the above analysis, the tiger switches to the action-mode and selects a subsequent action (see the flow diagram of figure 21-02b). In our example, this selection is also mechanical and automatic. However, in the more sophisticated reality of a living mosaic, the various options feed initially into further logical networks to determine the timing, intensity, number of repetitions of actions and so on. It is even possible to activate simultaneously two or more actions, even if they are incompatible. For example, the student may have watched a movie where the character wanted to hold his/her breath in total surprise, while trying to scream in horror at the same time.

Of course, living mosaics can and do override all this mechanical and automatic decision-making. For instance, in humans, the pre-frontal cortex decides ultimately whether a person's automatic, instinctive reactions will be carried out.

From Information towards Intelligence: The Implementation of Flow Charts

The passageways of the information flow inside the living mosaics are manifestations of numerous flow-charts, whose components are the implementations of logical functions. They determine, where, and along how many simultaneous paths the carriers of information will stream, and how their passage will alters their content. Hence, the practical realization of flow-charts and their underlying logical functions is a crucial step in trying to understand the intelligence of living mosaics.

In figures 18-01 to 18-05 we have already encountered a processing flow-chart. It created cyclical 'almost'-repetitions. Starting with one input 4-key-seed, its rule of logical progression was to maximize the overlap of the sequential 4-key-seeds.

This particular rule of progression is not a basic logical element in itself. However, the requirement of overlap maximization (or minimization) can easily be implemented as a series of '∧' and '~' that operate on the tiles.

In general, flow-charts may be realized as mosaics of key-seeds (= input information) that are guided from one stop to the next. At each stop the key-seeds are altered by manipulations of their tiles and holes.

We have described these basic manipulations in chapter 18 as replacements, transpositions, insertions, and deletions. Each of these, in turn, can be understood as a series of implementations of '∧' and '~'.

CHAPTER TWENTY-TWO

A CRITICAL VIEW OF MOSAICS

BEING COMPOSITE OBJECTS, LIVING MOSAICS INHERITED MATHEMATICAL PROBLEMS, AMBIGUITIES, AND PARADOXES. WHEN THERE ARE NO EXPLICIT MATHEMATICAL FORMULAS TO PREDICT EXPERIMENTAL DATA, COMPUTER SIMULATIONS MAY TAKE THEIR PLACE. YET, MANY MYSTERIES REMAIN

The book began with a seemingly simple definition of mosaics as collections of discrete tiles that fit perfectly into their frames. However, upon closer examination they did not remain simple. On the contrary, they revealed a dizzying variety of different and complex realizations.

The very ability of mosaics to appear in so many different sizes and guises seems to qualify them at least as metaphors of the vast hierarchy and variability of living things. However, mosaics may provide more than metaphors, namely a calculus and the vocabulary, syntax and semantics of a 'language' for a size-invariant biological analysis.

In order to master the 'language' of mosaics, we must face the inherent problems of every language, namely the ambiguous functions and interpretations of its 'words', i.e. of the meaning of the tiles of living mosaics. In particular, we must realize how much their meaning depends on their context.

What is their context? Obviously, the mosaic! *The context of each mosaic's tiles consists of the other tiles, their own mosaic, and the mosaics with which they interact.* Whatever meaning a mosaic's tiles may have had in isolation, the other tiles must change it. Yet, in doing so, each of them also changed their own meaning.

It is a disturbingly circular situation, but being self-referential is the problem of every 'context', and surely not unheard of among living things! It is not the only unsettling quality of living mosaics. Therefore, in this last chapter, we will try to face some of them.

Tangible vs. Symbolic Mosaics; an Artificial Divide

Initially, I tried to persuade the student to make a clear distinction between body markings, bird songs, languages, texts, or computer programs on one hand, and livers, eyes, skeletons, eggs, termite nests, and cathedrals on the other. The first kind, which we called symbolic mosaics, although carried by material tiles and frames, had no task or function unless a living mosaic read and interpreted them. The other kind, which we called tangible, carried out tasks and functions regardless whether a living mosaic noticed and interpreted them or not.

Now, after the above study of living mosaics, such a distinction seems no longer clear, nor even justifiable. Experience shows that the signal processing devices of all living mosaics read every tangible mosaic in front of them, whether they are supposed to be read or not. Of course, we do not know what animals feel, but we can be certain that at least the human animal can never perceive tangible mosaics as carrying no message whatsoever.

In fact, not only artists have testified to the powerful expressiveness of all tangible bodies and their tangible tools like eyes and hands, but many scientists accept painful living and working conditions, as long as they are allowed to see the expressive beauty of the tangible objects of their work. Hence, not only poets, but also quite pragmatic scientists cannot help but project meanings into tangible mosaics.

The converse is also true. The tangible carriers of symbolic mosaics often acquire global functions and tasks much larger than their initial symbolic meaning. Examples are cathedrals, sculptures, the orchestration of a musical melody, but also the pioneering power of computer codes. In these cases the majesty, texture, or uniqueness of the tangible carriers of symbols dwarf the significance of their symbolic meaning.

In short, no living mosaic is either symbolic or tangible. Each has both qualities. The interplay between them creates the emergence of beauty, power, perfection, and even emotion and love.

Unpredictability; the Need of Simulation Methods

A major reason for the difficulty of predicting the behaviour of mosaics is the lack of suitable mathematical methods. As we mentioned several

times before, the successful mathematical descriptions of the universe as developed by physicists, employ mostly continuous and even differentiable functions, while mosaics are no good targets for this approach, because the discontinuities of tiles and frames do not lend themselves to descriptions by differentiable functions for the following reasons.
1. The interactions between individual tiles are far too local and inhomogeneous. Most of them pose mathematical problems that have no explicit solutions.
2. The tiles change in *unpredictable* ways. In practice, we can pretend that they change in *random* ways, but this is unsatisfactory. *Unpredictability* only reflects our insufficient knowledge of the dynamic causes, while *randomness* claims that there are no causes.
3. Even if the interactions between any two tiles were relatively simple, there may be far too many of them for the human mind to keep the necessary numerical overview.

An alternative solution for the problem became available with the advent of modern computers. Even if no explicit mathematical theories were available, computers could *simulate and model* the behaviour of mosaics and, thus predict it at least in practice.

Two simulation methods have become most widely used. One method, called **'cellular automata'** ignores all detailed mechanisms of the tile-tile or tile-frame interactions altogether, and replaces them with a list of logical rules that only formulate their outcomes. The most famous example is the so-called 'Game of Life' by John Horton Conway.

Another most successful method is the so-called **'Monte-Carlo-method'**, developed by Stanislav Ulam (1909-1984). At the time of his discovery, he was a member of the 'Manhattan-project'. The practical problem he had to solve was to predict the parameters of a successful nuclear chain reaction. In this case, the number of variables, which contributed to the chain of individual events, exceeded human abilities to determine their values. Worse, attempts to find them experimentally could lead to a nuclear catastrophe.

In order to solve the problem, Ulam suggested letting random number generators determine the values of the variables, and to use them to calculate a single chain of events over and over again, thus effectively simulating the thermal diffusion of neutrons, instead of offering an explicit formula for the interaction between neutrons and Uranium-235.

Each simulation is a mathematical 'gamble' whose detailed result is always irreproducible. In reference to the name of the gambling capitol of his time, Ulam named the method the 'Monte-Carlo' method.

These two methods are not mutually exclusive, but complement each other. They are attractive approaches because they only require a bare minimum of knowledge of the underlying mechanisms. Unfortunately, they are also in danger of oversimplifying them. As usual in science, a great deal of sound judgement is required, before simulations become meaningful.

Let us illustrate them with a simple example.

Example Simulation

Assume that thousands of cells of type 'X' are migrating on a flat surface. We assume that they are attracted to each other, i.e. once they come close enough, they will move towards each other even closer. All the while, their numbers increase through mitosis. Let us also assume that there is a second cell population of type 'Y' that may be attracted to the first, but not to each other. They, too undergo mitosis. The question is, what tissue pattern will the cells create?

The first pre-condition of every simulation is a clear definition of its specific goals. In our case we may wish to find out how many Y-cells end up in clusters of X-cells, depending on the radius and the strength of the attraction between the two cell types. Hence the goals are extensive statistical evaluations of the cellular distributions.

The second pre-condition is the selection of suitable digital representations of the objects. In our case the X-cells are shown as gray circles; the Y-cells as black circles (figure 22-01).

The third pre-condition is the choice of the initial configuration of all component parts (figure 22-01a). Clearly, the outcome of the simulation depends critically on the starting situation. In fact, testing different initial conditions can become the goal of the simulation in itself, because one often knows experimentally the results, but tries to find out the initial conditions that created them.

Given the initial conditions, the simulation program places the digitized objects into a certain space, such as the volume inside the frame of the mosaic, and subsequently keeps track of their locations while they move.

In our example, the locations of the X-cells are random, while the Y-cells form an approximate straight line (figure 22-01a. The frame of this mosaic is larger than the panels, and not shown.)

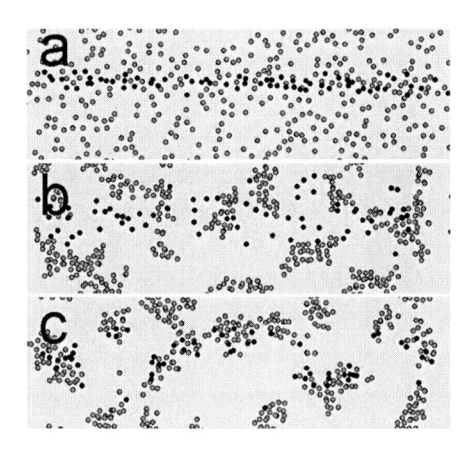

Fig. 22-01. Example of a typical Monte-Carlo simulation of the behavior of a living mosaic. The model mosaic is a 'loose' one, in order to offer sufficient space for the movements, that the illustration simulates. The gray dots represent cells of type X in tissue culture seeded randomly on top of a flat substrate. In addition, a number of cells of a different type Y (black dots) are plated out along a line. The cells are able to migrate randomly about. (The frame of this mosaic is larger than the panels, and not shown.)
(a) Initial configuration of the cells. (b) Result of 64 rounds of simulation if the A-cells are attracted to each other within a given radius. (c) Result of 64 rounds of simulation if the A-cells are attracted to each other and the B-cells are attracted to the A-cells within a certain radius.

The actual simulation consists of a large number of repetitions of single 'rounds of simulation'. Round by round, each cell (tile) is allowed to move a minute amount away from its present position. In order to determine the direction of the displacement, the program consults a random number generator of a random angle between 0° and 360°. As to the distance of the displacement, the program can use another random number generator, or apply certain logical rules that examine the configuration of the neighbouring cells. Of course, it can also do both.

In our example, it calculates how close the nearest X-cell is and computes a next position closer to it by a small amount. In doing so, it takes into account the distances and positions of all the nearest neighbours.

After having determined the next displacement of the cell, the program does not displace it by this amount, but only commits the computed value to memory. Only after each single cell has received its own next displacement, the simulation program *moves them all simultaneously to their new positions*.

This order of computations is most important. Only then, the procedure guarantees a proper simulation of reality for the following reason.

Obviously, the cells only can respond to where all others are now, but not to where they will be in the future. Hence, no cell must have moved already to its future position while another cell is still 'computing' where to go next.

The detailed results of the simulation are always irreproducible. That is the very reason it is legitimate to apply statistical methods in order to obtain highly reproducible statistical results.

In addition, there are general qualitative results. For example, the general result shows reproducibly that the Y-cells move independent of the clustering of the X-cells, unless they are attracted to them. In this case, they are able to integrate into the X-cell clusters and even bridge between them.

Of course, there are more details to take into account. For example, the simulation could look at the effects of other relevant parameters, such as initial cell density, cell size, mitotic rates, etc.

As convenient and powerful as simulations can be, they have limitations. Practical considerations must limit the numbers of components, logical conditions, and rounds that are involved, in order to be completed in a reasonable amount of time, Furthermore, each round only considers certain local interactions between groups of components, but largely ignores global effects, which the entire mosaic and the environment may have on each of the tiles.

Most questionable, however, is the limited accuracy of each computation. It may be due to numerical errors which are caused by the 'graininess' of the simulation: At each round, the tiles move a very small, but not infinitely small amount. No matter how small the individual computational errors may be, if they happen to accumulate over thousands of tiles and hundreds of thousands of rounds, they may yield an entirely erroneous result. As an example, think of the enormous difficulties to predict the exact path of hurricanes by simulating the influence of the entire Caribbean and the South-eastern coast of the United States.

Therefore, the final step of every simulation is a careful comparison with the experimental observations – here the problems of consistency proofs enter again – but also a realistic estimate of all numerical errors is indispensible.

Digital Mosaics; the Imitators and Universal Memory

A new form of tool-mosaic has appeared among us even during my own lifetime. Its most prominent property is its uniformity. What I have in mind is the 'digital' mosaic.

Digital mosaics are infinitely compatible with pictures, music, texts, programs, and the mechanics of logical operations. They are undoubtedly mosaics. Their tiles consist of 'bits' and 'pixels'. Their frames are the formatting and transmission protocols of the data. Their manifestations are electronic image displays, printers, or loud speakers. Their driving force is the perfection of minutely detailed imitations and simulations of real mosaics.

Since all their tiles are strings of bits, digital mosaics need to obey practically no rules of *local* fitting. Every bit or pixel fits every other in the most trivial way. Once everything is a pixel, there is no difficulty for a pixel to follow another one in space, time, or along a string. The original objects, which the digital mosaic represents, depicts, or imitates, fulfils already all other requirements of *global* fitting, no matter how demanding they may be.

In this way, digital mosaics store information in a completely neutral mode. The ease of storage and retrieval is entirely unaffected by the digitized contents. Therefore, they are a very effective carrier of memory.

The digital transformations of paintings, symphonies, books, and photographs are shear wonders of life-likeness, simplicity, and have a degree of portability that allows them to travel across continents and oceans in fractions of seconds. On the other hand, their 'life-expectancy' is surprisingly short and exceeds hardly a few decades.

A major reason for their short lifetime is the endless strife of engineers for perfection. The digital mosaics, utterly dependent on conventions about rules of formatting, high-speed processing technology, and the proper amplitude and AC frequency of the power sources, are constantly being outdated in the course of 'improvements'. As a result, many such digital recordings, hardly a decade old, are already obsolete. Today, no one can read or use Betamax tapes or the Windows95 operating system any more. Worse, even without the drive for updating, their material carriers such as memory chips, CDs, or magnetic tapes decay physically with a half-life time of 10 to 20 years, all by themselves. Neither they nor their successors will survive for millennia like the caves of Altamira, the pyramids, or the Parthenon. Even if future archaeologists are lucky to find un-decayed remnants of digital records, without knowing the proper formatting and the correct power supplies, how can they make sense of them?

Mosaic Noise and Noise Mosaics; Friend and Foe

Often scientists treat noise as a fast, random, and ubiquitous natural phenomenon that is effectively a nuisance because it invades and spoils the results of experiments. No system is entirely free of it. Even at temperatures of absolute zero, physicists have to ascribe the zero-point energy ('Nullpunktsenergie') to the vacuum.

Yet, the interpretation of noise as a perturbation for experimenters is misleading. It is also an 'enabler' of experiments. For example, the so-called 'noise analysis' gives scientists the opportunity to carry out experiments even though they have no control of the input signals, as in the cases of the second-by-second ion fluxes through membrane channels in living cells, the molecular transport through nuclear pores, or the saltatory movements of mitochondria in the cytoplasm of cells. Yet, the omnipresence of noise and its ability to affect the experimental systems offer practical handles on them. As long as the experimenter is able to observe the noise and its effects in detail, then s/he can leave it up to the noise to deliver the required input signals for the test. S/he only needs to wait, and observe the unknown mechanism's responses to the noise stimuli.

In one way or another, all living mosaics are experimenters who experience noise in both roles, namely as perturbation but also as 'enabler'.

Let us first look at noise as a perturbation of living mosaics. Naturally, thermal noise bounces their vital molecular components around. To see what powerful, destructive enemy it is, just put a mammalian cell into plain water and puncture its membrane under the microscope. The osmolar shock destroys instantly its order and purposeful movements. In their place, countless tiny granules bounce around in violent Brownian motion inside an expanding membrane bubble.

Naturally, cells have evolved numerous mechanisms to protect themselves. Thermal noise suppression increases with increasing mass, and is the more complete the larger the object is. Therefore, membranes, molecular complexes, and polymers are the main weapons of cells with which they combat thermal noise. Compared to molecules, these are effectively macroscopic objects, which are much less affected by thermal noise. Thus, the cells may enclose sensitive molecules inside membrane-protected compartments such as vesicles. Large numbers of molecules may combine to form massive complexes such as ribosomes. Even more effective, they may insert themselves into huge membranes, or chain themselves to huge polymers. The most important of the latter are the microfilaments, intermediate filaments, and microtubules that span the entire cell body. Together, these polymers carry the name of the 'cytoskeleton'. Unfortunately, the name is misleading, as these polymers, far from behaving like a rigid skeleton, are quite dynamic.

On the other hand, noise can be of great help for living mosaics. Whenever they rely on diffusion or molecular equilibration, they rely critically on the actions of thermal noise.

Similar to the way human experimenters use noise analysis, animals can use noise signals as substitutes for a missing signal from another animal. The method may sound absurd, but it is very effective for the noise-dependent phenomenon of 'stochastic resonance', which allows e.g. the paddle fish in the muddy waters of the Mississippi to detect extraordinarily weak signals of a distant prey [21].

Of course, there are other kinds of noise, including self-created noise, which is due to the openings and closings of the contacts between a living mosaic's functional tiles, sub-tiles, sub-sub-tiles.... (In fact, even thermal noise can be interpreted in this way, namely as the result of the openings and closing between the 'tiles' of a thermodynamic mosaic, i.e. between the collisions between its molecules, molecular clusters, polymers etc.)

These kinds of 'contact-jitters' are unpredictable for the same reason that statistical mechanics deems molecular collisions unpredictable, namely because there are too many of them, and because the exact values of the relevant parameters are unknowable. In addition, we must take into account the notorious individuality of the tiles. It adds individual circumstances and constraints to every contact variation, which, in turn, adds to their unpredictability.

The amplitudes of noise are not necessarily microscopically small. Living mosaics also create large amplitude noise in the form of optical, acoustic, or olfactory signals coming from their bodies. Just listen to the ubiquitous cacophony of animal calls and the humming of insects! Undetectable by humans, dogs sample thousands of odour marks; bats and insects are engaged in a battle of spontaneous ultrasound signals humans cannot hear; vipers see – or better 'smell' – a landscape illuminated by infrared light invisible to humans.

Noise not only is influencing living mosaics. Arguably, noise is a functional mosaic itself. After all, it is a composite of many independent signals and sources of perturbations. Especially, if the carrier of the noise consists of discrete particles or – more generally speaking - of tiles, it becomes the unpredictable barrage of discrete events, and thus a spatial and temporal mosaic. Since discrete, stochastic events usually follow a Poisson distribution, we may describe it as 'Poisson-noise'. Examples are the sounds of cicadas in a tree, or the trains of electrical pulses traveling along the optic nerve. It is not restricted to living mosaics. Examples from the inanimate world are the shot-noise, or the radioactive decay.

Complementarity; Necessity and Paradox

No matter what task a living mosaic performs, it must be able to start the task but also to terminate it. Inevitably then, it has to implement for every action an effective counter-action. How else could it stop it?

In general, control of a task requires more than starting and stopping it. It needs to determine the amplitude (extent, strength, speed, direction,…) of the actions. Hence, whatever is capable to increase the amplitude, there must be an opposite mechanism that is capable of decreasing it. For example, the heart must be able to compress *and* dilate. The body must fight bacteria *and* cultivate them as symbionts for its digestion. The student will have no problem to find countless other examples of the simultaneous presence of opposing forces and actions in the control of all living things.

The same applies to tools. Whenever living mosaics develop instruments and tools for actions and products, they too must be controllable and, hence have contradictory properties. For example, a car must be able to accelerate *and* brake, to steer left *and* right. A nail must slide smoothly into wood *and* hold.

These complementary requirements provide another justification that living things are best build as mosaics: Whatever property a tile may have, since its neighbour is always abruptly different, it is possible to endow it with the opposite property. Thus, the whole mosaic acquires contradictory properties.

Obviously, the coexistence of opposite properties, must not lead to paralysis or other forms of permanent mutual inhibition. Hence, *all living mosaics are engineered compromises that enable the co-existence of opposites.*

True, this kind of co-existence must weaken the attacks and jeopardize the defences of living mosaics, but it is vital, nevertheless. For example, without a predator's weakness, how could its prey ever escape? But the chances of escape and defeat are both important for the balance and vitality of ecologies. Without an Achilles heel of the prey, how could the predator ever succeed? If there was an all powerful, invincible living mosaic, evolution would long have eliminated it, as its unopposed success would have depleted its own resources.

The coexistence of opposing properties of living mosaics makes it also impossible to characterize them by a single quality. Due to their Janus-character, one always has to include the opposing quality in their every description. On the other hand, the simultaneous presence of a property and its opposite is a frequent recipe for logical problems. Paradoxes may appear all too readily and threaten to thwart the expectation of scientists for consistency and simplicity.

Emergence I: A Subjective Type (e.g. a Heap of Sand)

We have mentioned multiple times the phenomenon of emergence that follows whenever tiles incorporate into a mosaic and interact with each other: New properties appear that no single tile had before. Atoms turn into chemicals, chemicals turn into genomes, cells turn into animals, notes turn into symphonies, pigment spots turn into paintings, letters turn into novels, and symbols turn into computer programs.

The ancient Greek characterized it by *sorites*, which means 'heap'. They formulated it as the question, what number of grains is sufficiently large, to be called a heap? Or more generally, when does a local phenomenon turn into a different global one?

In many cases, the apparent discrepancy between local and global properties is merely an artefact of our brain's data processing methods. In fact, the transition from a handful of grains to a sand heap is one of them. Nothing actually changes fundamentally as the number of grains increases. However, at certain poorly defined points the economy of communication with other people causes us to stop counting the number of grains, and start using simplifying terms like 'many' or 'heap', instead.

Often, our arbitrary values determine the transition point. Twenty grains of sand will hardly cause a neighbour to boast with possessing a heap of sand whereas s/he might describe a collection of twenty gold doubloons as a 'treasure'.

Another such case may be the arms of galaxies. Actually, galaxies do not have spiral arms. The stars, which constitute any one of these 'arms' never rush along spiral trajectories towards the centre of the galaxy or away from it. On the contrary, they all run conventional elliptical orbits safely around that centre.

Yet, we all see these spirals quite clearly. Instead of describing every star individually, it is much more economical for our brains to package the accidentally denser areas of stars into single objects, and then line them up into larger structures. Since the majority of the stars have a common direction of rotation, and since Newton's Laws force the stars and dust clouds to orbit the faster, the closer they are to the centre of the galaxy, our brain simplifies the resulting distribution of the density patches of stars into 'spirals'. Naturally, the spiral arms disappear as we look at galaxies at higher and higher magnification.

The existence of such subjective types of emergence may call into question our very concept of mosaics. If we cannot be sure when grains end and heaps begin, or how to define galaxies unambiguously as mosaics containing spirals and dust lanes as their tiles, then it all may seem a subjective matter of perspective and level of resolution.

Indeed, it is not the first time we encounter this problem. Earlier, we had to come to terms with the notion that mosaics can be tiles, frames can be mosaics, tiles can be mosaics, mosaics can be frames, and so forth. How can mosaics and modules function as fundamental concepts of biology, if their definitions seem so arbitrary?

Admittedly, there are borderline cases, but most living mosaics are not fabrications of our mind. One can identify their frames and modules in objective ways, even if they contain other mosaics, tiles and frames: Russian dolls do not cease to be dolls because they contain smaller dolls inside. Similarly, liver and triceps are distinct and isolatable mosaics, while also acting as the tiles of the meta-mosaic of the body. Their discreteness and semi-autonomy are not fabrications of our mind.

This will be clearer in the next sections about cases of emergence that are not products of our mind.

Emergence II; an Objective Type (e.g. Irreversibility)

A heap of sand can also express emergent properties that are objective. For example, in a gravitational field it can form flanks with specific slopes. However, there are even more intriguing emergent phenomena that no one can explain away as creations of our minds. Perhaps, the most notorious among them is the phenomenon of irreversibility.

Single electrons, atoms, or tiles act quite reversibly, as long as they act in isolation. In contrast, some mosaics with huge numbers of tiles may change irreversibly. Take a sugar cube! After its approx. $4 \cdot 10^{21}$ sucrose molecules dissolved in water, they can never come back out of solution by themselves. In general, as the component numbers of mosaics grow, irreversibility seems to latch onto them as if they were highjacking them. However, it would be absurd to blame our own brain for it.

What is the difference between these cases and the previous ones, which our mind created itself? Thermodynamics associates the concept of irreversibility with increases in entropy, which only is meaningful if the mosaics concerned are composed of very large numbers of thermally moving atoms and molecules, i.e. if they are thermodynamic systems.

Numerous other irreversible changes of mosaics are not the direct consequences of entropy changes, i.e. of altered temperatures, pressures, molarities, or inner energy. They do not even involve large numbers of tiles. For example, in order to leave mosaics in configurations that cannot return to their original state by themselves, it may suffice to alter tile shapes or their spatial order. These changes are irreversible, but not for thermodynamic reasons.

Other examples include the logical impossibility to carry out some of the sub-task of mosaics in the reverse order. Once a task has modified the starting materials, the products cannot return to their native state without carrying out further tasks.

Also the fitting of the tiles is a source of irreversibility. The more sophisticated the original fitting procedure was, the less blind actions of natural forces are able to restore a disrupted mosaic. No amount of swirling, rattling, or heating will restore even simple jigsaw puzzles, once somebody has disordered them. There are far too many possible alternative configurations of the pieces to allow the puzzle to fall back into its singular fitted states by blind accident.

The irreversibility of thermodynamic systems, i.e. of mosaics with typically 10^{23} molecules may seem to be a quite different phenomenon than the irreversibility of smaller mosaics with (say) $10^3 - 10^6$ tiles. Yet, they have a common source.

In the cases of thermodynamic systems, their recovery from an irreversible event requires to lower their entropy, i.e. to add negative entropy ('negentropy'). In cases of the other kinds of mosaics, such as the macroscopic living mosaics, one also needs to add energy and information in order to restore their order and functions. Therefore, scientists like Erwin Schrödinger (1887-1961) and Leon Brillouin (1889-1969) have generalized the definition of entropy by identifying negentropy with information.

In other words, if the interactions between growing numbers of tiles destroy information as they turn into mosaics, then they create irreversible phenomena. On the other hand, if growing numbers of tiles loose information because our brain is too 'lazy' to keep it - as in the case of no longer counting individual grains of sand, but conveniently calling it a 'heap' – the corresponding phenomena remain reversible.

Optimal Values Linked to Actual Irreversibility

Irreversibility is also characterized by the tendency of certain parameters to assume extreme values. As mentioned before, the prime driving force of irreversibility of thermodynamic systems is entropy, which assumes its relatively *largest value* in the most disordered state.

Living mosaics also assume extreme values in their fitted state. We have already mentioned the extremely low probability that the tiles in a fitted state arrange themselves by accident. This probability is very much smaller than the probability to find them in a disordered state. Conceptually speaking, it corresponds directly to entropy.

However, there are additional kinds of extreme parameter values of fitted mosaics. For example, if the tiles have fixed and solid shapes like the pentominoes, they cover the smallest area (volume), once they fit the model-frame exactly. Obviously, any kind of shift or disorientation among them would create holes between the tiles, which would increase the occupied area. Hence, the successfully fitted state occupies the relative *smallest possible* space.

If the tile interactions of mosaics are very dynamic, many of their non-spatial quantities are optimal. If perturbed, their changes may become irreversible in other ways. For example, the resilience of well-fitted ecologies tends to rely on the optimization of their diversity, while the stability of well-balanced economies tends to minimize risk factors. In short, disrupting the fitting of mosaics and even reducing the perfection of their fitting can be formulated as irreversible changes, which are also accompanied by the loss of optimized properties.

Emergence III; Extra Dimensions Beyond Space and Time

Other emergent properties of mosaics that are not a fabrication of our minds are their dimensions. The most abstract definition of 'dimension' is that of linear algebra, namely that *the dimension of a vector space is defined as its largest number of linear independent vectors*. For anyone who has studied linear algebra, it is a quite profound and powerful definition.

In daily life, however, the concept of 'dimension' is less cryptic. In practice, the word usually indicates the three degrees of freedom in space, but not the degree of freedom in time, because its unidirectional flow is perceived as inexorable tyranny.

We have various ways of 'sensing' space and time. In order to determine time we read our inner clocks, but also use outside clocks like the positions of objects in the sky, running water, and others. As to space, we 'sense' distances by optical methods such as blueness, parallaxes, and relative movement. We 'sense' the directions of accelerations by our vestibular system, and use our short-term, medium-term, and long-term memory to order chains of events in space and time. The methods are fallible, but amazingly effective and responsive, regardless.

However, 'dimensions' need not mean space or time. More generally speaking, they mean 'degrees of freedom of something or somebody'. Besides space and time, many other quantities are free to change within a continuous range of values.

Still, our brain treats them as something very different from the dimensions of space and time, probably because their range does not seem to be infinite, and because they are not pre-conditions of all others, although they can also be 'sensed'. For example, pressure, temperature, brightness, colour, pitch and volume of sound, taste and odour of chemical concentrations, strength of forces, etc. are such 'sensed' dimensions. Similar to the parameters of space and time, one can plot their values on the axes of a graph, and use them as parameters for composite objects.

For some of these dimensions we have objective perception. Most of us are able to tell colour, i.e. the wavelength of visible light. Indeed, objects deviating from their expected colours, such as a green stop sign or a blue-shifted copy of the 'Mona Lisa' in the Louvre, would startle most of us. Nevertheless, there are a number of well-known optical delusions that demonstrate that our colour perception is the result of very complex, adaptive computations of our brains, which can be fooled by the colours of the surroundings.

In contrast, most of us would not even notice if a band played the National Anthem in a different key. Nevertheless, some people do have perfect pitch, i.e. that they are able to tell absolutely the frequency of sound. These people feel disturbed as much by a shift of a key-signature as others are by a shift of colour.

The perception of 'sensed' dimensions does not have to be true to their physical properties. For example, vipers and other snakes use their facial pits to detect the infrared emissions of their warm-blooded prey. Although their infrared image lacks details because the facial pits are not exactly optical lenses, the brain of the animals detects the directions and distances of the prey with amazing accuracy. Surprisingly, the nervous connection between the facial pits and the brain passes first through the olfactory bulb. It suggests that these animals do not 'see' an infrared image of the prey, but literally 'smell' its location and warm temperature.

A great many of the dimensions that are important for living mosaics are emergent. The earlier mentioned fractal dimensions (chapters 8, 27) are an obvious example. They emerge from the infinite nested repetition of a k-dimensional motif, which create an object that seems to have a spatial dimension somewhere between k and k + 1. We will call these and other emergent properties '**extra dimensions**'.

In the case of living mosaics, a major source of extra dimensions is the fitting process. Examples of mosaics with emergent extra dimensions are melodies, hues, figures, patterns, novel functions and skills, efficiency, productivity, functionality, impact, information, information linkage, meaning, and many more. They are composite objects, which highlight the emergent qualities of a mosaic that its tiles did not have before they were fitted together. Nevertheless, they are real quantities, which scanning and other physical methods are able to detect and measure. One can order their data along a scale of magnitude, which then becomes the extra dimension.

Since the tiles of living mosaics are usually living mosaics, too, the tiles could possibly express extra dimensions themselves. And so can the tiles of *their* tiles and so forth.

Emergence IV; the Ultimate Emergent Phenomenon: Consciousness

The question 'what is the true nature of something' is hardly scientific. However, when discussing consciousness, it is irresistible to ask that very question. Ranging between the ancient Indian text of the 'Upanishads' [24] to modern studies of neurological dysfunctions, there have been numerous attempts of defining and understanding the '*I*', the '*Self*', or whatever name we have given our inner certainty, that there is inside us a thing that is continuous in space and in time, and that watches the world from inside us, and that makes our decisions.

Consciousness is certainly an emergent property of the brain, but it does not involve the entire brain. Numerous parts of the brain can become dysfunctional by accident, stroke, or disease without loss of consciousness. For example, blindness, deafness, language disabilities, or paralysis do not suppress consciousness. On the other hand, sleep, anaesthesia, trauma, drugs, and certain neurological diseases can cause its temporary or permanent loss. The degree of its alertness can be modified by the neurotransmitter system, but only if consciousness is already present. It all may suggest is that we are closing in on the problem, but the main breakthrough is still missing.

What kind of breakthrough do we need? At this stage of our understanding, we only know with certainty a number of *necessary conditions* of consciousness, such as the functionality of specific parts of the brain, the right body temperature, the right composition of the blood and interstitial fluid, and so forth. Without them, there is no consciousness. Yet, if a mosaic would fulfil all these conditions, consciousness would still not magically appear. In fact, many unconscious patients fulfil these conditions. What we need, is at least one *sufficient condition*.

Promising approaches to find sufficient conditions are studies of certain neurological dysfunctions, where the body responds normally to stimuli, whereas the patient should be aware of them, but is not. In itself, the phenomenon would not be very remarkable. In fact, we are all unaware of most every action of the body; except in these special pathological cases we are certain, that the patient *should* be aware of the stimuli. By pinpointing the location of the neurological defect that prevented the expected awareness, we may at least find the portal where physical signals, of which we are normally aware, and their corresponding neuronal actions begin their transformation into consciousness.

For example, the eyes of a patient with so-called 'blindsight' [18][26], can follow accurately a moving light point, while the patient insists that s/he is entirely blind. Without the neurological defect, the patient would surely be fully aware of the moving light.

Can the concepts of living mosaics help to characterize consciousness? Arguably [107], it is composed of many loosely connected and incoherent fragments of processed and memorized experiences. Using the terminology of this book, we may therefore describe it *as a loose, symbolic mosaic*. Our minds do not seem to store it anywhere specifically, but make a new copy and update it whenever we 'become aware of something'. Like all other self-replicated living mosaics (chapter 19), it re-assembles its tiles every time *de novo*. Hence, the resulting replicas will not be faithful, but like most other *de novo* assembled living mosaics, will vary from copy to copy, although they produce 'almost'-copies of each other. In special situations they may vary more significantly and even become 'conveniently' edited by our intentions and impulses.

Considering the disordered, fallible, and fragmented state of our conscious mind, why then, do we perceive it as a single and unique 'oneness' inside us? John Gray [107] explains the oneness as an illusion that results from our ability to view this 'jumble of symbolic tiles' from the outside as if seen from a distance. We may call the product of this pseudo-distant view our symbolic 'self-reality'.

We produce many other symbolic representations. There is a much larger, much more detailed, and much more accurate representation of the world around us and of our body as 'seen' from the inside and as reported by our sensory organs and nerve endings. We may it call our 'ambient-reality'.

Curiously, we always patch our self-reality into the centre of this ambient one, and thus make it the reference point and 'meaning'-dispenser of everything else.

Detour: The Homunculus Conundrum

Gray's explanation begs the question, who or what is doing the postulated kind of pseudo-distant viewing of the self-reality? Let us assume for the moment that an 'inner, neuronal screen' displays our self-reality together with the ambient reality. In order to view it, somebody or something inside us must 'sit in front of it', monitor it, and report to us the results.

This homunculus must *consciously* watch the screen. Filming the screen with a mindless movie camera would not do much good, unless somebody who is conscious is watching the movie later on. Therefore, in order to be a conscious observer, the homunculus must have its own inner conscious homunculus, who watches the experiences of the first homunculus. The latter homunculus, of course, also needs its own conscious sub-homunculus inside, and so forth. In short, we face an infinite regress of sub-sub-...-homunculi, which never reaches an answer.

But wait! The postulate of a conscious homunculus as the creator of consciousness is either not applicable to biology, or else is a useless tautology: If we carry the above regress to infinity, the postulate is not applicable to biology, because nothing in biology is infinite. Alternatively, if we would stop the regress after the kth iteration and settled for the result that consciousness is the work of the kth conscious sub-homunculus, then we only have arrived at the tautology that *consciousness is the result of consciousness on mind level 'k'*. Therefore, we should dismiss the homunculus argument as a meaningless play with words. Let us consider another way.

Emergence V; a Second Ultimate Emergent Phenomenon: 'Will'

If consciousness is an enduring illusion, and if we know this to be true, why do we continue re-creating and trusting it? Even more discouraging should be the recognition how small a role consciousness plays in our lives.

During sleep we are unconscious almost one third of every day. During the wake state, we experience visual, auditory, olfactory and tactile sensations far too rapidly to be conscious of their complex processing. We control our upright walk, the balance of our bodies, and our reactions to the movements of others with equally rapid, hence subconscious neuronal routines. Although we can consciously intend to speak a word, our motor control for moving lips, tongue, and vocal chords operates on 'autopilot', too. Artists, instrumentalists and even chess-players seem to rely on gut-feelings much more than conscious decision making, lest they stutter, tumble, and get confused. In life-and-death emergencies, we instinctively turn off consciousness and act by instinct alone. In moments of ecstasy, we also silence our consciousness, as it spoils and trivializes the experience. So, what little consciousness we actually experience, we seem to dislike and avoid it. Hence, let me repeat the question. Why do we not ignore consciousness?

Actually, we have no choice. There is an overpowering drive inside us to create representations of reality. Only deep sleep, drugs, and major traumata seem to be able to suppress it. For lack of a better word, we may call it an inherent 'will'.

Quite obviously, like consciousness, this 'will' is not the property of single neurons, but an emergent property of large brain domains. It powers the assembly of our self-representation and its remoteness. However, it does not only operate with various strengths in humans, but in all living mosaics, because *it is partly what powers the defining 'tasks' that all living mosaics have.*

Hence, no meaningless series of conscious sub-homunculi exists that can explain consciousness. Instead, *there may be an **irrepressible will** to create symbolic representations of the world and of ourselves, and to scan and to re-create them ceaselessly.*

In terms of this notion, **the 'feeling of awareness' is not a stable state of contemplating a map of our ambient and internal world. It is rather the incessant effort of creating, re-creating, and updating a 'fluid' map of it all**. It is the feeling that streams of details flow in and out of our temporary memory while building new map portions and deleting old ones. Some map portions will eventually be selected to become permanent; others will disappear forever. As maintaining all this data traffic is strenuous, we seem eager to avoid the labour. Yet, somehow we feel gratified for every minute that we were able to satisfy our instinctive urge of keeping the map alive.

Of course, this mysterious 'will' remains unexplained. Still, it may present a step forward, because it seems more accessible to scientific exploration for the following reasons.

In the first place, in our age of intelligent machines, explaining the programmed commands for a 'will' inside us may be easier than the very subjective 'feeling' of consciousness. More importantly, in view of the minor importance of consciousness in our daily lives, it is hard to see the evolutionary advantage of 'feeling aware'. In contrast, the instinctive 'will' to create a 'reality map' of us and our ambient world makes much more sense. Considering the countless dangers around us, scanning and updating the map must improve our chances of survival significantly. Perhaps, studying this heritable, programmed 'will' along the evolutionary tree of living mosaics may help us discover its mechanisms.

The idea that the combination of mental representations and an underlying 'will' are responsible for our conscious perception of the world is far from new. Almost 200 years ago, one of the most influential European philosophers, Arthur Schopenhauer (1788 - 1860) suggested it in his work 'Die Welt als Wille und Vorstellung' (The World as Will and Mental Representation).

Still, Schopenhauer was not the first, either. More than 2,500 years earlier, the ancient Indian text of the *Kena Upanishads* [24] *formulated the paradoxical union between mental representations and an irrepressible 'will' in this most elegant and profound way.*

'That which makes the mind think, but cannot be
Thought by the mind, that is the Self, indeed.'

****

After having presented the above, mosaic-coloured views of consciousness at length, I hasten to admit that there are entirely different interpretations of the nature of consciousness. Raymond Tallis [22] has argued an almost diametrically opposed view, which maintains that consciousness is a metaphysical concept. Trying to explain it by merely physical mechanisms such as neuronal activity and natural selection cannot do it justice, but must trivialize and ultimately fail it.

We must stop here because the present book is not the place to debate philosophy. However, it should remind the students who gravitate naturally toward scientific data and experiment-based explanations, that nobody who strives earnestly to understand reality will be able to escape philosophy.

Uncertainties Caused by Observing Mosaics

The famous uncertainty relations of quantum mechanics state that it is impossible to observe certain *pairs of physical variables* simultaneously. The more accurately we determine the value of one, the more 'smeared out' the data of the other become. In other words, our very attempts to pin down one of them perturb the other in unpredictable ways.

The most frequently quoted example concerns the simultaneous measurements of the location and speed of an electron.

The usual formulation of uncertainty relations between two variables X and Y more accurately, is

equ. 22-01
$$\Delta X \cdot \Delta Y \geq L,$$

where ΔX and ΔY are the numerical variance of the variables X and Y, while L is a positive number that indicates that the product $\Delta X \bullet \Delta Y$ can never be made smaller than L. In particular, neither ΔX nor ΔY can ever be reduced to zero.

Quantum mechanics knows several such pairs of variables. However, disregarding the hypothetical case of 'Schrödinger's cat', these quantum mechanical uncertainties only are concerned with the observation of atoms and sub-atomic particles.

The uncertainty relations limit the observations of sub-atomic particles and, therefore, they seem to question their credibility and relevance. Although this sounds like a negative, the uncertainty relations are by no means merely a justification to doubt the accuracy of observations.

On the contrary, the uncertainty relations have very important *constructive* consequences. As we have reminded the student in chapter 21, the existence and strength of chemical bonds is a consequence of the uncertainty relations.

Either way, such considerations seem to have little relevance for living mosaics. In the first place, their tiles are much larger than sub-atomic particles, and biological actions rarely require the simultaneous knowledge of the location and speed of electrons or photons with absolute precision. Even the photo-receptors of the retinal cilia only respond to the energy of the photons, but they need not 'know' their location with any more accuracy than the diameter of a retinal cell.

Still, as we shall see, observing living mosaics may face obstacles of its own, and hinder measurements in fundamental ways. Here are some examples where uncertainties call biological data into question.

Inevitable Artefacts and Perturbations of the Living State

Sometimes the situation in biology is worse than in physics, because the biological uncertainties do not even require *pairs* of incompatible variables;

already *single variables* can be perturbed in unknowable ways by the observations. For example, as demonstrated by the work of many behavioural scientists, the observation of animals can be perturbed to unknowable degrees by the mere presence of the observer, even though the observer and the animals are far larger than atoms and electrons. Here, the uncertainties arise from the animal's perception of the observer, even if s/he exerts actively no perturbing forces.

In other cases, the observer remains uncertain about the data because the applied methods must distort the specimens in unknowable ways. Here the uncertainty results ultimately from the absence of independent control observations.

Obviously, it causes a gross distortion of reality if the observation of a living mosaic requires killing it. Yet, many times, biologists have no other choice. Especially, if we want to observe extremely small details, we must accept that the preparation method not only may have taken away all living functions, but also have altered the specimen in additional, yet unknown ways.

For example, living cells and their components are always immersed in water and electrolytes, even if their (say) land living organism is not. The logical approach to arrest these components for observation would be to freeze the specimen extremely fast. However, even if the rapid freezing would not kill the specimen, it alters it in other ways, regardless.

The speed of freezing is necessarily slower than the thermal speed of the molecules. The warm molecules must first collide with very cold (= slow) ones before they can slow down themselves. Hence, freezing temperatures advance predominantly across the specimen by energy transfers during molecular collisions, although energy loss by infrared photon emission plays an important role, too. Hence, by molecular standards the molecules have plenty of time to change their location and shapes, before the cold could finally stop them.

Furthermore, the natural electrolytes around and inside cells are much less soluble in freezing than in warm water. Hence, the freezing water forces some of them out of solution. The result is a 'wall' of precipitating salts racing ahead of the freezing front. Inevitably, before the electrolytes come out of solution, their concentrations must grow to very high levels. As a result, they are likely to denature all macromolecules before the surrounding ice has locked them in place. In short, rapid freezing of living system to 'capture a molecular snap-shot', creates molecular uncertainty, instead.

How can we know how much our observations of living mosaics have distorted and altered them? In order to determine the extent of the distortions, we would need a set of perfect controls, i.e. experimental methods that do not alter the specimen. However, they do not exist. What should we do?

Fortunately, some of the above artefacts do not have the character of principle barriers. So far, year by year, a growing number of technical improvements added credibility to our methods of preservation. In some cases it was even possible to find model systems that happened to be immune to one or the other artefact.

Nevertheless, uncertainties remain, which no amount of technical advancement can avoid. The answer to the problem is always the same in science. We must avoid all wishful thinking, use the knowledge derived from other fields, and try to estimate the extent of the artefacts, while surrounding the observations with a network of testable hypotheses. (At this point I cannot resist referring the student to Hempel's famous 'Philosophy of Natural Science' [109]).

Self-Reference and Progressive Catastrophes

Perhaps, the student has participated in the following little experiment. One person sits down on a chair; a second person sits on the lap of the first; a third on the lap of the second, and so forth (figure 22-02a).

The resulting line of sitting people is, of course, perfectly stable, unless somebody pulls the chair out from under the first, in which case the first person will fall and, through a chain reaction, make everybody else tumble, too.

There is, however, a way of removing the chair without causing a collective collapse, by seating the people in a circle, The first person does not sit on a chair, but on the lap of the last (figure 22-02b). It takes a bit of skills to set up such a ring. Nevertheless, once established, the arrangement is a perfectly self-stabilizing mosaic.

I believe, it is more than that, namely a metaphor for an essential property of the biosphere, of all ecologies and, indeed, of every living thing. They all exist in stable configurations, because a number of other living systems exist, which support them. Paradoxically, these 'supporters', in turn, only can exist, because their 'supportee' and several others exist.

Such a seemingly vicious circle not only is self-stabilizing but also self-referential, because its components derive their purpose, function, and definition from each other. Pulling yourself up by your own bootstraps certainly violates Newton's law, but it is commonplace among living things!

Countless cases of symbiosis operate because of this kind of self-reference. Food chains form self-supporting circles. Flowering plants and insects are locked into each other's lives. The function of lungs, heart, and digestive system depend critically on each other, and so forth.

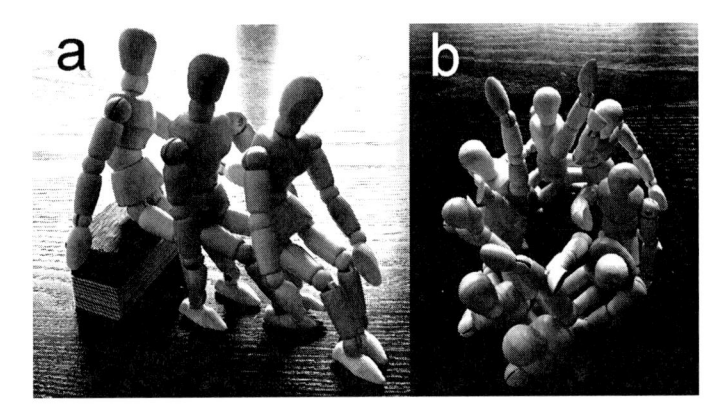

Fig. 22-02. Forward support and Self-reference. (a) Forward transmission of the mechanical support provided by the bench on the left side. (b) Self-stabilization through the circular seating arrangement with no support by a bench.

Of course, setting up these kinds of inter-dependence for the first time can be quite difficult and delicate, and may need long times to evolve. Just consider how long it took primitive plants and micro-organisms to deliver more and more oxygen into Earth's atmosphere while more and more oxygen-tolerant and oxygen-dependent organisms evolved, which, in turn, became indispensable for the cultivation, replication, or as food of the oxygen-producers!

Although the system of inter-dependent circles can spread quite far, it must end, because all life forms depend ultimately on the energy supply from the sun or the heat of Earth's interior. All circular dependences end right there, because obviously these two ultimate energy sources are truly autonomous as they do not depend on the life forms they support.

The weakness of all self-referential systems is their local, point-like vulnerability. As in the case of the circle of seated people, only linked to their own parts, they are not sufficiently anchored to their surroundings. Hence, the disruption of only one component must trigger the disassembly of all others. It does not matter which particular member of the circle falls first. All others will follow.

Therefore, self-referential mosaics, like living objects, are in danger of disintegrating by chain reactions. They may suffer from propagating catastrophes, because even during their collapse, the self-referential mosaics are self-feeding the progression of their own destruction.

It does not take huge energies to trigger the progressive catastrophes of a mosaic. Often small changes acting over millennia and eons may build up damages within its circle of dependences that will ultimately destroy them.

Whatever the detailed causes of progressive catastrophes, they can be categorized as either damage to the tiles, flawed fitting, faulty key-seeds, disrupted frames, or missing linkages with other living mosaics. Of course, in practice, they may be called by different specific names such as overpopulation, mono-cultures, destruction of resources, pollution, over-specialization, etc.

In order to protect themselves from this danger, the inter-dependent circles evolved to share their members with the members of other circles. As members of one ring they could survive its collapse, if they remained supported by the members of a second ring, as long as the second ring stayed intact.

Definition of Living Mosaics; How Autonomous is Autonomy?

Our initial definition of living mosaics stated that their tiles are modules and, therefore, are 'semi-autonomous'. We continued to use this term although it sounds like an oxymoron. How can something be half-autonomous? It should be either autonomous or not.

The argument is purely academic. In fact, 'incomplete autonomy' is what we encounter everywhere. It is 'complete autonomy' that seems unrealistic, if we look at it closer.

Autonomy does not mean the freedom to act independent of every other object around. Even apparent autonomous systems depend on supplies of material, energy, action spaces ('Spiel-Raum'), and tools by others. Autonomy is simply a matter of the degree of the freedom to make decisions.

For example, let us consider a predator stalking a prey and a prey using all its powers to escape. Are any of their mutually responsive actions actually autonomous? If each move is dictated by the attack or defence of the other, and ultimately by the compulsion to feed or to survive, the freedom of action and choice of action that autonomy requires is severely impaired. In addition to the constraints that living mosaics place on each other, there are constraints imposed by the inanimate world. Of course, the gravitational field of the Earth, her rotation, her oceans, mountains and her atmosphere create more than obvious limitations on the actions of living mosaics. Much less trivial are the constraints that are rooted in accidents of the evolutionary history and the individual fate of each living mosaics.

The actions of living mosaics are subject to entirely indirect and obscure limitations. Startling examples are freely moving protozoa such as *Chlamydomonas*. They cannot help but form elaborate, highly structured colonies depending on the locations of the surrounding light and oxygen sources. The indirect and obscure cause behind these colonies, however, is the physical separation between their centres of gravity and of buoyancy of their own bodies [10]. There is no intent or other biological motivation behind the structures of their colonies. Yet, the protozoa cannot escape these inanimate influences, because they are the pre-condition for their survival.

The cells inside organisms are naturally even more constrained. The histology of muscle, nerve, and connective tissue make it immediately clear how much freedom and autonomy the tissue cells must have surrendered compared to free-living protozoa in return for their narrowly regulated, but protected and bountiful life inside the organisms.

Surprisingly, the loss of autonomy of single cells goes even further. In the interest of the whole organism, they have to forgo even their right to self-defence and -preservation. Here is an example.

One of the most ancient and universal mechanisms of self-preservation is the so-called 'heat-shock' or stress response of cells. After temperature increases or exposure to certain toxins, free cells shut down their normal protein synthesis and switch to a 'rescue' mode that ultimately results in the synthesis of ancient and universal proteins ('chaperones') that are able to fix proteins that may have been misfolded by the trauma.

As discovered by Veena Prahlad et.al. [17], once they are part of an organism, the somatic cells have to hand over even this power to certain nerve cells in the organism that decide on their stress response. This profound kind of 'cell-non-autonomy' has since been confirmed in many organisms.

In view of the myriads of interdependences of all living mosaics and their tiles, and sub-tiles,..., their tasks, sub-tasks, and sub-sub-tasks,...we need to re-define autonomy to include the many influences and factors that enter into their actions. Going back to the original Greek meaning of autos (αυτoσ) = 'self' and nomos (νoμoσ) = 'law', it means literally to make one's own laws. Therefore, *the autonomy of a mosaic, tile, or frame in a given situation means that it can choose an action and makes the final decision whether, where, and when to carry it out, after weighing the preferences of other mosaics or tiles. In contrast, semi-autonomy means that the mosaic, tile, or frame cannot carry out certain actions without the decisions and actions of other mosaics, tiles, or frames.*

Definition of Living Mosaics; how Discrete is Discreteness?

The second major condition of our initial definition of mosaics was the discreteness of their tiles. More specifically, the discreteness was tacitly understood as spatial and/or functional separations between them without any gradual transition. When we introduced iterations, cycles and other kinds of temporal mosaics, we extended the condition of discreteness of tiles into the time dimension.

However, during our discourse, we weakened the condition of discreteness. When we accepted fractal mosaics, we allowed the spatial discreteness of the tiles to vanish altogether as they approached their infinite limits of the nesting of smaller tiles inside larger tiles.

There are more compelling reasons to qualify the conditions of discreteness of tiles and mosaics. In the first place, the interactions between adjacent tiles require that forces, energy, or information flow from one to the other. This flow between discrete tiles may be continuous in space and/or time.

Furthermore, seemingly disconnected objects may be connected, nevertheless, although the connection follows along paths that we cannot observe.

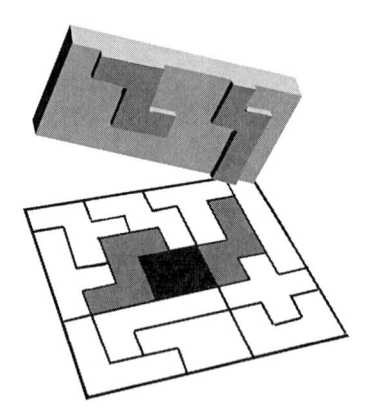

Fig. 22-03. Interpretation of the 'Z'- and the 'N'-pentominoes as the 2-dimensional 'prints' of one and the same 3-dimensional object. In this way, 2-dimensional tiles may no longer be discrete, but actually continuous along a path across a higher dimension.

The argument is another version of the age-old Hindu metaphor of the blind men who touch the same elephant, and thus describe the animal as quite different objects. One blind man describes the elephant as a trunk, another as ears, yet another as tusks, legs, a tail, etc. Yet, no matter how different and irreconcilable their descriptions are, the actual elephant connects them in a completely continuous and meaningful way.

'Flatland' enthusiasts among the students are quite familiar with another example of hidden connections between seemingly isolated objects through higher dimensions. In 'Flatland' one can easily construct situations where 2- or more-disconnected 2-dimensional objects are actually the 'footprints' of the same 3-dimensional object. Figure 22-03 illustrated such a 'Flatland' example, where two different and separated 2-dimensional pentominoes are merely 'prints' of a 3-dimensional 'stamp' that joins them across the third dimension.

In general terms, it is always conceivable that the discrete tiles of an n-dimensional mosaic are merely the n-dimensional 'footprints' of (n + 1)- or higher-dimensional objects.

The linking dimensions need not be spatial. As long as the connection between the tiles has a measurable size and a direction, it can be considered a dimension and form a temporary, oscillatory, or permanent bridge between them. Hence, the dimensions of tiles and their connections can be any combination of spatial, temporal, or symbolic parameters.

*In other words, the condition of discreteness of the tiles cannot be absolute. Our definition of the discreteness of tiles only can require the existence of certain components and directions where some of the properties of mosaics, tiles, or frames end abruptly, thus creating a 'border'. It does not require that **all** their other properties end abruptly at the same border. On the contrary, a number of additional parameters and properties of the tiles may pass through their border and mediate the connections and linkages with other mosaics, tiles, or frames.*

Definition of Living Mosaics; the Blessing and Threat of Modularity

We had introduced the concept of modularity as an engineering principle to guarantee successful maintenance and service. However, modularity not only is a required strategy for the actions and maintenance of living things. Reformulated, it states that the parts of living things must be well separated from each other, while closely cooperating at the same time.

In view of the above re-definitions of autonomy and discreteness, this is not a contradiction. In fact, it is a powerful weapon against the blind and merciless attacks of the inanimate world. The discreteness of tiles guarantees that most damages remain confined to the tile under attack. Moreover, semi-autonomy of the tiles means that some of their functions *are* autonomous. That, in turn, guarantees that these functions are able to protect, defend, and even attack without the help from others.

Indeed, the stronger the tiles of a living mosaic express individuality while cooperating the most with each other, the more successful and victorious the mosaic will be. As the highly individual members of the most successfully co-operating species on Earth, I believe, we all know the truth of it all too well.

Unfortunately, there is a flipside. The recipe of overcoming inanimate nature is also a recipe to destroy other living things, including one's fellow co-mosaics. A great deal of wisdom and compromise is needed to defuse the dangers of this most powerful principle.

Is the Terrestrial Principle of Modularity Actually a Universal One?

Look again at the arguments in chapter 21 about the lumpiness of the world! They are all based on principles of physics that are actually cosmic universals. Hence, people have called 'lumpiness' a universal cosmic principle.

Now pair the cosmic principle of lumpiness with the terrestrial principle of evolution! One can justify this, because evolution is hardly an exclusively terrestrial phenomenon. Its specific results may be unique for planet Earth, but the principle of evolution should be the way in which hugely complex life forms originated everywhere. That includes cases where life forms on a planet did not actually start there, but grew from seeding materials that originated elsewhere. The seeds ultimately evolved in that other place, nevertheless.

Therefore, we can be sure that evolution everywhere in the universe has a chance of transforming the locally present lumps. In some instances, they may have evolved into modules with distinct borders and functions. In other cases, they may have remained inanimate lumps. Which of the two occurred, has actually nothing to do with planet Earth.

Therefore, modularity of life is most likely a cosmic principle. And if this argument does not convince the student, there is always the justification of modularity as a necessity of successful engineering.

Why should such a generally valid principle be restricted to Earth?

A Portal for Creativity in Living Mosaics?

Nature's creativity is unquestioned, but that does not explain the creativity of individual living mosaics. In order to be living, mosaics carry out many tasks, including many routines that occupy their time almost completely.

For example, while painting the ceiling of the Sistine Chapel, Michelangelo most of his time was busy breathing, keeping balance high on a scaffold, mixing pigments, blowing coal dust through the holes of his cartoons, and so forth. When did he find the opportunity to become creative?

When and where is the portal through which creativity enters? We only ask this very limited question about creativity, because there are ultimately no explanations for creativity. There is not even a consensus about what it means.

Therefore, we are not asking what it is, but only now, at the end of the book, risk the question where in the midst of their normal routines living mosaics face the need to resort to creativity.

We may think of seeking help from the digital mosaics that we mentioned earlier. Could their power of analysis and imitation help us to pinpoint where creativity shows up?

Of course, there have been numerous attempts to empower digital machines with creativity, but they remain unconvincing, because they are all held captive by the assumption that random number generators can impersonate the sources of creativity's unpredictability.

For example, after recording the sound spectra of several Cremona violins, a synthesizer can imitate the sound perfectly. However, a circuit boards full of electronic oscillators and random number generators would never have invented the characteristic and enticing sound spectra of these violins.

After digitizing all of Bach's works, computers may be able to use the statistics of his music to simulate a composition that sounds like him, at least for a handful of measures, and which can create the sound of polyphony for listeners who have never played Bach's works. In order to deeply feel the difference, one must have struggled with and marvelled at the sparkling, stunning, unpredictable, yet in retrospect completely logical, alienations by which Bach creates and prepares a new musical thought many measures downstream. Without the template of his compositions, would a digital composing machine ever have come up with the musical universes of the Brandenburg concertos, the 'Matthäus Passion', the Partitas for solo violin, or the Art of the Fugue? If that was possible, a digital composer should be able to predict the music of the next future Bach, or perhaps even become itself a next composer of this magnitude.

Considering that such a programmer would need to be a composer of Bach's calibre, I doubt that this is a very likely scenario. Why would a programmer with Bach's creativity use machines to compose? Bach himself never composed even from the keyboard, which was the only 'music-machine' available to him in his time. In fact, he mocked people who did so as 'Klavier-Husaren' (keyboard-hussars).

As mentioned before, performing a composition, i.e. assembling a living mosaic, is never a reproduction, but always a re-creation from scratch. There is no digitization of the spontaneity and context-alienation of the human creative process and its enormous breadth from which it draws its inspiration. It has no rules or algorithms that we have ever identified, nor is it random. The monotonous and content-remote routines of digitization are entirely antagonistic to the profound unpredictability of enthusiasm and deep dreaming. Surely, a future Mozart may write for synthesizers or other electronic instruments, but it would mean nothing more than composing for a novel kind of a 'fiddle'.

Occasionally, one should remember the dangers of overestimating the abilities of machines. The mysteries of living mosaics will always reveal machines as shadows of living things, as Loren Eiseley argued so passionately in 'The Bird and the Machine' [105].

No, trying to formulate an algorithm for creativity is a hopeless enterprise. Still, can we at least try to identify and characterize the parts of a living mosaic, where creativity erupts into view?

Of course, creativity's hallmark is change. Can we at least ask where a living mosaic expresses typically an unexpected, yet convincing change?

A powerful mechanism of change is the mutation of mosaics, but this is exclusively nature's genetic mechanism of creativity. The particular kind of creativity of individuals that I have in mind is a unique occurrence, not a heritable one.

Other candidates could be the mechanisms that cause the 'almost'-repetitiveness of living mosaics. Could not one or the other 'almost'-repetitions become a point of departure that leads to innovation?

I doubt it. Creativity is most often a 'life-saver', born out of a crisis, and the rather tame deviations of the 'almost'-repetitions may not cause sufficiently drastic breaks from normal routines.

So, we should ask, what precipitates crises in individuals? Of course, this question opens the flood gates for answers by modern psychiatry, neuroscience, and psychology. However, if we focus on known creative people, I believe, that the critical trauma is often the encounter with a contradiction that concerns their fate and work, and resists resolution. Indeed, the foregoing sections have pointed out a number of inevitable contradictions from which the concepts of living mosaics suffer. Are we to hypothesize they are the portals of creativity?

After a lifetime of loving and admiring painting and music, I deduce my criteria of creativity from my most admired artists, Shakespeare, Goethe, Hölderlin, Dostoevsky, Bosch, Leonardo, Michelangelo, Rembrandt, Cezanne, van Gogh, Klee, Bach, Mozart, Beethoven, and Tchaikovsky, to name a few.

Naturally, I perceive all their works as *living* mosaics. Based on this understanding, I believe creative works do not begin with an explosion of novelty. On the contrary, they usually emerge almost stealthily from their conventional precursors. Yet, when they do, they are completely unexpected, although in retrospect their appearance always seems surprisingly logical in their self-created context.

To be sure, the conventional materials of their origin are never simple. At least, they contain several problematic, even paradoxical elements. It seemed that the artist's effort to reconcile them produced the opportunity and the necessity to deviate from the norm. Still, the conventional materials of the creative nuclei must at least be fertile, because they have to support their subsequent growth.

Hence, it seems to me that the crucial factors are the following.
1. Conventional, yet fertile materials.
2. Local conflicts between tiles that need to be reconciled.
3. Impossibility of reconciliation without eliminating or damaging the conflicted tiles.
4. Relinquishment of any reconciliation.
5. Replacement of the conflicting loci with novel, never before used elements, derived nevertheless from the normal materials.

Earlier, when we described the conjunction of two key-seeds, we encountered this very situation. Figure 22-04 tries to illustrate that the conjunction of two 4-key-seeds fulfils all the above criteria.
1. In the first place, the materials used for their conjunction are conventional, as they contain, of course, conventional pentominoes and frames. Yet, like all other key-seeds, they are 'fertile' objects from which solutions can grow.
2. The conjunction of the two key-seeds leads to a conflict between their two 'L'-shaped tiles. In their present orientation, it is impossible for both 'L's to enter the conjunction intact. Their overlap area only is a single square (labelled white in figure 22-04).

3. Normally, the conjunction would simply eliminate both tiles.
4. However, let us assume that the 'author', who decided to seek a conjunction between two different, fertile elements, cannot accept this course of action.
5. In this case, s/he can place a novel tile in the overlap area. It could be another pentomino, a door to the outside, or even an entirely different tile.

To be sure, this only is the moment when a creative process may begin. It is the crucial moment where the 'strange' choice of an added tile may send the mosaic on its path towards an entirely new object.

The resulting mosaic is likely to appear both surprising and logical. It may appear surprising, because the initial change of both key-seeds was small, yet novel. Yet, it may appear logical, because it followed the logic of conjunctions, albeit with a deviation. Furthermore, the subsequent resolution of the conflicted conjunction incorporates conventional, logical rules.

Of course, whether the result appears ultimately creative, depends entirely on how novel, yet how convincing the treatment of the conflicted tiles was. Furthermore, this method of placing creative nuclei is hardly the only one. Otherwise, creativity would obey a rule of nucleation, whereas creativity never obeys any rules. Still, I believe the described strategy is one of the more frequent beginnings of creative mosaics.

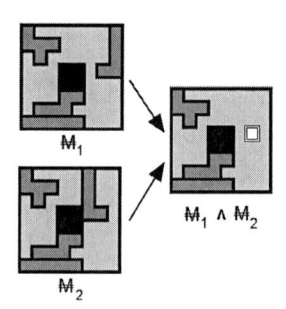

Fig. 22-04. Incompatible, overlapping tiles in the conjunction of conventional key-seeds. The overlap area is marked white. Are such conflicting areas the portals for creativity?

Climbing a 'Wittgenstein Ladder', and Tying up a Few Loose Ends

Teachers and authors face many times the dilemma that Ludwig Wittgenstein (1889 – 1951) described famously at the end of his 'Tractatus Logico-Philosophicus' *[28]*

'My propositions are elucidatory in this way: He who understands me finally recognizes them as senseless, when he has climbed out through them, on them, over them. (He must so to speak throw away the ladder, after he has climbed up on it.) He must transcend these propositions, and then he will see the world aright.'

In the preceding chapters, I had asked the student repeatedly to climb such 'Wittgenstein-ladders' towards a more comprehensive understanding of the core concepts of living mosaics. Some of the ladders were more wobbly than others, although I tried to construct them stable enough for at least one climb.

I hope, the student, who has accompanied me along these paths, eventually acquired the confidence to test the concepts of living mosaics and their modularity in his/her real world. I believe, they are able to depict biology as a unified, stable edifice of thought, and paint a clearer and less intimidating picture of life on Earth.

To be sure, the concepts cannot explain biology, nor what humans are, let alone what consciousness is. They cannot predict unknown biological mechanisms, either. On the contrary, these concepts only seem to predict an overwhelming diversity of biological mechanisms. Yet, at the same time, they may provide a common format to think about this very diversity, and help order, orient, and guide our minds in their attempts to conquer it.

****

Presenting the concepts of mosaics and modularity as uniform, size-independent ways of describing living systems based on engineering principles does not mean that we will always find an engineering justification for the observed actions and properties of biological systems. Some may remain beyond our understanding for some time, because our own level of engineering has not yet reached theirs. Imagine, trying to understand the engineering of the nervous system in the year 1700 when we had never heard of electrical pulses and their use in data processing! Other actions and properties of living mosaics may still be primitive and have failed to evolve yet to a perfect level of engineering. Nevertheless, applying the principles of engineering may point our research in new directions and help our insights into a common inner logic of living things and the amazing speed of evolution.

****

Let me close with a note about my apparently conflicted beliefs. On one hand, I am sure that it is self-evident that both living and inanimate objects are aggregates of discrete components.

On the other hand, as a former physicist, I am also aware that physics has been spectacularly successful by ascribing lesser importance to the discreteness of things, and instead seeking their underlying laws in differentiable quantities that are quite the opposite of mosaics.

A contradiction? Is the universe of biology in truth also continuous, while our minds, possibly terrified of infinity, create the delusion of its fragmentation?

Perhaps. However, encouraged by Niels Bohr's validation of antinomies as quoted at the beginning of the book, I believe that the mosaic nature of things and especially the modular structure of all living things may very well be also true.

REFERENCES/SEARCH TERMS

AT THE AGE OF POWERFUL SEARCH ENGINES, ADDING REFERENCES TO A BOOK SEEMS RATHER SUPERFLUOUS. BY OFFERING STARTING REFERENCES AND SEARCH-TERMS, THE FOLLOWING FEW REFERENCES MAY AID THE READER'S ELECTRONIC SEARCHES AND SELF-STUDY

[01] Albrecht-Buehler, G., (1976) Numerical evaluation of the validity of experimental proofs in biology. Synthese 33, 283-312.
[02] Albrecht-Buehler, G. (1998): http://www.basic.northwestern.edu/g-buehler/cellint0.htm.
[03] Albrecht-Buehler, G.,(1990), In defence of non-molecular' cell biology. International Review of Cytology 120:191-241
[04] Clune, J, Mouret, JB, and Lipson, H. (2013) The evolutionary origins of modularity. Proceedings of the Royal Society B. 280: 20122863.
[05] Eiseley, Loren (1978) The Bird and the Machine. In 'The Star Thrower', Times Books Inc., New York, NY.
[06] Golomb Solomon W. (1965) Polyominoes. Charles Scribner's Sons, New York.
[07] Gray, John. (2002) Straw Dogs. Thoughts on Humans and Other Animals. Farrar, Strauss, and Giroux, New York.
[08] Harari, Yuval Noah (2017) Homo Deus, A brief history of tomorrow. Harper Collins Publisher, New York, NY.
[09] Hempel, Carl Gustav (1966) Philosophy of Natural Science. Prentice-Hall, Inc., Englewood Cliffs, N.J.
[10] Kessler, J.O. (1985),Contemp. Phys.26: 147-166.

[11] Koyaanisqatsi, Indie film/Experimental film, (1982) (Director: Godfrey Reggio).
[12] Levin, Beth.(1993), English Verb Classes and Alternation, A Preliminary Investigation. The University of Chicago Press.
[13] National Research Council. (2008) The Role of Theory in Advancing 21st-Century Biology: Catalyzing Transformative Research. Washington, DC: The National Academies Press. https://doi.org/10.17226/12026.
[14] Patrusky, Ben (1981) Why is the Cosmos Lumpy? Science 81: 96
[15] Peitgen, H.-O., H.Juergens, D. Saupe (1992) Chaos and Fractals. New Frontiers of Science.
[16] Piaget, Jean (1954) The construction of reality in the child. Basic Books, New York, NY.
[17] Prahlad V, Cornelius T, Morimoto RI (2008) Regulation of the cellular heat shock response in Caenorhabditis elegans by thermosensory neurons. Science 320(5877):811-814. doi: 10.1126/science.1156093.
[18] Ramachandran, Vilayanur S. (2004). A Brief Tour of Human Consciousness. Pi Press, New York.
[19] Rizzolatti, G., Fadiga, L., Fogassi, L., Gallese, V. (1999) . Archives italiennes de biologie 137(2-3):85-100.
[20] Russell, Bertrand. (1922) The Analysis of Matter. London, George Allen and Unwin Limited.
[21] Russell David F., Wilkens Lon A., Moss Frank. (1999) Use of behavioural stochastic resonance by paddle fish for feeding. Nature 402: 291-294.
[22] Tallis, Raymond (2016), Aping Mankind. Routledge, New York, NY
[23] Thompson, D'Arcy Wentworth. (1945) On Growth and Form. Cambridge University Press; New York: Macmillan.
[24] (The) Upanishads. (© 1987) Nilgiri Press, Tomales, CA.
[25] Wagner, GP. (1996) Homologues, Natural Kinds and the Evolution of Modularity. American Zoologist. 36:36-43.
[26] Weiskrantz, Larry. (1986) Blindsight. Oxford University Press, Oxford.
[27] West, Geoffrey. (2017) Scale. Penguin Press, New York.
[28] Wittgenstein, Ludwig. (1922) Tractatus Logico-Philosophicus. Harcourt, Brace, and Company, Inc., New York, London.
[29] Soll, David. R. (2009) Why does *Candida albicans* switch? FEMS Yeast Res., 1-17.

[30] Shapiro, James A. (1988) Bacteria as Multicellular Organisms. Scientific American (June 1988) p. 82-89.
[31] Fawcett, Don W. (1981) The Cell. W.B. Saunders, Philadelphia, London, Toronto.

GLOSSARY

General

A mosaic is a 'frame' filled completely and accurately with a set of 'tiles', which are discrete objects with finite sizes. Each tile is fitted to another tile and/or the frame. The transition between any two tiles is abrupt. There is no continuity between them. If there are signals and forces exchanged between the tiles, they travel along paths and channels that are tiles, as well.

Depending on the 'substance' of the tiles, mosaics may be tangible or symbolic. The tiles of tangible mosaics may be spatial, temporal, or functional. The tiles of symbolic mosaics are non-material objects that only exist in the minds of living things, although they may be attached to material carriers.

Depending on the amount of information the minds of living things have to contribute to symbolic mosaics, the symbolic mosaics may be information-rich (i.e. the recipient mind contributes no information, but only receives it), or information-free (i.e. the recipient mind contributes all relevant information, but receives none).

The frames of mosaics are mosaics themselves.

By turning themselves into tiles, mosaics may create meta-mosaics of unrestricted sizes.

Individual basic terms

Mosaic: Composite object.
Tiles: Discrete, semi-autonomous components of a mosaic.
Frame: The spatial, temporal, and/or functional constraints under which the tiles of a mosaic interact.
Fitting: Optimization of the interactions of the tiles under the constraints of the frame (see chapter 5).
Autonomy: Autonomous mosaics, tiles, or frames choose their actions and make the final decision whether, when, and where to act.

Semi-autonomy: Mosaics, tiles, or frames are semi-autonomous with respect to certain actions that they cannot carry out without the inputs of other mosaics, tiles, or frames.

Discreteness: The existence of certain properties and directions of where mosaics, tiles, or frames end abruptly, thus creating a 'border' It does not require that all other properties of these mosaics, tiles, or frames end abruptly at the same border. On the contrary, a number of additional parameters and properties of the tiles may pass through their border and mediate the connections and linkages with other mosaics, tiles, or frames.

Task: A mosaic whose modules are instructions and 'names' for processes and functions that are placed in a specific order within a certain temporal and/or spatial frame.

Living mosaic: Biological object. The combination of a mosaic and a task, which was, is, or will be associated with the mosaic.

Module: Tile of a living mosaic whose functions are predominantly autonomous.

Solution: A mosaic that presents a perfect solution of the problem of fitting the tiles under the constraints of the frame.

Sibling-solution: Another mosaic with the same tiles, frame and fitting procedure.

Seed: Incomplete mosaic.

Key-seed: A seed that yields a uniquely defined solution, if completed by the fitting procedure.

Tile dissection: Mosaic in the shape of a tile and composed of smaller tiles.

Iterative sub-tiles: The tiles that make up a dissection.

Fragmentation factor: Number γ of iterative sub-tiles of a dissection.

Solution: Completely fitted mosaic using all available tiles.

Sibling-solution: Alternate solution with the same tiles and frame.

F,I,L,N,P,T,U,V,W,X,Y,Z : Pentomino names.

Other terms in the order of appearance

Mosaic assembly
Assembly algorithm: Algorithm or computer program to find all solutions for a certain frame and a set of tiles.
Search-code: String of letters (tiles) and numbers (orientations) to be tested by an assembly algorithm.

Seeds of mosaics
Seed: Group of initial tiles that can be completed into a solution.
n-key-seed: Group of 'n' initial tiles whose completion leads to a solution *with necessity*.
^{n}M̶: Symbol for the n-key-seed of mosaic M.
Rank 'n' of a key-seed: The number 'n' in ^{n}M̶.
Complementary key-seeds: 2 key-seeds ^{m}M̶$_1$ and ^{n}M̶$_2$ whose sizes 'm' and 'n' add up to the size N = m + n of the mosaic. They have no tiles in common and together they fill the frame precisely to constitute the mosaic.
M̶* Symbol for complementary mosaic of the key-seed M̶. It contains the tiles of M that fit the holes of M̶ in the same order and orientation as M. If M contains symmetrical tile groups, M̶* may not be a key-seed. (see figure 11-05a).
'Inner seam': Given 2 complementary key-seeds of a mosaic, its inner seam consists of all the tile-to-tile contacts, where a tile of the first key-seed touches a tile from the second.

Inner structure of mosaics
v_w : Volume of tile #w.
s_w : Surface of tile #w.
V_Φ: Total volume of the frame.
S_Φ : Total surface of the frame.
S_T : Total surface of all tiles in a mosaic.
S_P : Total surface of all tiles in a mosaic if all tiles are pentominoes.
S_M : Total surface of a mosaic.
Ξ : Total inner surface of a mosaic.

Fractal mosaics
Fractal mosaics: Mosaics whose tiles are mosaics of smaller tiles *ad infinitum*.
Level independent tiles: Every level uses the same tile dissections for recursive replacements.
Level dependent tiles: Different levels using different tile dissections for recursive replacements.

Iteration driven mosaic cycles
'natural' iteration series: The 'next' key-seed has a maximal size of overlap
τ_R time interval between successive replications
τ_{Td} time it takes to duplicate and assemble the new tiles
τ_{Ip} interphase time interval between the completion of a replication and the onset of the next.

Replication of mosaics
'dimension of a living mosaic' : The number of independent directions which the mosaic uses in order to arrange the tiles without overlap.

'dimension+1' rule: The mechanical separation of 2 complementary key-seeds cannot happen along the mosaic's own dimension(s). The mechanism requires an additional independent direction at right angles to those.

'embryo':   The set of tiles or their precursors together with the information of their placements and growth that is needed to replicate a 3-dimensional mosaic.

Partial replication
'mother-mosaic' The complete mosaic from which a part is selected for partial replication.

coded mosaic: Information-rich mosaic whose tiles represent the letters of a code.

'transcription': The partial replication of code-bearing sub-sets of the tiles of an information-rich mosaic.

'translation': Implementation of the specific function of different partial replicas of a symbolic mosaic

'expression': The combined actions of transcription and translation of an information-rich mosaic.

Living mosaics and their 'tasks'
'living mosaics'   Discontinuous, composite objects whose components are other mosaics.

'material'   A portion (sub-set) of the world including the mental world.

'process'   Every transformation that changes one kind of material into another.

'starting materials'   The initial input material of a process

'products'   The final output materials of a process.

'task'   Ordered set of processes (mapping) to turn starting materials into products.'

'tools' Materials that enter a process at various stages and exit it again at a later stage, while their composition and shapes remain essentially unchanged. They constitute no part of any final product.

A task-based taxonomy of living mosaics
Tasks
    Performer ('Who or what carries out the task?')
    Action ('What is done to perform the task?')
    Target ('At which object is the action directed?')
    Means ('What procedures, instruments, or strategies are applied?')
    Manner ('What are the circumstances under which they are applied?')

A most important functional mosaic: The iteration
'Iteration-mosaics': Functional, single tile mosaics that can be applied to themselves. Their frames are abstract qualities. The condition of the 'fitting' is the requirement that both its input 'i' and input 'o' fit the frame (input and output must have the same abstract quality). Hence, **IT** can operate on its own output, i.e. if $o_n = \mathbf{IT}(i_n)$, then $i_{n+1c} = o_n$, i.e. $i_{n+1} = \mathbf{IT}(i_n)$. ($n = 0, 1, 2, \ldots$). Iteration mosaics are useful tools to formulate causality and the history concerning biological objects,

Fractal mosaics
Self-similarity: If magnified by a certain *reduction factor 'r'*, every enlarged portion of the set looks identical to the original. If 'r' is able to reproduce the initial set identically, then $r^2, r^3, \ldots, r^k$, etc. must be able to do the same.

Applying a scale change of r^k, means that the unit of length δ_0 is reduced to a length of $\delta_k = \delta_0/r^k$ in order to appear as large as the original unit length. The unit length δ_n is also called the *resolution unit*, and $1/\delta_n$ as the *resolution at n^{th} level*, while 'n' indicates the nth *level of refinement*.

Series of resolution units: $\delta_n > 0$, ($n = 0,1,2,\ldots$), with $\delta_{n+1} = \delta_n/r$, and r the reduction factor >1. It follows that $\delta_n = \delta_0/r^n$ ($n = 0,1,2,\ldots$).

Patches: Point sets with diameters of unit length that cover certain portions of the point set to be measured. The *number $N_n(\delta_n)$ of patches* required to cover at least the point set grows to infinity as the resolution unit δ_n shrinks to zero. A plot of the logarithm of the number of patches versus the logarithm of the resolution ($1/\delta_n$) at n^{th} level yields a straight line with good approximation.

$\log[L_n] = \eta \cdot \log[1/\delta_n] + \gamma$ (n = 0,1,2,...).

Fractal dimension η: The slope of the linear regression of '$\log[L_n]$ vs. $\log[1/\delta_n]$'.

Tile dissections: The process of replacing recursively each tile of a mosaic with a mosaic made of smaller tiles.

Iterative sub-tiles: 'Tiles-that-are-actually-mosaics-made-of-smaller-tiles' by the process of tile dissection.

Static sub-tiles: Some parts of the dissection reserve spaces for something other than the iterative sub-tiles. They are never replaced during the advancing rounds of recursion.

Level independent dissections: The tile dissections do not vary from round to round of replacement but use always the same shaped, albeit smaller mosaics.

Level dependent dissections: The tile dissections vary from round to round of replacement by using always differently shaped, albeit smaller mosaics.

The seeds of mosaics

Seed' of a mosaic: A mosaic with empty spaces that can be completed to a mosaic by fitting the appropriate tiles into the empty spaces.

Key-seeds M of a solution M: seeds of a mosaic that **only** can be completed in one way. They contain a subset of tiles that determine uniquely the final solution M. The assembly of a solution starting with a key-seed is entirely independent of the fitting-protocol, as long as all tiles are different and available and as long as this protocol is able to find the only tile that fits a hole with the exact same shape.

The m-key-seed ^{m}M of a solution M: Key-seeds M of a solution M that contains exactly m tiles. The integer 'm' of an m-key-seed ^{m}M will be called the 'rank' of the key-seed.

Complementary key-seeds: Two key-seeds ^{n}M$_1$ and ^{m}M$_2$ for the mosaic solution M are complementary to each other, if
 (a) Both are key-seeds of the same mosaic.
 (b) They have no tile in common.
 (c) Together they fill the mosaic precisely.
Symbolically:
 (a) ^{n}M$_1 \cap {}^m$M$_2$ = M.
 (b) ^{n}M$_1 \cup {}^m$M$_2 = \emptyset$, where $\emptyset$ symbolizes the empty set.

The replication of fitted mosaics
Mosaic replication:
- (a) Production of new starting materials such as frames and tiles,
- (b) Assembly the new tiles inside the new frame in precisely the location and orientation of the original, and
- (c) Growth and development of the mosaic to the stage of maturity where it can replicate again.

'Canonical' method of mosaic replication: Each pre-existing mosaic M is the union of 2 complementary n-key-seeds $^nM_1$ and $^nM_2$. It replicates by splitting into these seeds along their 'inner seam' and subsequently completes both of them.

The strategies of mosaic replication
'dimension' of a living mosaic: The number of independent directions which the mosaic uses in order to arrange the tiles without overlap.

'mosaic-dimension + 1' rule: The mechanical separation of 2 complementary key-seeds requires an additional independent direction at right angles.

Partial replication of fitted mosaics ('Coding' and 'Expression')
partial replication of a mosaic: Not all of its components are replicated, but only a selection of its tiles and/or frames.

'colour': A non-essential modifier of tiles. Partial replicas must have different 'colours' than the original tile sets.

transcription: The partial replication of an information-rich mosaic.

translation: Implementing the specific functions of different partial replicas of the information-rich mosaic.

expression: The combined actions of transcription and translation.

Iteration-driven mosaic cycles
natural iteration: Iteration where the 'next' key-seed in the series is the most similar, yet not identical one.

rank of key-seed: Number of tiles in the key-seed

basic key-seed changes: (same rank) replacements and transpositions.
(different rank) deletions and insertions.

replacement: One of the tiles of the seed is removed and replaced with another tiles.

transposition: One or several tiles of the seed are moved to new locations.

deletion: Changes the rank of the key-seeds by removing a tile.

insertion: Changes the rank of the key-seeds by adding a tile.

'key-seed pool': The set of key-seeds on which the iteration algorithm draws when it selects the next key-seed.

key-seed # or solution #: In order to determine more economically whether a particular key-seed or mosaic re-occurs during the iteration either randomly, all the time, only after regular periods, each key-seed and mosaic is assigned its own unique index number.

'rounds of iteration' and 'time points': The advancing rounds of iteration are assumed to occur in time. That does not mean that they correspond to equidistant time intervals. Yet, the terms 'rounds of iteration' and 'time points' are used interchangeably.

Regeneration period of key-seeds (srp): Every key-seed returns to the pool of available key-seeds after 'srp' rounds of iteration. After the srp is passed, the key-seed will not automatically enter the iterative series, but only the pool. The iteration mechanism is merely able to call it up, if and when it needs it again.

APPENDIX A

NUMERICAL METHODS FOR THE ASSEMBLY OF MOSAICS

Numerical Methods for the Fitted Assembly of Mosaics

The physical assembly of complex mosaics would be extremely difficult, if not impossible without representing them in the weightless, compact forms of the written or computer languages. Therefore, we will translate them into abstract, numerical form of search-codes.

Mosaic Search-Codes

There is more than one way to translate mosaics into abstract representations. The method presented here is illustrated using the example of pentomino mosaics of the model-frame. More details are described in Appendix C.

In order to find all sibling-solutions of a particular kind of mosaic, we assume that its common frame and all of its 'N' tiles #1, #2,...,#N, are known. Each tile #k can exist in a number O(k) of different orientations, which are expressed as indices ω_p^k, indicating that the kth tile is in its pth orientation.

Furthermore, there is a systematic way of scanning each sibling-solution. For example, if the siblings are 2-dimensional, one may scan them from top-to-bottom and left-to-right. The scanning method is arbitrary, and several equivalent methods may exist for a given type of mosaic. The only requirement of the scanner is that it never misses a tile, and that it probes every spot within the frame exactly once.

Whenever the scan encounters a tile for the first time, it determines the #k of the 'N' tiles and its orientation index ω_p^k. These data pairs, chained together in the order of the scan, will be called a 'search-code'.

A small, but important technicality needs to be mentioned: Each tile must have a 'handle', i.e. a point that represents the position of the tile. After all, tiles are usually extended objects, and one of their points will be designated as their handle. In other words, the tile is where its handle is.

equ. 24-01
Depending on the scanning method, one chooses as handle the first point that is encountered when one scans the tile in its particular orientation.

equ. 24-02
Mosaic search codes have the structure
$$\{\#k_1 / \omega_{p1}^{k1}, \#k_2 / \omega_{p2}^{k2}, \ldots, \#k_N / \omega_{pN}^{kN}\}$$

Since all the tiles of a mosaic must be different, the numbers k_1, k_2, ..., k_N have to be a permutation π of the numbers 1, 2,...,N. In contrast, the orientation indices ω_p^k have no such restrictions. Their only requirement is that the index p must not exceed the total number of different orientations $O(k)$ of the kth tile.

As shown in Appendix C, a typical search code of the sibling-solutions for the model-frame is
{Z/2,Y/5,W/0,N/6,L/2,F/0,I/0,X/0,P/7,V/3,T/2,U/1}. It indicates that the first pentomino that the scanner encountered was a 'Z' in orientation '2'. Next it hit a 'Y' in orientation '5', and so on.

The search-codes not only will record the codes of different sibling-solutions, but also all of their rotations or reflections. Obviously, there is no benefit in recording these trivial multiples. In order to exclude them, we may use a simple trick: We select a particular tile whose shape happens to be completely asymmetrical. If there is no such tile, one can create one by attaching specific markers to one of the tiles that make it completely asymmetrical.

Such an 'indicator'-tile allows us to exclude unwanted rotations and reflections by the following reasoning. We only accept sibling-solutions where the indicator-tile is oriented in its essentially different orientations. Should the assembly method yield a solution that contains this particular tile in any of *its* rotations, the solution can be recognized as the rotation of the entire sibling-solution, and will be rejected. In the following we will understand 'sibling-solutions' as a set of solutions that exclude mirror images and reflections of their entire mosaics.

Reconstructing Sibling-Solutions from their Search Codes

Once we have a search-code $\{\#k_1/\omega_{p1}^{k1}, \#k_2/\omega_{p2}^{k2},\ldots, \#k_N/\omega_{pN}^{kN}\}$, it allows us to re-construct its sibling-solutions in the following way. We take the empty frame and, following the path of the scanner, we place first tile $\#k_1$ in the orientation ω_{p1}^{k1}. After moving along the scanner path we arrive at the next empty spot in the frame, and place tile $\#k_2$ in the orientation ω_{p2}^{k2}. By virtue of the way we found the search-code, the second tile will fit exactly the first tile in this location and orientation. Similarly, the third tile will fit the second, and so forth until the last tile $\#k_N$ in the orientation ω_{pN}^{kN} completes the reconstruction of the sibling-solution.

Tree-Searches of Mosaic Search-Codes

Assume, all we know about the sibling-solutions of a class of mosaics are their N tiles and their possible orientations in a standardized order. In this case, equation 24-02 reflects the general structure of the search-codes.

If we want to obtain all the solutions, all we need to do is to write down all permutations of the N tiles together with the combinations of all their legitimate orientation codes, and construct the mosaics for each of the codes.

Of course, writing down all possible search-codes must be done in a systematic way, or else we may miss some of them. However, this is not our main problem. Unfortunately, the codes give us much more than just the solutions. The resulting strings of symbols contain the codes for all mosaics, fitted solutions and non-solutions, alike.

In order to identify the codes that lead to fitted sibling-solutions, we need to weed out the non-solutions. This must be done by testing one-by-one, which of the search-codes does not lead to a fitted mosaic.

The main road-map for this testing will be a search-tree. It is a natural consequence of our systematic way of constructing the search-codes. As there are always multiple choices of the next-level entries into a progressing search-code, representing these choices as branches will inevitably generate the search-tree.

Although it may sound simple in principle, it is not in practice. The astronomically large number of all potential search-codes presents a monstrous problem, as testing them one after another would take prohibitively long times. For example, as pointed out in Appendix C, finding all the sibling-solutions of the model-frame, would require the testing of $3 \cdot 10^{16}$ different codes. Even, if we were able to test 1000 codes per second, it would take 951,293 years to complete the test.

Therefore, we need to search this tree in a much more economical way. A widely used method is to perform a truncated tree-search. 'Truncation' means that we will not even look at codes that cannot possibly lead to a fitted solution, let alone test them.

How can we know in advance a code will not lead to a fitted solution? Take for example, our model-frame with its 12 pentomino tiles. Each search-code must contain a permutation of the numbers 1, 2, ..., 12. Say, the permutation of our next search-code to be tested is [ZYXNTFIWPVLU], but after placing the 'Z' and the 'Y', we find that it is impossible to place the 'X' at the next open spot of the scan. Therefore, we can be sure no permutation [ZYX....] that begins with 'ZYX' can ever produce a solution: Regardless, what pentominoes we place after the 'X', and in whatever orientation we place them, we will find no solution, because the 'X' already does not fit. Hence, we are allowed to ignore (truncate) all the branches of the search-tree that follow the path 'Z'→'Y' →'X'.

This amounts to a gigantic savings of search-codes that need not be tested. In our example, they include all 9! = 362880 permutations of the remaining tiles F, I, L, N, P, T, U, V, W, times the number K of all combinations of all their orientations, namely $K = 8 \cdot 2 \cdot 8 \cdot 8 \cdot 8 \cdot 4 \cdot 4 \cdot 4 = 524288$. So, the total comes to $362880 \cdot 524288 = 2 \cdot 10^{11}$ search-codes that can be skipped without testing them. Keep in mind that this only is the truncation of one of the branches. There are a great many others like that.

The actual implementation of the truncations of search-trees can require elaborate programming, but given our fast computers, it can always be done. The result is the quite efficient listing of all fitted solutions.

Monte-Carlo Methods

Still, even truncated tree-searches may fail to reduce the search times to manageable sizes. The creation of a search-tree begins with generating all permutations of the number N of all tiles. If N is large, even the setting up the list of all permutations may already exceed every practically available amount of time.

In such cases, we may have to give up our goal of generating *all* solutions. Instead, we may settle for as many solutions as the available time permits. Nevertheless, we want to create a representative, unbiased sample of sibling-solutions. In terms of our metaphor that describes sibling-solutions as isolated, extremely rare islands that are scattered throughout an ocean of non-solutions, we want to sample randomly as many spots all over the ocean as possible, hoping to catch some of the islands.

Instead of generating *all* search-codes for the N tiles, we settle for creating a large number of random ones by using the Monte-Carlo methods in order to generate their permutations and orientation codes.

Random numbers can create permutations in the following way. Beginning with the original order of the N numbers we select randomly 2 positions 'r' and 's' between 1 and N, and swap the numbers at these positions. Then we choose other numbers 'r' and 's' and repeat the procedure, and so forth.

Let me illustrate it with the example of N = 6. Starting with the original series [1,2,3,4,5,6], we begin by (say) selecting r = 3 and s = 5, and consequently exchanging the third number with the fifth. It yields [1,2,**3**,4,**5**,6] → [1,2,**5**,4,**3**,6]. Next we select randomly r = 1 and s = 2 and obtain [**1**,**2**,5,4,3,6] → [**2**,**1**,5,4,3,6]. Then we choose r = 2 and s = 6 and obtain: [1,**2**,5,4,3,**6**] → [1,**6**,5,4,3,**2**], and so forth. We will do this N times with each permutation as an initial one. In this way it is guaranteed we never duplicate any of the numbers, but merely scramble up their order.

In short, Monte-Carlo method can create permutations by choosing random numbers r = random(N) and s = random(N) to determine the next pair-exchange.

equ. 24-03

Starting with an initial permutation of the numbers 1,2,...,N, a random permutation can be created by repeating N times the swap of the number at the rth position with the number at the sth position. Each such act is called an 'inversion'.

$$\begin{array}{cc}(r) & (s) \\ \downarrow & \downarrow\end{array} \quad \rightarrow \quad \begin{array}{cc}(r) & (s) \\ \downarrow & \downarrow\end{array}$$
[1,2,...,x,...,y,...,N] → [1,2,...,y,...,x,...,N],

with r = random(N), and s = random(N).

Repeating this procedure thousands of times, we obtain thousands of permutations. There is a draw-back, though. Although we can guarantee that they are all permutations, some of them may be duplicated and many of them will be missing, unless we go on *ad infinitum*.

As to the orientation codes, each tile #k has O(k) different orientations. Again we use Monte-Carlo methods to determine which particular orientation ω_p^k each tile in the search code will have.

equ. 24-04 The Monte-Carlo orientation codes ω_p^k, which express that tile #k is in its pth orientation, are

ω_p^k = random(O(k)),

where O(k) is the number of all possible orientations of the tile #k.

equ. 24-05 Monte Carlo search codes: If the mth permutation π_m = [k$_{m1}$,k$_{m2}$,...,k$_{mN}$] of the numbers 1,2,...,N, is obtained by a random number of successive random pair exchanges, and if each tile #j has O(j) different orientations, then the corresponding Monte-Carlo search-code is

{k$_{m1}$/random(O(k$_{m1}$)), k$_{m2}$/random(O(k$_{m2}$)) ..., k$_{mN}$/random(O(k$_{mN}$))}

Monte-Carlo Method of Setting Up Search-Trees for Testing

Regardless how little systematic our Monte-Carlo method of creating search-codes is, and how incomplete the yield will be, we can still sort them into a search tree, by the first, second, third, ... digit of their permutation. We can also apply the logic of truncation to the process, even before we search for solutions.

Testing for solutions does not even require that individual search-codes are completed. Preliminary testing of their initial tiles and orientations suffices to mark them as failures. Consequently, we can flag their locations in the search-tree, and stop the generation and testing of every new search code, whenever it begins like the code for an already known non-solution.

Of course, this method will still take very long processing times regardless of truncations. Yet, we do not have to wait until all solutions are found. We may stop the procedure and start evaluating the data, whenever we have reason to consider the sample size of siblings to be sufficient for a statistical treatment. The approach may yield sufficiently many solutions to determine some of their common properties for testable hypotheses about their structures and behaviours.

APPENDIX B

FITTING AS A MATRIX OPERATION

What do the perfectly fitted sibling-solutions have in common? Or, to use our earlier metaphor, what distinguishes a string of tiny islands of sibling-solutions from the rest of the surrounding ocean of non-solutions?

Perhaps an answer can be found, if we develop a common mathematical formulation for each of them. Their abstract structure may reveal their common features much better than verbal arguments can.

There may be other benefits of such a formulation. For example, it may suggest how a fitted solution may be able to transition into another one, without first labouring through astronomical numbers of failed intermediates. In this way, the mathematical description may suggest ways in which evolution can move from fitted solution to fitted solution with high speed.

This mainly mathematical chapter is placed as an Appendix, because it is not needed for the understanding of the narrative. Nevertheless, I believe that it has practical use for the analysis of mosaics.

Spatial Fitting

The Solution Codes

We illustrate our approach with the example of the 64 different 'essential' solutions to fit the 12 pentominoes into the model-frame like figure 3-03. (The term 'essential' means that we do not count rotations and mirror images of the entire mosaic).

As shown in more detail in Appendix C, each of the fitted solutions can be encoded by a 'solution-code'. The latter is obtained by scanning the solution from top-to-bottom and left-to-right (see figure 25-01a). Whenever the scan encounters a pentomino for the first time, the solution code is incremented by the letter of the pentomino followed by an orientation code number that specifies in which way it is rotated, as illustrated in figure 25-01b.

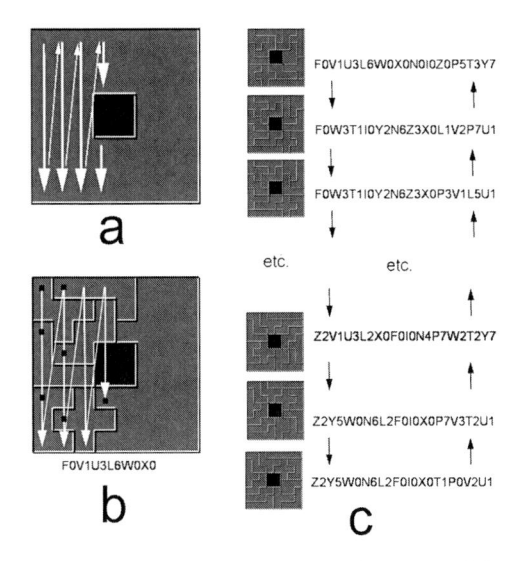

Basic Biology for Born Engineers: Living Mosaics 339

Fig. 25-01 Construction of the solution-code for each of the 64 essential solutions of the model-frame (see figure 3-03.
(a) top-to-bottom and left-to-right scanning of each solution.
(b) The letter and index of rotation is added to the solution code after each first encounter of the scan with a pentomino, The black spot indicates the 'handle' of each tile, which is the first of the 5 squares that a 'top-to-bottom' and 'left-to-right' scan would encounter (see also (equation 24-01 and figure 25-02).
(c) List of the solutions together with their solution-codes. (Further explanation sees text).

The following list shows the first and last 4 solution codes for the pentomino solutions of the model-frame.

equ. 25-01
#1: F0V1U3L6W0X0N0I0Z0P5T3Y7
#2: F0W3T1I0Y2N6Z3X0L1V2P7U1
#3: F0W3T1I0Y2N6Z3X0P3V1L5U1
#4: F0W3T1I0Y2N6Z3X0V0P5L3U1
..................etc....................
#61: Z2Y5W0N6L2F0I0X0P7T0V2U1
#62: Z2Y5W0N6L2F0I0X0P7V3T2U1
#63: Z2Y5W0N6L2F0I0X0T1P0V2U1

#64: Z2Y5W0N6L2F0I0X0T1P0V2U1

How do we find explicit representations of these operations that transform a solution code #x into a solution code #y? Let us take the example of solution codes #1 and #2.

Reformulations of the Solution Codes

At first we rewrite the codes by separating the letters and the numbers into 2 brackets:

#1: [FVULWXNIZPTY] [013600000537]
#2: [FWTIYNZXLVPU] [031026301271]
etc.   etc.

The left-hand brackets of symbols, such as [**FVULWXNIZPTY**], list the 12 pentominoes in the specific order found by scanning their corresponding solutions. Since each pentomino must occur exactly once in each solution, the 12 symbols in this bracket must be a permutation of the 12 letters '**FILNPTUVWXYZ**'.

Of course, each tile (pentomino) in each solution must be oriented in a certain way, in order to fit. The symbols in the right-hand brackets, such as [**013600000537**], indicate symbolically the orientation of each pentomino tile relative to an 'initial' orientation. As an example, figure 25-02 shows the 8 actual orientations of the 'P'-pentomino and their symbolic code numbers as they are used throughout the text.

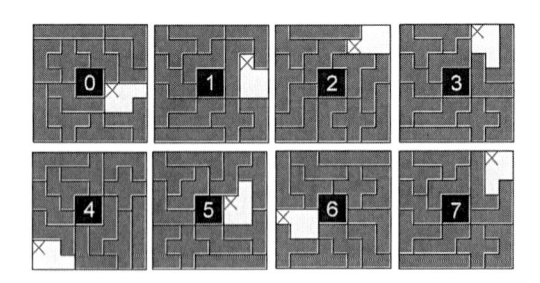

Fig. 25-02. Examples of the 8 possible orientations of the 'P'-pentomino. The white letters indicate the code numbers that symbolize them throughout the text. The 'x' indicates the handles (see equation 24-01) of each of the 8 orientations.

The orientation code numbers are assigned as follows. One begins with an initial orientation, called #0. Subsequently, one rotates and reflects the tile shape in a systematic way, and assigns a new code number to each new orientation. The choice of the 'initial' state of each tile is arbitrary. However, once chosen, it is binding for all solutions of the mosaic.

In the above example, the orientation code numbers #1 through #3 indicate successive counter-clockwise rotations of the initial 'P'-pentomino by 90°. Next, the series of reflections begin. The code # '4' is the up-down mirror image of the initial orientation. Then the orientation code numbers #5 through #7 indicate the same series of counter-clockwise rotations starting with #4 as a new initial orientation.

Since the pentomino orientations were determined by the same scan that found the order of the tiles, the right-hand brackets list the orientation codes in the identical order as the tiles of the left-hand bracket. In our example it means that 'F' has the orientation code #0; 'V' has #1; 'U' has #3; and so forth.

Actually, there is no need to keep the codes of the right-hand brackets in this order, because the left-hand bracket recorded already the specific order of the pentomino tiles in the solution. Therefore, we lose no information if we re-order the right-hand brackets by listing the orientations of the pentominoes in the standard order **'FILNPTUVWXYZ'**. In this way, the right-hand brackets turn effectively into look-up tables for the orientation of the pentominoes in the corresponding left–hand brackets. Hence, we can rewrite the right-hand bracket as follows.

FVULWXNIZPTY(permutation)→ FILNPTUVWXYZ(std order)
[**013600000537**] → [**006053310070**]

FWTIYNZXLVPU(permutation)→ FILNPTUVWXYZ(std order)
[**031026301271**] → [**001671123023**]

The results of all these rearrangements are codes that list the pentomino tiles in their solution-specific permutation and add a look-up table for their orientations in a standard order:

[**FVULWXNIZPTY**][**006053310070**]
[**FWTIYNZXLVPU**][**001671123023**]

The Representation of the Permutations

We will now use the standardized form of the solution codes to construct a suitable representation for the operations that turn any solution into any other. We begin with the permutations matrices.

First we define (arbitrarily) the initial order of the tiles by a vector

$\pi_0 = [\mathbf{F,I,L,N,P,T,U,V,W,X,Y,Z}]^T$

(The 'T' indicates the matrix is transposed, i.e. it is supposed to be a column instead of a row; but they are easier to handle as text, if we write them as rows). Next, we replace the letters with integers that are easier to permutate. The result is

$\pi_0 = [\mathbf{1,2,3,4,5,6,7,8,9,10,11,12}]^T$.

Using as example permutation

$\pi_x = [\mathbf{Z,Y,W,N,L,F,I,X,T,P,V,U}]^T$,

it transforms into

$\pi_x = [\mathbf{12,11,9,4,3,1,2,10,6,5,8,7}]^T$.

Like any other permutation, π_x can be represented by the product of a particular permutation matrix $\Pi_{(0 \to x)}$ and the standard vector π_0.

| equ. 25-02 |

$\pi_x = \Pi_{(0 \to x)} \otimes \pi_0$.

Whenever there is no possibility of misunderstandings, we will write Π_x instead of $\Pi_{(0 \to x)}$. The permutation Π_x is represented by a matrix whose elements are all '0', except a single '1' is placed at the intersection between the kth column and the nth row, if the permutation π_x has transposed the kth tile to the nth place.

Explicitly, the matrix is

equ. 25-03

$$\begin{bmatrix} 12 \\ 11 \\ 9 \\ 4 \\ 3 \\ 1 \\ 2 \\ 10 \\ 6 \\ 5 \\ 8 \\ 7 \end{bmatrix} = \begin{bmatrix} 0 & 0 & 0 & 0 & 0 & 0 & 0 & 0 & 0 & 0 & 0 & 1 \\ 0 & 0 & 0 & 0 & 0 & 0 & 0 & 0 & 0 & 0 & 1 & 0 \\ 0 & 0 & 0 & 0 & 0 & 0 & 0 & 0 & 1 & 0 & 0 & 0 \\ 0 & 0 & 0 & 1 & 0 & 0 & 0 & 0 & 0 & 0 & 0 & 0 \\ 0 & 0 & 1 & 0 & 0 & 0 & 0 & 0 & 0 & 0 & 0 & 0 \\ 1 & 0 & 0 & 0 & 0 & 0 & 0 & 0 & 0 & 0 & 0 & 0 \\ 0 & 1 & 0 & 0 & 0 & 0 & 0 & 0 & 0 & 0 & 0 & 0 \\ 0 & 0 & 0 & 0 & 0 & 0 & 0 & 0 & 1 & 0 & 0 & 0 \\ 0 & 0 & 0 & 0 & 0 & 1 & 0 & 0 & 0 & 0 & 0 & 0 \\ 0 & 0 & 0 & 0 & 1 & 0 & 0 & 0 & 0 & 0 & 0 & 0 \\ 0 & 0 & 0 & 0 & 0 & 0 & 0 & 1 & 0 & 0 & 0 & 0 \\ 0 & 0 & 0 & 0 & 0 & 0 & 1 & 0 & 0 & 0 & 0 & 0 \end{bmatrix} \otimes \begin{bmatrix} 1 \\ 2 \\ 3 \\ 4 \\ 5 \\ 6 \\ 7 \\ 8 \\ 9 \\ 10 \\ 11 \\ 12 \end{bmatrix}$$

There is a particular permutation 'E' that does not change the initial order, or the order of any other vector π_x. It is called the neutral element.

equ. 25-04
$\pi_x = E \otimes \pi_x$, for every π_x, with

$$E^{(12)} = \begin{bmatrix} 1 & 0 & 0 & 0 & 0 & 0 & 0 & 0 & 0 & 0 & 0 & 0 \\ 0 & 1 & 0 & 0 & 0 & 0 & 0 & 0 & 0 & 0 & 0 & 0 \\ 0 & 0 & 1 & 0 & 0 & 0 & 0 & 0 & 0 & 0 & 0 & 0 \\ 0 & 0 & 0 & 1 & 0 & 0 & 0 & 0 & 0 & 0 & 0 & 0 \\ 0 & 0 & 0 & 0 & 1 & 0 & 0 & 0 & 0 & 0 & 0 & 0 \\ 0 & 0 & 0 & 0 & 0 & 1 & 0 & 0 & 0 & 0 & 0 & 0 \\ 0 & 0 & 0 & 0 & 0 & 0 & 1 & 0 & 0 & 0 & 0 & 0 \\ 0 & 0 & 0 & 0 & 0 & 0 & 0 & 1 & 0 & 0 & 0 & 0 \\ 0 & 0 & 0 & 0 & 0 & 0 & 0 & 0 & 1 & 0 & 0 & 0 \\ 0 & 0 & 0 & 0 & 0 & 0 & 0 & 0 & 0 & 1 & 0 & 0 \\ 0 & 0 & 0 & 0 & 0 & 0 & 0 & 0 & 0 & 0 & 1 & 0 \\ 0 & 0 & 0 & 0 & 0 & 0 & 0 & 0 & 0 & 0 & 0 & 1 \end{bmatrix}$$

The mathematical theory behind the set of all permutations and many other sets of operations is called Group-Theory. It shows that each permutation matrix has an inverse matrix Π^{-1}_x that turns the permuted vector π_x back into the standard vector π_0. This follows common sense. Whatever series of symbol exchanges created the permutation π_x can, of course, be applied in reverse order to it.

equ. 25-05
If
$\Pi_x := \Pi_{(0 \to x)}$, then
$\Pi^{-1}_x = \Pi_{(x \to 0)}$, with
$\pi_0 = \Pi^{-1}_x \otimes \pi_x$, and
$\Pi^{-1}_x \otimes \Pi_x = \Pi_x \otimes \Pi^{-1}_x = E$.

Therefore, one can bypass the standard vector and go directly from any permuted vector π_x to any other permuted vector π_y.

If $\pi_x = \Pi_x \otimes \pi_0$,   and
       $\pi_y = \Pi_y \otimes \pi_0$.   Then
       $\pi_0 = \Pi^{-1}_x \otimes \pi_x$.   Hence,
       $\pi_y = \Pi_y \otimes \pi_0 = \Pi_y \otimes \Pi^{-1}_x \otimes \pi_x$.

Therefore, the matrix that mediates the transition from a permutation π_x to another permutation π_y is given by

equ. 25-06
$$\Pi_{(x \to y)} = \Pi_y \otimes \Pi^{-1}_x .$$

Of course, a similar procedure can be done with any other number and any other kinds of tiles, by using appropriately larger matrices and vectors.

The Representation of the Orientation Codes

Like the letters that stand for the pentomino tiles, the orientation codes are also arbitrary symbols with which one cannot perform mathematical operations. Therefore, we have to replace them, too, by matrices that represent the corresponding rotations and reflections of the pentominoes. In our example, the tiles are 2-dimensional, and, consequently, these operations are represented by the rotational matrices ρ in 2 dimensions.

equ. 25-07
$$\rho = \begin{bmatrix} \cos(\alpha) & -\sin(\alpha) \\ \sin(\alpha) & \cos(\alpha) \end{bmatrix},$$

where α is the angle of rotation. Mirror reflections are expressed by flipping the coordinates of one of the axes. If the object is mirrored about the horizontal axis (i.e. $y \to -y$), the matrix ρ becomes ρ_{horiz}

equ. 25-08
$$\rho_{horiz} = \rho \otimes \begin{bmatrix} 1 & 0 \\ 0 & -1 \end{bmatrix} = \begin{bmatrix} \cos(\alpha) & \sin(\alpha) \\ \sin(\alpha) & -\cos(\alpha) \end{bmatrix}$$

Alternatively, if the object is mirrored about the vertical axis (i.e. $x \to -x$), the matrix ρ becomes ρ_{vert}

equ. 25-09

$$\rho_{vert} = \rho \otimes \begin{bmatrix} -1 & 0 \\ 0 & 1 \end{bmatrix} = \begin{bmatrix} -\cos(\alpha) & -\sin(\alpha) \\ -\sin(\alpha) & \cos(\alpha) \end{bmatrix}.$$

For example, Figure 25-02 shows the angles of rotation for the 8 different orientations of the 'P'-pentomino.

equ. 25-10

$\alpha_0 = 0°$, $\alpha_1 = 90°$, $\alpha_2 = 180°$, $\alpha_3 = 270°$, $\alpha_4 = 0°$, $\alpha_5 = 90°$, $\alpha_6 = 180°$, $\alpha_7 = 270°$.

The corresponding matrices are $\rho_0, \rho_1, ..., \rho_7$. They are to be used in the following way. If a pentomino has the orientation code 'k' then each point $p = [x_p, y_p]^T$ of this particular tile must be transformed by the matrix ρ_k.

equ. 25-11

$p' = \rho_k \otimes p$.

Here are the explicit forms of these matrices.

equ. 25-12a

$$\rho_0 = \begin{bmatrix} \cos(\alpha_0) & -\sin(\alpha_0) \\ \sin(\alpha_0) & \cos(\alpha_0) \end{bmatrix} = \begin{bmatrix} 1 & 0 \\ 0 & 1 \end{bmatrix};$$

$$\rho_1 = \begin{bmatrix} \cos(\alpha_1) & -\sin(\alpha_1) \\ \sin(\alpha_1) & \cos(\alpha_1) \end{bmatrix} = \begin{bmatrix} 0 & -1 \\ 1 & 0 \end{bmatrix};$$

$$\rho_2 = \begin{bmatrix} \cos(\alpha_2) & -\sin(\alpha_2) \\ \sin(\alpha_2) & \cos(\alpha_2) \end{bmatrix} = \begin{bmatrix} -1 & 0 \\ 0 & -1 \end{bmatrix};$$

$$\rho_3 = \begin{bmatrix} \cos(\alpha_3) & -\sin(\alpha_3) \\ \sin(\alpha_3) & \cos(\alpha_3) \end{bmatrix} = \begin{bmatrix} 0 & 1 \\ -1 & 0 \end{bmatrix};$$

As illustrated in figure 25-02, the matrices ρ_4 and ρ_6 are ρ_0 and ρ_2 mirrored around the horizontal axis, whereas ρ_5 and ρ_7 are ρ_1 and ρ_3 mirrored around the vertical axis. Hence,

equ. 25-12b

$$\rho_4 = \rho_0 \otimes \begin{bmatrix} 1 & 0 \\ 0 & -1 \end{bmatrix} = \begin{bmatrix} 1 & 0 \\ 0 & -1 \end{bmatrix};$$

$$\rho_5 = \rho_1 \otimes \begin{bmatrix} -1 & 0 \\ 0 & 1 \end{bmatrix} = \begin{bmatrix} 0 & -1 \\ -1 & 0 \end{bmatrix};$$

$$\rho_6 = \rho_2 \otimes \begin{bmatrix} 1 & 0 \\ 0 & -1 \end{bmatrix} = \begin{bmatrix} -1 & 0 \\ 0 & 1 \end{bmatrix};$$

$$\rho_7 = \rho_3 \otimes \begin{bmatrix} -1 & 0 \\ 0 & 1 \end{bmatrix} = \begin{bmatrix} 0 & 1 \\ 1 & 0 \end{bmatrix} \quad \Box$$

Like the permutations, the rotations and reflections have corresponding inverse operations. Again, common sense demands that the clock-wise rotation of an object by a certain angle can be undone, by afterwards rotating it counter clock-wise by the same angle. Formally,

equ. 25-13

$$\rho_k \otimes \rho_k^{-1} = \rho_k^{-1} \otimes \rho_k = E = \begin{bmatrix} 1 & 0 \\ 0 & 1 \end{bmatrix}, \text{ for } k = 0,\ldots,7.$$

Using equ. 25-10, the student can easily verify that

equ. 25-14

$\rho_0^{-1} = \rho_0;\ \rho_1^{-1} = \rho_3;\ \rho_2^{-1} = \rho_2;\ \rho_3^{-1} = \rho_1;$
$\rho_4^{-1} = \rho_4;\ \rho_5^{-1} = \rho_5;\ \rho_6^{-1} = \rho_6;\ \rho_7^{-1} = \rho_7.$

Please note: after each rotation and reflection, the location of the tile's handle (equ. 24-01) may have to be re-adjusted. We omit this step here, because it only confuses the formulas. The student who implements these transformations will have no problem, subtracting the handle-vector from the tile before the operations, and adding the appropriate new handle-vector back afterwards.

Now we have formulated all the matrices that express explicitly the orientation codes, we are able to transform the brackets of standardized orientation codes like [**001671123023**] as a collection of matrices Ω

equ. 25-15
[**001671123023**]→ Ω = [$\rho_0,\rho_0,\rho_1,\rho_6,\rho_7,\rho_1,\rho_1,\rho_2,\rho_3,\rho_0,\rho_2,\rho_3$],

where the matrices ρ_k are defined by equation 25-12a,b.

The Matrix Representation of Mapping Fitted Solutions onto Others

In summary, we can now write each operation that transforms a fitted solution Σ_x into another fitted one, Σ_y by the following matrix operation that is illustrated by the example of transforming solution code # (x = 2) into solution code # (y = 62).

equ. 25-16
(a) Solution codes
#x: **F0W3T1I0Y2N6Z3X0L1V2P7U1**
#y: **Z2Y5W0N6L2F0I0X0P7V3T2U1**

(b) Standardized orientation codes
#x: [**FWTIYNZXLVPU**] [**001671123023**]
#y: [**ZYWNLFIXPVTU**] [**002672130052**]

(c) Standardized permutation codes
#x: [**1,9,6,2,11,4,12,10,3,8,5,7**] [**001671123023**]
#y: [**12,11,9,4,3,1,2,10,5,8,6,7**] [**002672130052**]
:
(d) Definition of permutations
#x: [**1,9,6,2,11,4,12,10,3,8,5,7**] [**001671123023**]
#y: [**12,11,9,4,3,1,2,10,5,8,6,7**] [**002672130052**]
$\pi_0 = [1,2,3,4,5,6,7,8,9,10,11,12]^T$.
$\pi_x = [1,9,6,2,11,4,12,10,3,8,5,7]^T$.
$\pi_y = [12,11,9,4,3,1,2,10,5,8,6,7]^T$.

(e) Definition of permutation matrices based on
$\Pi_x : \pi_x = \Pi_x \otimes \pi_0$.
$\Pi_y : \pi_y = \Pi_y \otimes \pi_0$.

(f) Definition of orientation matrices
$\Omega_x = [\rho_0,\rho_0,\rho_1,\rho_6,\rho_7,\rho_1,\rho_1,\rho_2,\rho_3,\rho_0,\rho_2,\rho_3]$.
$\Omega_y = [\rho_0,\rho_0,\rho_2,\rho_6,\rho_7,\rho_2,\rho_1,\rho_3,\rho_0,\rho_0,\rho_5,\rho_2]$.
(g) 1. Transforming standard initial values into solution #x, or #y:
#x: $\{\pi_0; \pi_x = \Pi_x \otimes \pi_0; \Omega_x\}$, with
$\Omega_x = [\rho_0,\rho_0,\rho_1,\rho_6,\rho_7,\rho_1,\rho_1,\rho_2,\rho_3,\rho_0,\rho_2,\rho_3]$.
#y: $\{\pi_0; \pi_y = \Pi_y \otimes \pi_0; \Omega_y\}$, with
$\Omega_y = [\rho_0,\rho_0,\rho_2,\rho_6,\rho_7,\rho_2,\rho_1,\rho_3,\rho_0,\rho_0,\rho_5,\rho_2]$.

(h) 2. Transforming solution # x into solution # y:
In this case the permutation code $_x$ has to be returned to the standard order π_0, and subsequently turned into the permutation π_y (See 25-06). Similarly, the orientation of every pentomino tile of #x has to be returned to the initial orientations of every pentomino tile in the standard order by applying a vector
$\Omega_{-x} = [\rho_0^{-1},\rho_0^{-1},\rho_1^{-1},\rho_6^{-1},\rho_7^{-1},\rho_1^{-1},\rho_1^{-1},\rho_2^{-1},\rho_3^{-1},\rho_0^{-1},\rho_2^{-1},\rho_3^{-1}]$.

Subsequently, all the initial values have to be turned into the orientations of #y, by applying Ω_y. The result is

#x→#y: $\{\pi_x, \pi_y; \pi_y = \Pi_y \otimes \Pi^{-1}_x \otimes \pi_x; \Omega\}$, with

$\Omega = [\rho_0 \otimes \rho_0^{-1}, \rho_0 \otimes \rho_0^{-1}, \rho_2 \otimes \rho_1^{-1}, \rho_6 \otimes \rho_6^{-1}, \rho_7 \otimes \rho_7^{-1}, \rho_2 \otimes \rho_1^{-1},$
$\rho_1 \otimes \rho_1^{-1}, \rho_3 \otimes \rho_2^{-1}, \rho_0 \otimes \rho_3^{-1}, \rho_0 \otimes \rho_0^{-1}, \rho_5 \otimes \rho_2^{-1}, \rho_2 \otimes \rho_3^{-1}]$.

Therefore, the matrix that mediates the transition from a permutation π_x to another permutation π_y is given by equation 25-06.

Generalization of Spatial Fitting

How do the above formulae apply to living mosaics? In the first place, we need to be able to scan the mosaics. It requires that we actually know their tiles, at least one fitted solution, and preferably several sibling-solutions.

These conditions are generally fulfilled in the case of structural biological mosaics. Although we may not know all sibling-solutions, we usually have access to sufficiently many of them, in order to be able to assess the tile permutations and their basic orientations. Based on such preliminary analyses, the above formulae may help us to formulate and subsequently find additional sibling-solutions.

Of course, the 2-dimensionality of the tiles in the above example is not really limiting. The rotational matrices of 3- and higher dimensions are well-established in group-theory, and can be used appropriately.

Therefore, the above example can be generalized to apply to a number of other cases of structural mosaics. In contrast, it is not obvious how the above formulae apply to living mosaics, if they are predominantly functional mosaics.

Functional Fitting

If we look at the ways biologists have traditionally depicted the actions and tasks of functional mosaics, we find that they use circuit diagrams, flow-charts and similar graphs. Examples are the metabolic pathways, the flowcharts that depict the cell types and interactions of the immune system, computer graphics of biochemical or molecular biological mechanisms, and so forth. Effectively, the representations are graphs, where the vertices represent the various functions, and the edges represent the interactions between them.

However, these are not simple graphs, not even in their most abstract form. Most of their edges are directional. Some of the vertices represent the inner or outer environment, or a conduit with a circulating medium. Since most functions must have access to all of them, these special vertices must be connected directly and directionally to most other vertices of the graph.

Many of the edges lead to logical elements, such as conditional ('IF'-elements) or inversion ('NOT-) elements. Whenever a directional edge leads to a conditional element, it branches into 2 directional edges.

Inevitably, like all functional mosaics, the living mosaics have an overall processing direction that renders some of the vertices as 'early' and others as 'late' with respect to its underlying process. In addition, the process usually requires that some vertices are synchronized with each other. Therefore, a 'clock'-element must be present and, like the inner and outer environment and the circulatory medium, it must also be connected with all other vertices.

Furthermore, these functional mosaics being living mosaics, are capable of self-control and self-regulation. Hence, their graphs contain numerous directed edges, which lead back to earlier vertices in the form of feed-back loops, which are likely to create unwieldy Gordian knots.

Hence, the graphs of the functional mosaics of living system have to solve very difficult topological problems. Just ask the engineers of integrated circuits, what kind of tantalizing problems can arise in their designs!

Therefore, we will take an easier and more transparent approach. As will be shown below, each such graph can be formally depicted by a better readable matrix that lists which tile is interacting with which other one, and in which direction.

The Interaction Matrix I

Regardless of their specific details, all functional mosaics consist of a set of N interacting tiles τ_k (k = 1,...,N) and a list of the interactions between tile # i and tile #k, or tile #i and the frame. The set of interacting tiles may include one or several common 'embedding' or 'circulating' media. Their spatial fitting is not required, although the tiles of most living mosaics are well-fitted in 3 dimensions.

In chapter 20, we had introduced the contact-matrix, which listed the physical contacts between tiles (see figure 20-09). Quite similarly, we can formulate an 'interaction'-matrix, which lists the functional linkages between the tiles of functional mosaics, instead.

However, there will be an important difference. The contact-matrix $\Omega = \{\omega_{i,k}\}$ was symmetrical: Whenever tile #i touched tile #k, then inevitably the tile #k touched tile #i in the same places (see equation 20-10). Using conventional notation, this symmetry was expressed by the equation $\Omega = \Omega^T$.

The same is not true for interactions. In contrast to the contacts between tiles, their actions and interactions are directional: Tile #i may acts upon tile #k without the latter acting upon the former at the same time, or same strength, or actually acting upon it at all. Calling the interaction matrix I = $\{\iota_{ik}: i, k=1, 2,..., N, N + 1\}$ (the index N + 1 indicates the frame) we formulate it as follows.

equ. 25-17

In general, the interaction matrix of functional mosaics is not symmetrical:

$I \neq I^T$.

This difference between Ω and I has further implications. Contacts are binary, i.e. yes-or-no quantities. In contrast, the in- and outputs of interactions may depend on a great many parameters, because interactions have timing or strength that need to be considered. Of course, contacts, too, can form or break in time, the strength of adhesion at the point of a contact is an important and variable quantity, and contacts may transmit material and information. However, this means that they are no longer just contacts, but that, in addition, there are interactions linked to them.

Furthermore, the use of the interaction matrix will be quite different from the use of the contact matrix. We had used the contact matrix merely as a practical scheme to express the inner structure of a mosaic.

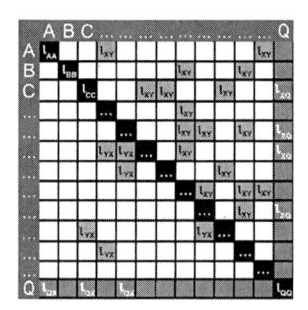

Fig. 25-03. Schematic of an interaction matrix. Assuming functional tiles A, B, C,... and a frame Q, and using the same colouring scheme of figure 24-09, their interactions are expressed by the components ι_{XY} (tile#X acts upon tile#Y) and ι_{YX} (tile#Y acts upon tile#X), which are different and, therefore render the matrix asymmetrical. The tiles may act upon themselves, which is expressed by the non-trivial diagonal components ι_{XX} (tile#X acts upon tile#X). The frame interactions are expressed by ι_{XQ} (tile#X acts upon the frame Q) and ι_{QX} (frame Q acts upon tile#X)

However, it was not used as an operator that acts upon a target. Naturally, the interaction matrix expresses actions and their consequences. Hence, it acquires the character of an operator, and its elements ι_{ik} become functions of time and other parameters.

More specifically, the interaction matrix mediates between a set of inputs and the outputs that result from the interactions between the tiles. Interactions are even able to generate outputs where there were no inputs. For example, tiles may act upon themselves without input from other tiles. Hence, the diagonal elements will no longer vanish and be ignored. We write this situation as the matrix product

equ. 25-18

Given an input matrix $i = \{i_1, i_2, ..., i_N, i_f\}$, an output matrix $o = \{o_1, o_2, ..., o_N, o_f\}$, and a 'bias' output $b = \{b_1, b_2, ..., b_N, b_f\}$ where i_k are the inputs, o_k the outputs, and b_f the biases of tile #k of a functional mosaic (the index 'f' indicates the corresponding values referring to the frame), then the interaction between the tiles can be expressed as $o = I \otimes i + b$.

It should be emphasized this equation only represents a first approximation. It is 'linear' in the sense that a λ-fold input $i \rightarrow \lambda \cdot i$ yields a λ-fold (bias-reduced) output $o \rightarrow \lambda \cdot (o-b)$. There is no compelling reason that higher order interactions should not exist. If they occur, the simple matrix product $I \otimes I$ would have to be replaced with an input-dependent matrix $I(i)$.

APPENDIX C

ALGORITHMS TO FIND ALL STANDARDIZED SOLUTIONS OF PENTOMINO MOSAICS

The Problem of Frame Symmetries (Standardization)

Like any other square, the model-frame has 8 symmetries. Therefore, after the assembly algorithm has found a solution, it will find 7 more that are merely mirror images and rotations of the first. As they add no new information about the fitting problem, they may be ignored.

The elimination of redundant solutions could be accomplished by generating all solutions, identifying the rotations and mirror images of each one, and deleting all but one of them. However, this would be much too slow and error prone. Technically, the simplest way to eliminate mirror images and rotations of entire solutions is to restrict the mirror images and rotations of only a single tile, namely of one of the most asymmetrical tiles.

This is illustrated in Figure 26-01). The 'F' is one of the most asymmetrical tiles among the pentominoes, because it has 8 different orientations. All 8 orientations can be created by mirroring or rotating an entire solution (highlighted white in figure 26-01 a).

Now let us run the assembly protocol while excluding all but one orientation of the 'F' (highlighted in dark gray in panel a). As a result of this kind of **standardization** none of the solutions can turns out to be the rotation or mirror image of another: If it was, it should contain the corresponding rotations and mirror images of the 'F' pentomino. However, the exclusion rule prevented that such an orientation of the 'F' occurred during assembly.

If the frame has lower symmetries, like a rectangle, rotations and mirror images only can create half of the 8 orientations of the 'F' (figure 26-01 b). Hence, the standardization of the solutions consists of permitting 2 of the orientations of the 'F' are not mirror images of each other.

In the extreme case of a frame that has no symmetries at all, we have to admit all 8 orientations of the 'F' (figure 26-01c). Otherwise, the assembly protocol would not be able to find all solutions.

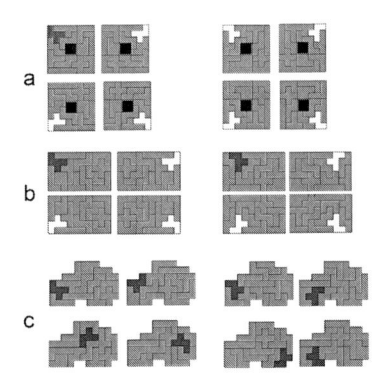

Basic Biology for Born Engineers: Living Mosaics 355

Fig. 26-01. Prevention of the co-assembly of mirror images or rotations of each solution. Depending on the symmetry properties of the frame, specific orientations of one of the most asymmetrical tiles (e.g. the 'F' pentomino) are excluded. The permitted orientations are highlighted in dark; the others in bright.

(a) The 8-fold symmetrical model-frame. Each solution could be assembled along with 7 mirror images and 90° rotations. However, the ones marked in white cannot occur, if their orientations of the 'F' pentomino are excluded.

(b) A 4-fold symmetrical rectangular frame. Each solution could be assembled along with 4 mirror images. However, the ones marked in white cannot occur, if their orientations of the 'F' pentomino are excluded.

(c) An asymmetrical frame. No orientations of the 'F' pentomino are excluded. The assembly of all shown solutions is permitted.

Growing the Solution from a Single Tile (Truncated Search)

Sooner or later we will need a method to construct *all* possible pentomino mosaics for a given frame. Should we try to construct them manually? For many years I have watched (gleefully) fellow scientists with exceptional intelligence and 3-dimensional imagination struggle for hours and sometimes days to find a single solution for the problem of fitting the 12 pentominoes into a 6·10 rectangular frame. There are 2339 such solutions. Finding them manually would take forever. No, we need a program and a fast computer.

However, as we will see later, even a fast computer may take thousands of years to find all pentomino solutions for a given problem. Very specific algorithms are required, before the computing time shrinks down to hours and minutes. I have written several such programs in the past using TurboC, C++, and Visual C++ as languages. The following describes some of the details that were needed to find all solutions in a reasonable time.

First we will have to agree upon a number of standardizations. Most important is the protocol by which the pentomino tiles are placed successively into the frame. Our standard protocol will be called 'growing the solution alphabetically from a corner'. It uses the following standardizations:

1. The tiles are ordered alphabetically by their names. In this case the order would be F, I, L, N, P, T, U, V, W, X, Y, Z.

2. Each tile has up to 8 possible orientations which are ordered by the numbers 0, 2,...,7. For example, the 'F' has 8 orientation, the 'I' only has two, and the 'X' only has one. Tiles and their orientations are written as F0, F1, F2, F3, F4, F5, F6, F7 and I0, I1, etc., respectively.

3. The first configuration of tiles and orientations to be tested is always 'F0I0L0N0P0T0U0V0W0X0Y0Z0'. Other test configurations may look like 'L7U3I0F7N4X0Y4W0T3P4Z0V2'. Let us call this kind of string the 'search-code'. Since all tiles only can occur once in a solution, each search-code contains a permutation of the 12 pentomino letters.

For instance, the initial permutation is FILNPTUVWXYZ, and the above example contains the permutation LUIFNXYWTPZV. (The latter is actually a solution).

4. Testing consists of determining the next empty spot ('test-point') inside the frame and trying whether the next letter of the search-code can be placed there without creating isolated squares, double squares,....or other violations of the fitting requirements.

5. The search for the next test-point begins at the left upper corner of the frame and then scans the remaining frame in a 'top-to-bottom-then-left-to-right' fashion. The next test-point is the first empty square found in this way (figure 26-02 a).

6. There is an additional standardization left to do, which was explained earlier. It is repeated here for the sake of completeness: If the frame is mirror symmetrical, the search protocol not only would find every solution, but also all of its mirror images.

For example, if we would use a rectangular frame of 10x6 units, the protocol would find for each solution 3 additional ones that are nothing but its left-right and up-down mirror images.

In order to exclude these, we would restrict the 8 orientations of the (say) 'F' to only 2 orientations that are not mirror images of each other. That eliminates the mirror image solutions for the following reason: Assume the protocol has found a solution S, but at some later time it also found its mirror image S^. Consequently, the 'F' in the mirror image solution S^ would be the mirror image of the 'F' in S. After applying our exclusionary rule, however, is impossible, because no mirror image of 'F' ever occurred in the search-code.

7. Now we may begin to test the possible search-codes one after another.

Or can we?

In fact, we would be quite naïve trying to test all possible search-codes, as it would take thousands of years to complete the search. It is mandatory to 'truncate' the search, i.e. we must not waste time testing the millions of search-codes which have no chance of being a solution.

The following is a more detailed explanation of the way one recognizes countless non-solutions and skips them. Although working through this explanation may be a bit dry and tedious, I add it here because, in practice, the strategy of finding the fitting solutions of mosaics lies at the core of the present book. Nevertheless, at first reading one can safely skip it.

The Search Algorithm for Solution-Codes

Since all parts only can occur once on a solution, each solution-code represents a permutation of the tile symbols followed by a list of the code symbols for their different orientations. More specifically, we may use the following convention.

1. We number the tiles 1,...,N. For example, we number the pentominoes F = **1**, I = **2**, L = **3**, N = **4**,..., Z = **12**.
2. We number the orientations of each piece in a systematic way. In the case of pentominoes, we numbered the orientations 0 -7.
3. We set up a list of all permutations of N numbers.
4. We set up a list of all combinations of the orientations.
5. We combine each permutation with all combinations of the orientations in order to generate all potential solution-codes (**search-codes**).

The Ascending Order of Permutations of Pieces

Next we need a systematic way to list these codes. A natural way to order them is to order them by ascending numbers.

For the sake of simplicity, the following illustration will not use 12 pieces like the pentominoes, but only four. Therefore, consider the following list of all permutations of the numbers **1, 2, 3, 4**.

1234, 1243, 1324, 1342, 1423, 1432, 2134, 2143, 2314, 2341, 2413, 2431, 3124, 3142, 3214, 3241, 3412, 3421, 4123, 4132, 4213, 4231, 4312, 4321.

If we interpret each permutation as a decimal numbers, e.g. **3214** as the number three thousand two hundred and fourteen, it is obvious that the above list is ordered in ascending decimal numbers.

If there are more than 10 pieces, one cannot use decimal numbers. For instance, the 12 pentominoes force us to generate number sequences such as **1-3-11-9-10-2-4-5-7-6-12-8** which must not be read as the decimal number 131191924576128, but has to be understood as a duodecimal number, i.e. a number representation that allows in each place 12 different digits.

In general, if there are N pieces, the permutations may be read as N-ary number representations before they can be listed in ascending order. In the previous text we used capital letters such as FLYW[7]NPUTZV instead of numbers in order to facilitate reading of the search-codes. However, the notations are equivalent because the order of ascending numbers was translated into the more readable alphabetical order.

The Complete List of the Combinations of Orientations

In contrast to the parts of a mosaic that have to be different from each other, the orientations of different parts may be the same. Therefore, the list of possible orientations consists of the list of all combinations of the orientation symbols.

Example: Assume that the 4 different pieces in the above example can each exist in two orientations. It means e.g. that in the permutation **3214** the piece **3** could exist in orientation 0 or 1, the piece **2** in the orientations 0 or 1, and so forth. Consequently, the permutation could exist in each of the following 16 combination of orientations.

0000, 0001, 0010, 0011, 0100, 0101, 0110, 0111, 1000, 1001, 1010, 1011, 1100, 1101, 1110, 1111

Basic Biology for Born Engineers: Living Mosaics   359

In a similar way one can construct the complete list of all orientations if there are more orientations and/or more pieces. It is also possible for some pieces a larger number of different orientations exist than for others. For example, the X-pentomino only has one orientation, whereas the F pentomino has 8. In such cases, one can construct the complete list of orientations as if all pieces had the maximal number of 8 orientations and afterwards erase all cases that assign more than one orientation to the X.

The List of Possible Search-Codes

By combining every permutation with every combination of the 2 orientations 0 and 1, one can easily generate the list of 'search-codes'

12340000, 12430000, 13240000, 13420000, 14230000, 14320000,
21340000, 21430000, 23140000, 23410000, 24130000, 24310000,
31240000, 31420000, 32140000, 32410000, 34120000, 34210000,
41230000, 41320000, 42130000, 42310000, 43120000, 43210000.
and
12340001, 12430001, 13240001, 13420001, 14230001, 14320001,
21340001, 21430001, 23140001, 23410001, 24130001, 24310001,
31240001, 31420001, 32140001, 32410001, 34120001, 34210001,
41230001, 41320001, 42130001, 42310001, 43120001, 43210001.
and
12340010, 12430010, 13240010, 13420010, 14230010, 14320010,
21340010, 21430010, 23140010, 23410010, 24130010, 24310010,
31240010, 31420010, 32140010, 32410010, 34120010, 34210010,
41230010, 41320010, 42130010, 42310010, 43120010, 43210010.

and so forth.
The search-code **32140101** means that one should test the fitting of the pieces in the order 3, 2, 1, and 4, with the pieces 2 and 4 in orientation 1 and all others in orientation 0. If fitting is possible, then **32140101** is added to the list of solution-codes of the mosaic.

The reading of the search-codes is facilitated, if the orientation symbols follow immediately the piece symbols, e.g. to write 30211041 instead of 32140101. We used this format in the example of the solution-codes of pentomino puzzles, such as **TXUIFNLWYZVP**100016224227 which we simplified into the form T1X0U0I0F1N6L2W2Y4Z2V2P7.

360   Appendix C

Truncation of the Search Algorithm

As mentioned before, the method of testing all possible search-codes must fail for practical reasons if there are too many tiles and orientations. Even relatively small numbers, like the number 12 may actually be too large. The number of permutations of the 12 pentominoes is $12! = 4.79 \cdot 10^8$. As mentioned above, among the 12 pieces there are 5 with 8 orientations, 5 with 4 orientations, 1 with 2 orientations, and 1 with 1 orientations. Hence for each permutation we have to test a total of $85 \cdot 45 \cdot 2 \cdot 1 = 6.7 \cdot 10^7$ possible combinations of orientations. Thus the method requires to test $N = 12! \cdot 85 \cdot 45 \cdot 2 \cdot 1 = 3 \cdot 10^{16}$ search-codes, in order to determine the solution-codes of the pentomino mosaic.

Obviously, this number of required tests is far too large. Even if a single fitting test would require as little as 1 microsecond, the total test time for the pentomino mosaics would last about 10,000 years. Therefore, one must find ways to make the search more economical. A rather general and effective method to economize the search is called a **truncated tree-search**.

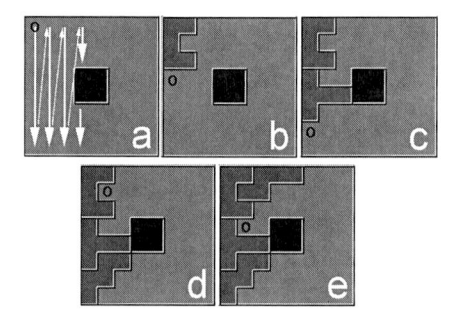

Fig. 26-02. Assembly protocol for the truncated tree-search.
(a). Standardized scanning mode (up→down + left→right)
(b)-(e). Sequential placement of fitting tiles ('o' indicates the position of the next tile.

For an illustration, let us assume that the search-code to be tested is 'U3T1W2N4F0I0L0P0V0X0Y0Z0'. Furthermore, assume that the test algorithm has successfully fitted the first four pieces (highlighted; also see figure 26-02, panels b,c,d,e).

The 5th piece in line is the F-pentomino in orientation 0. Hence, the testing method has to try to fit it into the space marked 'o'. Obviously, the F does not fit, no matter what its orientation is. Consequently, the method must strike the above search-code from the list and try the next.

However, striking this search-code from the list is literally a drop in the bucket. In fact, the method should strike many, many more search-codes from the list, because the space marked 'o' in figure 26-02 panel e only can be filled with a 'W' or an 'N'. However, neither is available since both are already used. Therefore, *no search-code that begins with U3T1W2N4 can ever become a solution-code, no matter in what order and orientation the remaining pieces FILPVXYZ are.*

How many such failing search-codes exist in the above case? There are 8! = 40,320 different permutations of the remaining letters, and they can exist in 8·2·8·8·4·1·8·4 = 131,072 different combinations of orientations. Since there is no point of testing any of them, the fitting algorithm should be able to automatically strike from the list as many as of 40,320·131,072 = 5,284,823,040 failing search-codes. Striking one code is a drop in the bucket, indeed.

But how can the algorithm find all these failure codes among the very large list of all search-codes?

Finding the 'Next' Test-Worthy Search-Code after a Failure to Fit One

Based on the above argument it is clear that the fitting algorithm has to truncate all the failure codes and leap to the next untested, yet test-worthy permutation. However, which is the 'next'?

Here is the rule to find it based on the **ascending order of permutations**. It is also the rule of truncation.

PERMUTATION-TRUNCATION RULE (PT-rule):
a. If the testing of a permutations fails at position 'n' with piece 'P', then move 'P' to the **end of the line of the larger, not yet used** pieces. Its orientations start again at 0.

b. If there is no larger number left, move the piece at position 'n-1 to the end of the line and replace it with the next of the remaining ones. Their orientations start again at 0.

As we shall see below, the PT-rule guarantees that a failed search automatically strikes all the appropriate failure codes and continues with the first, yet untested, but test-worthy search-code. Note that it will not require that the system memorizes astronomically long lists of search-codes. It turns out to be enough only memorizing the present configuration up to position n.

In order to illustrate the efficacy of the PT-rule let us return to the earlier example of the fitting of the search-code '**U3T1W2N4**F0I0L0P0V0X0Y0Z0' that failed in position n = 5 with trying to fit the 'F' (figure 26-02 panel e). In other words, after replacing F0 with F1, F2,.., F7 and testing them we found that the 'F' could fit in any of its orientations. Hence, the PT-rule applies:

We have to move the 'F' to the end of the line of the unused pieces ILPVXYZ. Furthermore, each of the pieces has to be tested first in the lowest number orientation. Therefore, we have to test as the new search-code **U3T1W2N4**I0L0P0V0X0Y0Z0*F0*.

The test consists of trying to fit the 2 possible orientations 'I' into the position marked 'o' in figure 26-02 panel e. That requires a total of 2 tests before we find out that the 'I' does not fit in position n = 5, either. So, we apply the PT-rule again, move the 'I' to the end of the line, and test the next search-code, namely **U3T1W2N4**L0P0V0X0Y0Z0*F0I0*.

Again we find that the 'L' does not fit in position n = 5. The test consists of trying to fit the 8 possible orientations 'L' into the position marked 'o' in figure 26-02 panel e. Hence, after only 2 + 8 = 10 tests we find out that the 'L' does not fit, either. So, we apply the PT-rule again, and so forth. None of the unused pieces FILPVXY will fit in position n = 5 until we find out that the even the 'Z' in the last tested code, namely **U3T1W2N4**Z0*F0I0L0P0V0X0Y0* yields no fit, either. Hence, even the 'Z' has to move to the end of the line, yielding as the next code **U3T1W2N4***F0I0L0P0V0X0Y0Z0*, which is how we started.

It only took us 2 + 8 + 8 + 4 + 1 + 8 + 4 = 35 tests, namely the sum of all orientations of the unused pieces ILPVXYZ, to find out that no search-code that begins with **U3T1W2N4** can ever yield a solution. As mentioned above, had we mechanically tested all possible search-codes, it would have required 5,284,823,040 tests to arrive at the same conclusion. Obviously, the PT-rule truncates the search quite effectively.

At this stage the second provision (b) of the PT-rule applies, because the first provision (a) fails: The next piece in line would again be the 'F', but it is not 'larger' than the 'Z'. In fact it is the smallest of all the failed pieces. Therefore, we have to go 1 position back to position n = 4, and place the 'N' at the end of the line.

Hence, the next search-code to be tested is **U3T1W2**F0I0L0P0V0X0Y0Z0*N0* and the described procedure begins again and continues until the list of search-codes is exhausted.

Finding the 'Next' Test-Worthy Search-Code after Finding a Solution

The PT-rule tells us the next search-code, after a failing one. However, which is the next search-code, after we found a solution? Obviously, at this point all pieces of the search-code have been placed successfully. None needs to be altered and tested.

For example, the search-code Z2Y5W0N6L2F0I0X0V1P0T3U1 codes for a solution (figure 26-03 a). *What is the next search-code after that?*

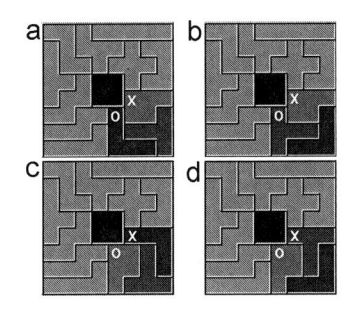

Fig. 26-03. Construction of the next search-code after completing a solution (PT-rule). Explanation see text.

As usual, first we need to try another orientation of the last piece, but this is obviously useless, because the last piece is fitting already and cannot possibly fit in any other orientation.

That is where provision (b) of the PT-rule comes in. Accordingly, we have to move the piece before last to the end of the line. In short, we will always have to exchange **at least the last 2** pieces. In most cases, however is not enough for fitting and the application of PT-rule (b) requires removing 3 or 4 and rearranging their order alphabetically and restarting their orientations.

Still, it may happen that it is actually enough to arrive at new solutions. An example is shown in figure 26-03. In this case the highlighted pentominoes form mirror symmetrical pairs and triplets. The alphabetic reordering and change of orientations results in flipping them around and only yields new solutions with new search-codes, where the terminal pieces have changed orientations. This can be seen clearer if we look at their search-codes:

a: Z2Y5W0N6L2F0I0X0**V1P0T3U1**
b: Z2Y5W0N6L2F0I0X0**T1P0V2U1**

c: Z2Y5W0N6L2F0I0X0**P7V3T2U1**
d: Z2Y5W0N6L2F0I0X0**P7T0V2U1**

Usually, however, after removing 3 or more pieces, none of the subsequent reordered search-codes permit fitting, and the application of the PT-rule (b) begins to move piece by piece backwards, until a new permutation leads to a new solution.

'Sibling' Solutions and Solutions with Internal Duplications

Using this algorithm, we found 64 different ways of fitting the 12 different pentominoes into the model-frame. Figure 26-04 shows them. The 'X' and the 'P' pentominoes are highlighted to make it easier to recognize that they are all different solutions of the same fitting problem.

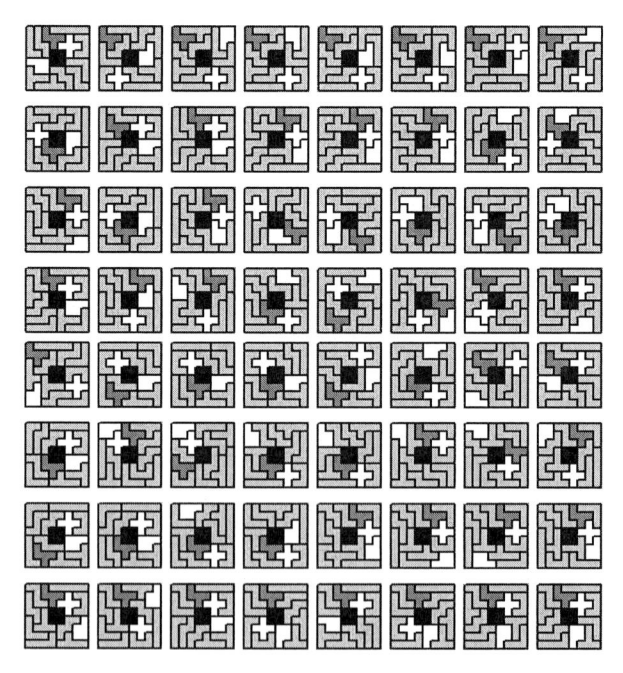

Fig. 26-04. All 64 sibling-solutions of the model-frame. Three pentomino tiles are highlighted in order to facilitate the differences between the solutions. The dark marked 'F' pentomino shows that no other orientation occurs in order to prevent the co-assembly of mirror images and rotations of each solution (see figure 26-01).

In the future, we will call the solutions that fit the same specific set of tiles into the same frame **'sibling-solutions'**.

In view of the 8-fold symmetry of the model-frame, all its sibling-solutions had to be standardized, i.e. the 'F' pentomino only occurred in one orientation (see the highlighted 'F' in figure 26-04).

In future applications we will also consider assemblies where one or the other pentomino is missing, while another one was duplicated in its place. Figure 26-05 shows some examples where one, a few, or all pentominoes were replaced with multiple copies of the 'P'. Obviously, such mosaics only can represent biological objects, whose viability tolerates the internal duplication of some of its tiles.

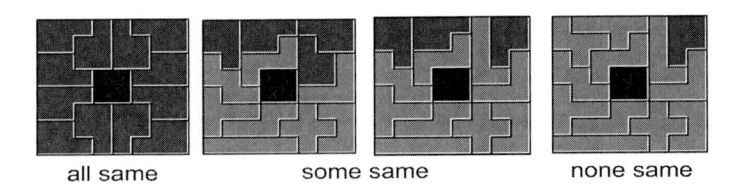

Fig. 26-05. Mosaic assemblies of the model-frame with internal duplications of the 'P' pentomino.

Needless to say there are many well documented examples of internal duplications of biological components. In some cases they represent simple aggregates, in others they may be eggs, embryos, and so forth.

APPENDIX D

COMPUTATION OF THE FRACTAL DIMENSION OF FRACTAL MOSAICS CREATED BY RECURSIVE REPLACEMENTS

1a. Level independent dissections; iterative sub-tiles without static sub-tiles.

In this case the number of iterative sub-tiles increases by the same fragmentation factor 'γ' (in panel (a) of the illustration, $\gamma = 4$) at every replacement, regardless whether the iterative sub-tiles have static sub-tiles or not.

equ. 27-01

$N^{(n)}{}_T = \gamma \cdot N^{(n-1)}{}_T = \gamma^2 \cdot N^{(n-2)}{}_T = \gamma^3 \cdot N^{(n-3)}{}_T = \ldots$ Hence,
$N^{(n)}{}_T = \gamma^n \cdot N^{(0)}{}_T$.

Since the number N_T cannot decrease, it follows that the fragmentation factor
$\gamma \geq 1$.

1b. Level dependent dissections; iterative sub-tiles without static sub-tiles (figure 8-09).

In this case, the increase in the total number of tiles changes from level to level by a variable fragmentation factor γ_n. Hence,

equ. 27-02

$N^{(n)}{}_T = \gamma_n \cdot N^{(n-1)}{}_T = \gamma_n \cdot \gamma_{n-1} \cdot N^{(n-2)}{}_T = \ldots, \gamma_n \cdot \gamma_{n-1} \cdots \gamma_1 \cdot N^{(0)}{}_T$,
or

$N^{(n)}{}_T = \prod_{1}^{n} \gamma_v \cdot N^{(0)}{}_T$.

Introducing the 'total fragmentation' at level 'n', $\Gamma^{(n)}$, we can also write

equ. 27-03
$N^{(n)}{}_T = \Gamma^{(n)} \cdot N^{(0)}{}_T$, with

$$\Gamma^{(n)} := \prod_1^n \gamma_v \geq 1, \qquad \text{since } \gamma_v \geq 1;\ (v = 1,\ldots,n,\ldots).$$

Since $\Gamma^{(n)} \geq 1$, we can also write in analogy with equation 27-01

equ. 27-04
$N^{(n)}{}_T = (g_n)^n \cdot N^{(0)}{}_T$, with $g_n = (\Gamma^{(n)})^{1/n} \geq 1$.

1c. Level independent dissections; iterative sub-tiles with static sub-tiles.

There is no need to compute the number of active tiles $N^{(n)}{}_T$ with growing 'n'. Based on the above considerations, we know already it will converge to zero.

1d. Level dependent dissections; iterative sub-tiles with static sub-tiles (figure 8-09c).
(See the remark to the case 1c).

2. Computation of the unit of resolution δ_n.
As mentioned before, the increasing numbers $N^{(n)}{}_T$ of iterative sub-tiles that are 'squeezed' into the initial frame volume V_Φ, inevitably shrinks the sizes of the tiles, and with them the resolution units δ_n. Let us estimate by how much.

Not all tiles can be squares, cubes, circles, or other shapes that can be characterized by a single number. Most need 2 or more parameters to describe their shapes. Therefore, we define a single parameter to describe their shrinking sizes:

equ. 27-05
We define as the *'characteristic linear expanse'* of a p-dimensional object with the p-dimensional volume V the quantity λ, such that $\lambda^p = V$.

Example: Take a brick with the sides x, y, and z. Its volume is $V = x \cdot y \cdot z$. Its characteristic linear expanse λ, is defined as a length λ such that $\lambda^3 = V$. Hence, we would calculate $\lambda = (V)^{(1/3)} = (x \cdot y \cdot z)^{(1/3)}$.

Let us assume that the topological dimension of the 'volume' of the mosaic is 'p'. At the nth round of the recursion the volume V_Φ of the initial frame has been fragmented by a total of $N^{(n)}{}_T$ iterative sub-tiles. Hence, on average, each sub-tile has available a partial volume of $V_\Phi N^{(n)}{}_T$. Then we define as the unit of resolution δ_n at n^{th} round of the recursion **the characteristic linear expanse of the average tile**

equ. 27-06
$\delta_n = [V_\Phi N^{(n)}{}_T]^{1/p}$.

Using equation 27-04,
$\delta_n = [V_\Phi [g_n]^n \cdot N^{(0)}{}_T]^{1/p}$.

For n = 0, we can write
$\delta_0 = [V_\Phi N^{(0)}{}_T]^{1/p}$,

Hence, equation 27-06 becomes

equ. 27-07
$\delta_n = [g_n{}^{1/p}]^{-n} \cdot \delta_0$

The fragmentation factors γ that apply to dissections are effectively the same as the reduction factors r that apply to the units of resolution. Defining $r_{(n)}$ as the reduction factor at level 'n'

equ. 27-08
$r_{(n)} = g_n{}^{1/p}$.

We can write

equ. 27-09
$\delta_n = [r_{(n)}]^{-n} \cdot \delta_0$.

In the special case of level independent dissections that contain γ iterative sub-tiles (no static sub-tiles), it follows that $\Gamma^{(n)} = \gamma^n$, and hence $g_n = \gamma$. Therefore,

equ. 27-10
$\delta_n = \delta_0 / \gamma^n$.

which is equ. 27-02.

3. Fractal dimension calculation

Re-writing equation 27-07 yields $[g_n]^n = [\delta_0/\delta_n]^p$

Hence, equation 27-04 becomes

equ. 27-11
$N^{(n)}{}_T = [\delta_0]^p \cdot N^{(0)}{}_T \cdot [1/\delta_n]^p$,

or in summary

equ. 27-12
Given a p-dimensional mosaic frame,
an initial unit of resolution δ_0,
an initial number of tiles $N^{(0)}{}_T$,
a series of dissections with γ_n tiles,
at most a finite number of dissections with static sub-tiles.

Also given the definitions $\Gamma^{(n)} := \prod_{1}^{n} \gamma_v$,

$g_n := (\Gamma^{(n)})^{1/n}$,
$\xi := \log[N^{(0)}{}_T \cdot [\delta_0]^p]$,

Then the recursive replacement of every tile with its dissections yields as unit of units of resolution at the nth round
$\delta_n = [g_n{}^{1/p}]^{-n} \cdot \delta_0$,

and as total number $N^{(n)}{}_T$ of tiles at the nth round

$\log[N^{(n)}{}_T] = p \cdot \log[1/\delta_n] + \xi$.

The equation describes a straight line with slope $\eta = p$ and an ordinate intersect of ξ. Hence, its fractal dimension is p.

****

Similar to the fractal mosaic of static sub-tiles, *the fractal mosaic of the iterative sub-tiles also has an integer 'p' for its fractal dimension,* without, however, being a topologically 'normal' set.

Example: The Fractal Dimension of Pentomino Mosaics

It may help the understanding of how integral fractal dimensions arise, to use our model system of pentomino mosaics. We proceed in a 'pedestrian' and detailed way the construction of a fractal mosaic and the calculation of the fractal dimension of its surface. As dissections we use the mosaics like the ones in figure 4-02. There are many similar ones. As listed in Table 4-01, there are multiple ways in which each pentomino can be composed of 9 smaller ones whose sides are 3 times shorter, i.e. the reduction factor $r = 3$.

0 level of refinement: $\delta_0 = r^0 = 1$ (left panel of figure 8-05).
According to equation 20-07, the total surface S_{M0} of our starting mosaic is

equ. 27-13
$$S^{(0)}_M = (S_{\Phi(0)} + S_{P(0)})/2,$$

where $S_{\Phi 0}$ is the surface of the original frame (= model-frame) and S_{P0} is the total surface of all 12 initial pentomino tiles. As was calculated earlier, in the case of pentominoes

equ. 27-14
$$S^{(0)}_\Phi = 40, \text{ and } S^{(0)}_P = 142.$$

Hence,
equ. 27-15
$$S^{(0)}_M = (40 + 142)/2 = 91.$$

We also need to introduce the total number $N^{(n)}_P$ of all pentominoes that are contained in the mosaic at each nth level of refinement. Since the mosaic is a solution of the model-frame, the starting number of pentominoes is

equ. 27-16
$$N^{(0)}_P = 12.$$

1st level of refinement: $\delta_1 = (1/3)$ (middle panel of figure 8-05).
As the first step of the iteration we fill each of the 12 pentomino tiles of the starting mosaic with one of the pentomino-dissections. In other words, we consider **the total surface $S^{(0)}_M$ of the starting mosaic as the frame $S^{(1)}_\Phi$ of the next mosaic**, i.e.

equ. 27-17
$S^{(1)}{}_\Phi = S^{(0)}{}_M.$

Since each of the 12 pentominoes will be replaced with a pentomino-dissection, the entire mosaic will be filled with the 3-times smaller pentominoes. Their total number is

equ. 27-18
$N^{(1)}{}_P = 9 \cdot N^{(0)}{}_P.$

Next we have to calculate the new total surface $S_{P(\square 1)}$ of all these small pentominoes. Unfortunately, there is no simple answer, because it depends on which particular pentomino-dissection we selected to substitute each of the original 12 pentominoes: Different pentomino-dissections #τ have different total surfaces s_τ

The reason is this: As mentioned earlier, the 'P' has a surface of 10, whereas all other pentominoes have a surface of 12. Therefore, different pentomino-dissections #τ have one of four different total surfaces $s^{\square 1)}{}_\tau$, depending on whether they are a 'P' or not, and on whether they contain a 'P' or not. (This problem did not arise earlier for any of the complete pentomino solutions we discussed, because they contained *all* pentominoes, and not just a selection of 9. Hence, they always contained a 'P'!).

Using equation 20-07 yields 4 possible values.

If the pentomino #τ is not a 'P', then

(a) $s^{1)}{}_{\tau\,max} = (12 + 9 \cdot 12/3)/2 = 72/3$, if its pentomino-dissection does not contain a 'P', or

(b) $s^{1)}{}_{\tau 1} = (12 + (8 \cdot 12 + 10)/3)/2 = 71/3$, if it does.

If the pentomino #τ is a 'P' itself

(c) $s^{1)}{}_{\tau 2} = (10 + 9 \cdot 12/3)/2 = 69/3$, if its pentomino-dissection does not contain a 'P', or

(d) $s^{1)}{}_{\tau min} = (10 + (8 \cdot 12 + 10)/3)/2 = 68/3$, if it does.

Hence, instead of computing a single value, we are forced to compute the upper and lower limits of S_{P1}.

equ. 27-19
$s^{(1)}{}_{\tau min} = 68/3.$
$s^{(1)}{}_{\tau max} = 72/3.$ Then

equ. 27-20
$N^{(1)}{}_P \cdot s^{(1)}{}_{\tau min} \leq S^{(1)}{}_P \leq N^{(1)}{}_P \cdot s^{(1)}{}_{\tau max}$.

Since, $S^{(1)}{}_M = (S^{(1)}{}_\Phi + S^{(1)}{}_P)/2 = (S^{(0)}{}_M + S^{(1)}{}_P)/2$,
it follows
$(S^{(0)}{}_M + N^{(1)}{}_P \cdot s^{(1)}{}_{\tau min})/2 \leq S^{(1)}{}_M \leq (S^{(0)}{}_M + N^{(1)}{}_P \cdot s^{(1)}{}_{\tau max})/2$,
and therefore,

equ. 27-21
$(S^{(0)}{}_M + 68 \cdot N^{(0)}{}_P \cdot 3)/2 \leq S^{(1)}{}_M \leq (S^{(0)}{}_M + 72 \cdot N^{(0)}{}_P \cdot 3)/2$.

the 2nd to n^{th} level of refinement: $\delta_2 = (1/3)^2$ to $\delta_n = (1/3)^n$ (right panels of figure 8-05).

From the second level on, we continue the same procedure recursively *ad infinitum*. In all cases, the new frame is the total surface of the previous mosaic. Therefore, we use iteratively the previous equations

$S^{(n)}{}_\Phi = S^{(n-1)}{}_M$ (equ. 27-17)
$N^{(n)}{}_P = 9 \cdot N^{(n-1)}{}_P = 9^n \cdot N^{(0)}{}_P$. (equ. 27-18)
$s^{(n)}{}_{\tau min} = s \, s^{(n-1)}{}_{\tau min}/3 = 68/3^n$. (equ. 27-19)
$s^{(1)}{}_{\tau max} = s^{(n-1)}{}_{\tau max}/3 = 72/3^n$.

Furthermore,
$N^{(n)}{}_P \cdot s^{(n)}{}_{\tau min} \leq S^{(n)}{}_P \leq N^{(n)}{}_P \cdot s^{(n)}{}_{\tau max}$, (equ. 27-20), or
$68 \cdot N^{(0)}{}_P \cdot 3^n \leq S^{(n)}{}_P \leq 72 \cdot N^{(0)}{}_P \cdot 3^n$.

Since
$S^{(n)}{}_M = (S^{(n)}{}_\Phi + S^{(n)}{}_P)/2 = (S^{(n-1)}{}_M + S^{(n)}{}_P)/2$, (equ. 27-21),

it follows
$(S^{(n-1)}{}_M + 68 \cdot N^{(0)}{}_P \cdot 3^n)/2 \leq S^{(n)}{}_M \leq (S^{(n-1)}{}_M + 72 \cdot N^{(0)}{}_P \cdot 3^n)/2$.

==

Basic Biology for Born Engineers: Living Mosaics 373

Summary: Case of Pentomino Tiles; Model-Frame

equ. 27-22

Given the parameters $\delta_0 = 1$, $S^{(0)}_M = 91$, $M^{(0)}_P = 12$, $r = 3$.
Then for $n = 0,1,2,\ldots,\infty$
(a) $\delta_n = \delta_0/r^n$.
(b) $(S^{(n-1)}_M + 68 \cdot N^{(0)}_P \cdot 3^n)/2 \leq S^{(n)}_M \leq (S^{(n-1)}_M + 72 \cdot N^{(0)}_P \cdot 3^n)/2$.

The equations show that the unit of resolution δ_n converges to zero, while the total surface $S_{M(n)}$ increases to infinity as the number of iterations 'n' increases to infinity.

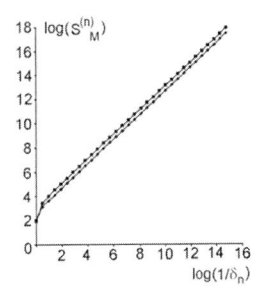

==

Fig. 27-01. Log-Log plot of the values of the total surface $S^{(n)}_M$ of a 'similar tile' as a function recursion as a function of the resolution $(1/\delta_n)$. The 2 curves depict the upper and lower limits of $\log[S^{(n)}_M]$. With the exception of a few initial points, their slope is precisely 1.0, i.e. their fractal dimension $\eta = 1.0$.

The equations show that the unit of resolution δ_n converges to zero, while the total surface $S^{(n)}_M$ increases to infinity as the number of iterations 'n' increases to infinity. Figure 27-01 shows a plot of the results of this iteration.

With the exception of the first few points, the data fall on straight lines with a slope $\eta = 1.0$. The data for the upper and lower limits are separated by a small parallel shift.

This result can be derived directly from equation 27-22 (b). Since the terms $68 \cdot N^{(0)}_P \cdot r^n$ and $72 \cdot N^{(0)}_P \cdot r^n$ increase rapidly with increasing 'n', the only stable solution $S^{(n)}_M$ of equation 27-22 (b) for large values of 'n' has to be proportional to r^n, too. Hence, we set for large 'n'.

equ. 27-23

$S^{(n)}{}_M = \gamma \cdot N^{(0)}{}_P \cdot r^n$, with a certain proportionality factor γ.
Hence, ((equ. 27-22, (b)))

equ. 27-24

$(\gamma \cdot N^{(0)}{}_P \cdot r^{n-1} + 68 \cdot N^{(0)}{}_P \cdot r^n)/2 \leq S^{(n)}{}_M \leq (\gamma \cdot N^{(0)}{}_P \cdot r^{n-1} + 72 \cdot N^{(0)}{}_P \cdot r^n)/2$,
    or
$N^{(0)}{}_P \cdot r^n \cdot (\gamma + 68 \cdot r)/2r \leq S^{(n)}{}_M \leq N^{(0)}{}_P \cdot r^n \cdot (\gamma + 72 \cdot r)/2r$.

Defining the constant values

equ. 27-25

$\xi_1 = N^{(0)}{}_P \cdot (\gamma + 68 \cdot r)/2r$.
$\xi_2 = N^{(0)}{}_P \cdot (\gamma + 72 \cdot r)/2r$.
$\alpha = \log[\xi_1 \cdot \delta_0)$.
$\beta = \log[\xi_2 \cdot \delta_0)$.

and considering that $\delta_n = \delta_0/r^n$, equ. 27-24 becomes

$\xi_1 \cdot \delta_0 \cdot (1/\delta_n) \leq S^{(n)}{}_M \leq \xi_2 \cdot \delta_0 \cdot (1/\delta_n)$.

Taking the logarithm of all terms yields finally

equ. 27-26

$\alpha + 1 \cdot \log[1/\delta_n] \leq \log[S^{(n)}{}_M] \leq \beta + 1 \cdot \log[1/\delta_n]$,

which states that a plot of $\log[S^{(n)}{}_M]$ vs. $\log[1/\delta_n]$ for large 'n' yields 2 different straight lines. Both have the same slope $\eta = 1.0$, but different ordinate intersects, which is precisely as in figure 27-01.

APPENDIX E

OUTLINE OF A TASK-BASED TAXONOMY OF LIVING MOSAICS

Note: Although a task-based taxonomy is about living things, the following will include the category of 'inanimate', as living mosaics may use, seek, and contain inanimate objects.

As mentioned in the main text, we may divide the elements of tasks as follows.

def. 28-01
Each task is described as a string of 5 elements.

Performer ('Who or what carries out the task?')
Action ('What is done to perform the task?')
Target ('At which object is the action directed?')
Means ('What procedures, instruments, or strategies are applied?')
Manner ('What are the circumstances under which they are applied?')

be described and illustrated in the following sections. Figure 28-03 will provide a summary illustration.

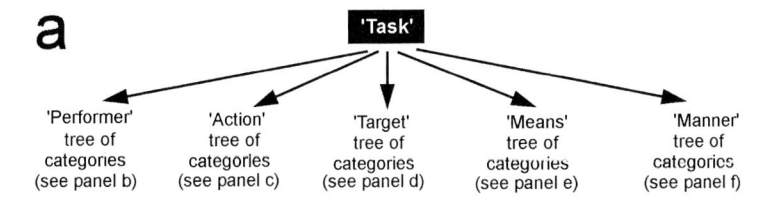

Fig. 28-01a. The basic flow diagram of the categories proposed here for the classification of 'tasks'.

376   Appendix E

1. Taxonomy of 'Performers'

The 'performers' have a tree of branching categories of their own.

Figure 28-01b illustrates this classification tree.

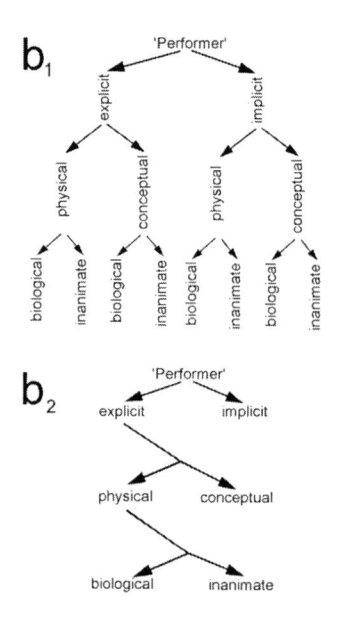

Fig. 28-01b. (b₁) The basic flow diagram of the categories for the classification of 'performers' in the explicit form. (b₂) An abbreviated version of the same diagram. *In the following all such flow diagrams will be shown in the abbreviated form.*

def. 28-02   *Explanation/ Examples:*

Performer *Which agent carries out the task?*
Implicit agent *[No performer is specified.](e.g. **Indirect, anonymous, conventional wisdom, prejudice, experience, faith, confidence, ...**)*
Explicit agent *[specific performer is named.]*
Physical agent
Biological agent *(e.g. **organs, limbs, hair, wings, claws, fins, wound, body, family, tribe, instinct, conviction, obsession, ...**)*

Inanimate agent. (e.g. text, music, work of art, car, air plane, train, robot, ...)

Conceptual agent
Biological agent   *(e.g. its name, person, animal, plant, microorganism, hive, university, school, army, company, government, behaviour, life, enemy, threats, war, disease,...)*
Inanimate agent   *(e.g. letter, tone, measure, frame, style, fashion, polyphony, harmony, computer language, computer program, operating system,...)*

2. Taxonomy of 'Actions'

Next, we need to define the element of 'Action' in more detail. We will not distinguish between an **action** and its **opposite action**. As negatives and positives of the same theme, they can be placed in the same category. This approach is similar to the way the Linnaean taxonomy classifies females and males of a bisexual species as identical, even in cases where the sexual dimorphism is extreme.

However, we will distinguish between actions such as mating, bleeding, breathing, etc. that only living mosaics can perform, in contrast to actions such as falling, swimming, heating, etc. that both biological and inanimate subjects can perform. They will be classified as 'biological' and 'general' respectively.

Explanation/ Examples:
A. Mutually Exclusive Categories of 'Actions'.

The result is a nested hierarchy of mutually exclusive categories.
1. It divides 'actions' into **'individual'** actions *(e.g. fly, add, signal)* and **'collective'** actions, *(e.g. swarm, mix, cooperate)*.
2. Each of these, in turn, can either be a **'physical'** action *(e.g. attack, kill, feed)*, or a **'conceptual'** action. *(e.g. decide, choose, plan)*
3. Each of these, in turn, can either be a **'biological'** action *(e.g. mate, bleed, breath)* or a **'general'** action.   *(e.g. fly, swim, sink)*
4. These, in turn, can either be actions concerning the **'self'** *(e.g. eat, flee, sing)* or all **'non-self'** objects *(e.g. feed, hunt, build)*.

5. These, in turn, can either be a newly **'initiated'** action *(e.g. conquer, camouflage, seek)* or the **'response'** to one *(e.g. defend, escape, liberate).*
6. These, in turn, will represent an action's **'continuation'** *(e.g. process, resume, direct)* or its **'interruption'** *(e.g. terminate, imbalance, surprise).*

Up to this point classifying an 'action' is simple and unambiguous, yet it would be quite crude. After all, numerous 'actions' describe drastically different activities, depending on their specific parameters. Therefore, we cannot avoid making the most relevant action-parameters an integral part of the classification.

B. The Parameters of 'Actions'

For example, if the action concerns…
(1)…the location of target or performer, it depends on a parameter such as **'location'** **(e.g. move, change, process, hold, maintain, freeze,…).**
(2)…the temporal aspect of the action, it depends on a parameter such as **'time'** **(e.g. preempt, speed up, postpone, delay,…).**
(3)…the physical direction of the action, it depends on a parameter such as **'direction'** **(e.g. seek, encourage, attack, avoid, discourage, defend,…).**
(4)…the size of target or performer, it depends on a parameter such as **'size'** **(e.g. add, enlarge, include, remove, reduce, exclude,…).**
(5)…the changes of contents of target or performer, it depends on a parameter such as **'content'** **(e.g. emit send, accept, receive,…).**
(6)…the pattern of the components of target / performer, it depends on a parameter such as **'pattern'** **(e.g. scramble, arrange, mix, link, fit, match, image, disrupt, select, unlink, dismantle,…).**
(7)…the effect/ consequence of the task, it depends on a parameter such as **'effect'** **(e.g. succeed, agree, support, fail, dispute, oppose,…).**

C. The Need for a 'Parameter-Code' of 'Actions'

No matter how important these parameters are for the characterization of an 'action', unfortunately, they are not mutually exclusive. For example, the action 'to exclude' can relate to the size *and* the contents of the performer/target. Similarly, 'to move' may need to be characterized by 'location', 'time', *and* 'direction'.

The complication that many 'actions' depend on multiple parameters cannot be avoided. Therefore, we have to specify for each 'action', on which combination of the parameters it depends.

For example, some 'movement' actions only depend on a single parameter, such as the 7 categories 'time', location', etc. were listed in the previous section B. Hence, here are maximal $\binom{7}{1} = 7$ such possibilities.

Other forms of 'movement' action may depend on combinations of 2 of the parameters, which add $\binom{7}{2} = 21$ different cases, as the order of the parameters does not matter.

Yet others may depend on combinations of 3, 4, 5, 6, or all 7 of the parameters. Hence, total number of variants of 'movement' actions, based on their parameter dependence alone, is $\binom{7}{1} + \binom{7}{2} + \ldots + \binom{7}{7} = 2^7 = 128$ different cases.

How can one differentiate between them? Do we need to invent 128 different names for each 'movement' action, but also for every other verb that qualifies for an 'action'?

Fortunately horrendous task is not required. The simplest way to handle the situation is to attach a 'parameter-code' to each 'action'. For example, if a particular 'flying' action like the seasonal bird migration depends on 'time' and 'direction' we can write it as '{t,d}-flying'. If it depends on 'location', 'time', 'content', and 'pattern', like the 'waggle dance' of bees, we can write it as '{l,t,c,p}-dance'. The notation is unambiguous, and covers all logical possibilities. Most importantly, the addition of a 'parameter-code' renders any overlapping 'actions' as mutually exclusive

The construction of 'parameter-codes' is not restricted to the mentioned 7 basic parameters. Should more parameters be needed in the future, one can easily expand their list and add their symbols to the corresponding parameter-codes.

In summary, the proposed classification scheme of 'actions' is a nested hierarchy of 6 mutually exclusive categories followed by the combination of a 'parameter-code' and the name of the action. The parameter-code is a combination of the first letters of the following categories: 'Location' / 'Time' / 'Direction' / 'Size' / 'Content' / 'Pattern' / 'Effect' / which is enclosed in { } brackets.

D. The Basic Nested Hierarchy of 'Actions'

The resulting nesting of the hierarchy of categories is shown in the form of a flow-diagram in Figure 28-01c.

def. 28-03
Action   What action is required to perform the task?
1. Individual action / Collective action
2. Physical action / Conceptual action
3. Biological action / General action
4. Concerning Self / Concerning Non-Self
5. Action initiated by performer/Performer responds to other action
6. Performer continues an action / Performer interrupts an action
7. {Parameter-code} & 'action' name.

The system is able to distinguish between $2 \times 2 \times 2 \times 2 \times 2 \times 2 \times 128 = 8192$ different actions, although it may happen one or the other of the sub-categories provided by the system cannot be filled with a suitable, known 'action'.

Regardless, the system can classify many more than 8192 different kinds of actions. The situation is similar to the Linnaean system, where the final category of the 'species' is a collection of a large number of different names of organisms. In a similar way, the above schematic can accommodate any number of different 'action' names and parameters in the final category.

Next we have to categorize the various possible 'targets' of a task.

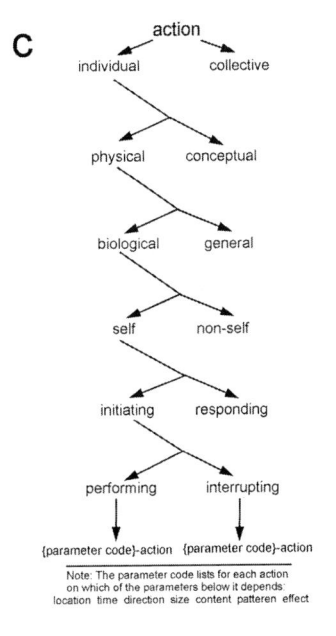

Basic Biology for Born Engineers: Living Mosaics 381

Fig. 28-01c. The nested hierarchy of sub-categories to classify 'actions'

3. Taxonomy of 'Targets'

Each 'performer' can also function as a 'target' or 'goal'. Therefore, the 'target'-category includes also the items of the 'performer' category. In contrast, some 'goals' cannot be performers of tasks, and are therefore missing from the 'performers' category.

def. 28-04 *Explanation/ Examples:*
Target *At what object is the action directed?*
Task-qualified. *The target itself is capable of performing a task. All Performers listed in definition 28-02) (e.g. all performers are capable, and thus can be targets.).*
Task-inept. *The target itself is not capable of performing a task.*

```
              d        'Targets'
                      ↙      ↘
              task-inept   task-qualified
                          (all 'Performers')
                    ↘    ↙
                  ↙        ↘
            biological    inanimate
                ↓             ↓
               aim           aim
                ↓             ↓
             material     material
                ↓             ↓
             territory    territory
```

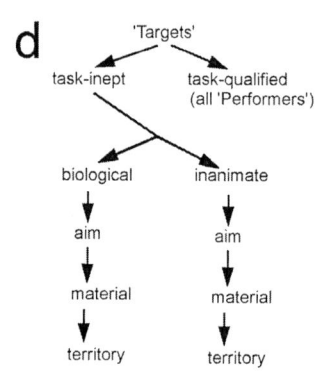

Fig. 28-01d. The basic flow diagram of the categories for the classification of 'targets'.

Biological
Aim. The targets of the task are biological concepts.*(e.g. function, freedom, importance, ignorance, value, understanding, injustice, miscommunication, symbol,...)*
Materials. The targets are materials with only biological meaning. *(e.g. starting materials, supplies, products, waste, food, clothes, money,...)*
Territory. The specific targets of the task are owned by biological objects. *(e.g. home, possessions,...)*
Inanimate
Aim. The targets of the task are concepts belonging to the inanimate world. *(e.g. weight, position, accident, danger, catastrophe, necessity, probability, climate, drought, flood, energy, tides,...)*
Materials. The specific targets of the task are inanimate materials *(e.g. substrate, heat, caves, shelter,...)*
Territory. The specific targets of the task are geological objects *(e.g. countries, mountains, oceans, river deltas, wetlands, sediments,...)*

The categories of 'Aim', 'Material', and 'Territory' are not meant to be mutually exclusive but supplemental. However, unlike the situation in the case of 'actions', they are not parameters of a 'target', but its variable compartments.

For example, the target of a task may involve the 'injustice' of 'waste' of somebody's 'possessions', or the 'danger' of 'caves' near 'wetlands'. In these cases not one, but all three items represent the 'target'. Consequently, all three will be added in brackets, e.g. task-inept / biological / ['injustice', 'waste', 'possessions'] or task-inept / inanimate /['danger', 'caves', 'wetlands'].

Next, we have to categorize the various possible tools, procedures, etc. the task needs to employ in order to accomplish its goal.

4. Taxonomy of 'Means'

These categories are to identify which particular 'means' the various 'actions' are employing in the course of their proceeding.

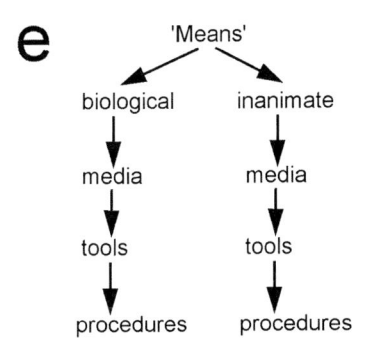

Fig. 28-01e. The basic flow diagram of the categories for the classification of 'means'.

def. 28-05 *Explanation/ Examples:*
Means *What procedures, instruments or strategies are applied by the action?*
Biological
Media. *The media of the method only have biological meaning (e.g.*
  probing, waiting for accidental encounter, searching,
  trapping, hunting, setting thresholds, keeping agents
  temporarily dysfunctional (e.g. e.g. immature, retaining
  products in storage,...)
Inanimate
Media. *The media of the method belong also to the inanimate*
  world(e.g. sound, light, electricity, gravity, chemical
  reaction,...)

Tools. The tools of the method belong also to the inanimate world (e.g. camera, X-ray, infrared, weapons, pen, fire, wheels,...)
Procedures. The procedures of the method belong also to the inanimate world (e.g. freezing, burning, submersing, dissolving,...)

Similar to the previous case of the 'Targets', the categories of 'Media', 'Tools', and 'Procedures' are not meant to be mutually exclusive but supplemental.

For example, the means of a task may include 'gesture', 'intelligence', and 'hunting', or 'light' of 'fire', and 'burning'. In these cases not one, but all three items represent the 'means'. Consequently, all three will be added in brackets, e.g. biological / ['gesture', 'intelligence', 'hunting'] or inanimate /['light', 'fire', 'burning'].

Finally, we need to categorize the manner and circumstances under which the various actions may apply instruments, tools, strategies, etc.

5. Taxonomy of 'Manners'

The ways in which the 'actions' apply their 'means' require several characterizations, such as…

def. 28-06 *Explanation/ Examples:*

Manners *In which way were the procedures, instruments or strategies are applied by the* action?
…time and place: *The methods are applied at a specific time and place (e.g. on New Year's Eve, at 4000 ft. altitude, every Monday, at my birthday, In the winter, in mid-air, periodically, post-mortem…)*

f 'Manner'
↓
time and place
↓
character
↓
arrangement

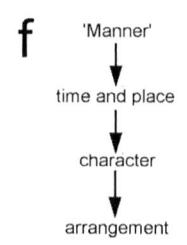

Fig. 28-01f. The basic flow diagram of the categories for the classification of 'manners'.

...*character:* *The methods have a common character, namely...(e.g. **American, agnostic, aerodynamic, population statistical, anatomical, fitted, ordered, fractal, random, sporadic, asymptotic, realistic, imaginary, temporary, symbolic, irreversible, improbable, frequent, ambitious** ...)*

...*arrangement:* *The performers or targets apply the methods in an arrangement of...(e.g. **single, multiple, together, sequential, simultaneous, iterated, independent, linked, serial, parallel, as a branching' tree', as 'inverse tree'** = **filter**,...)*

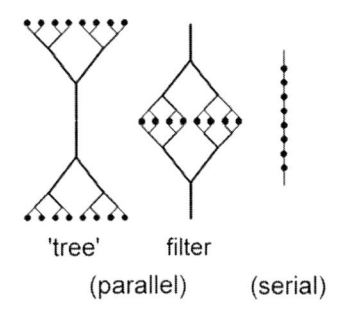

Fig. 28-02 The most frequent arrangements of performers or targets in biology, the parallel arrangements as in the case of a 'tree' or the inverted tree = 'filter', and the serial arrangement.

Although these characterizations may be considered mutually exclusive, most 'tasks' employ more than one of them at the same time. Therefore, it is not sufficient only to substantiate one of them. The situation is expressed in the flow diagram of Figure 28-01f.

Of course, each of these categories of the preceding schematic can be sub-classified even further.

Summary of the Proposed Taxonomy of 'Living Mosaics'

The proposed classification system is summarized graphically in Figure 28-03. Perhaps it is useful to remind the student that all the above efforts of classification were aimed at classifying 'living mosaics'. We only detailed the taxonomy of 'tasks' based on the equivalence between 'living mosaics' and 'tasks'. Nevertheless, as marked on the summary illustration, if a 'living mosaic' needs to be classified, the very first step is to identify the 'task', which it fulfils.

'Time' / 'Direction' / 'Size' / 'Content' / 'Pattern' / 'Effect' /

The next step is to use the above flow charts to classify this 'task'. Three examples are added in order to illustrate how to this may be done in practice.

We begin with our earlier example of Tom Sawyer's task 'to whitewash Aunt Polly's fence with paint and brush on a Saturday morning'

Based on the flow diagram of Figure 28-03 we begin with the identification of the *performer*, who is 'Tom Sawyer', which is the name of a biological object and, therefore, *Explicit / Physical / Biological /*.

Secondly, we determine the *'action'*, namely 'to whitewash', which is *Individual / Physical / General / Non-Self / Initiating / Performing /* {c,e}- to whitewash. The parameter code {c,e} indicates that whitewashing changes the content and effect of the fence.

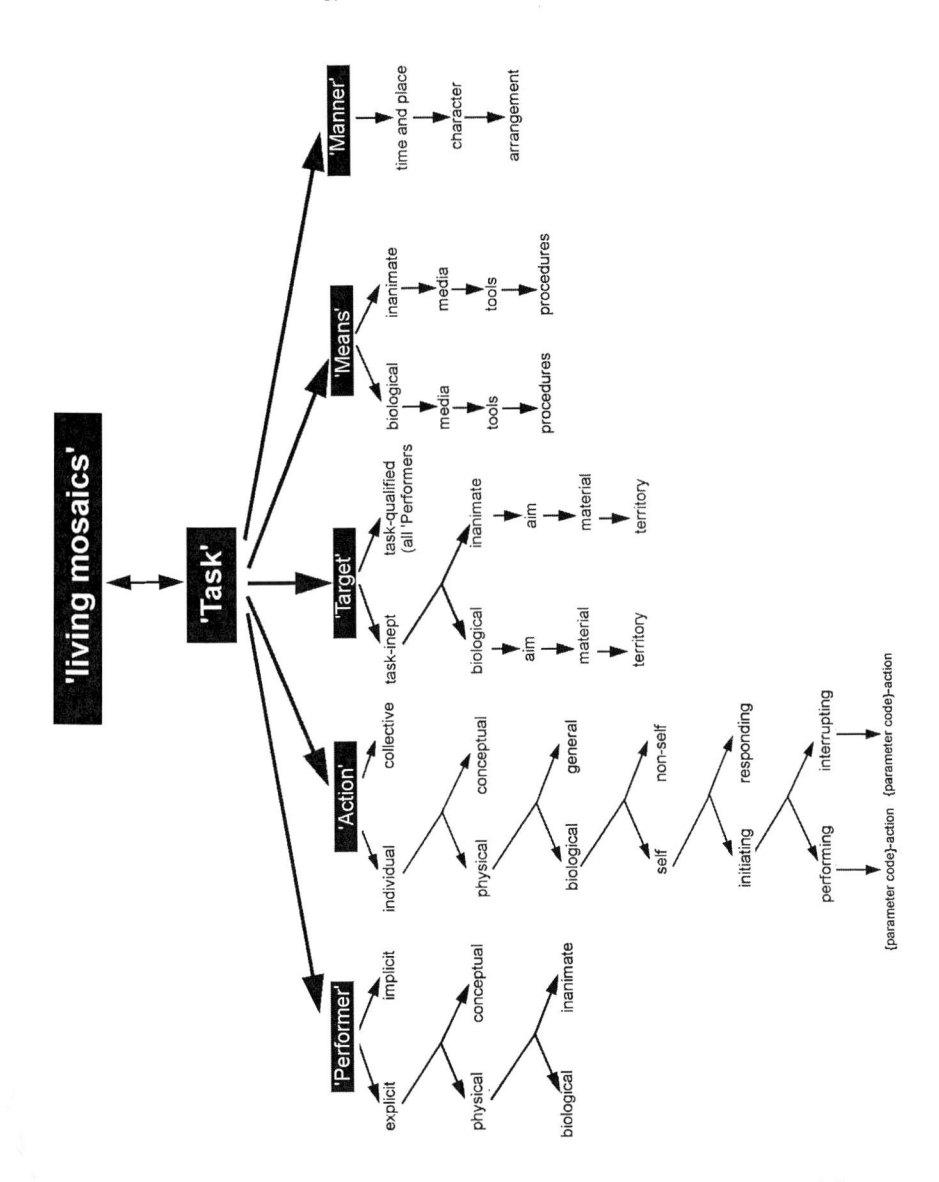

Basic Biology for Born Engineers: Living Mosaics 387

Fig. 28-03 Summary of the proposed schematic of categories used for taxonomy of 'living mosaics' The category {parameter code} is formulated from the initial letters of the parameters 'Location' /

The third step requires determining the *target* of the action, namely 'Aunt Polly's fence', which is *Incapable of performing a task / Inanimate Target / Material /*.

Next, we determine the media, tools, or procedures of the action, namely 'with paint and brush', and find as its classification *Inanimate / Tools /*.

And finally, we need to determine the circumstance, and find *Time and Place /* 'on a Saturday morning'; *Character /* 'permanently; *Arrangement /* 'single'.

The classification of the above task consists of all 5 sub-classifications, and can be summarized as follows.

Example 1
Living mosaic:' Tom Sawyer'
Task: **Tom Sawyer's task is to whitewash with paint and brush Aunt Polly's fence that runs along the street on a Saturday morning.**
1. Performer: *Explicit / Physical / Biological /* 'Tom Sawyer'.
2. Action: *Individual / Physical / General / Non-Self / Initiating / Performing /* 'to whitewash'.
3. Target: *Incapable of performing a task / Inanimate /*
Aim / 'Aunt Polly's Fence';
Material / 'Fence pickets' ;
Territory / 'along street'
4. Means: *Inanimate /*
Media / 'white paint';
Tools / 'brush';
Procedures / 'Manual brushing'.
5. Manner: *Time and Place /* 'on a Saturday morning';
Character / 'permanent;
Arrangement / 'single'.

Similarly, we classify

Example 2.
Living mosaic: Bee hive
Task: **The bees of the hive are to swarm this afternoon while surrounding the new queen.**
1. Performer: *Explicit / Physical / Biological /* 'The bees of the hive'.
2. Action: *Collective / Physical /*
Biological / Self /

Basic Biology for Born Engineers: Living Mosaics

                Initiating / Performing / 'to swarm'.
3. Target: *Task-inept / inanimate /*
                Aim / 'undetermined location';
                Material / 'tree';
                Territory / 'hollow space';
4. Means: *Biological /*
                Media / 'air';
                Tools / 'scout bees';
                Procedures / 'surrounding new queen'.
5. Manner:
                Time and Place / 'this afternoon';
                Character / 'random;
                Arrangement /   'dense'.

Example 3.
Living mosaic: Mammalian mitochondria
Task: **Mammalian mitochondria are to perform saltatory movements in the cytoplasm parallel to direction of microtubules.**
1. Performer: *Explicit/Physical/Biological/* 'Mammalian mitochondria'.
2. Action:
                Collective / Physical /
                Biological / Self /
                Initiating / Interrupting /do saltatory movements'.
3. Target:
                task-qualified / biological /
                Aim / 'unknown';
                Material / 'microtubules';
                Territory / 'cytoplasm'.
4. Means:
                Biological / Media / 'colloidal';
                Tools / 'unknown';
                Procedures / 'stop-and-go'.
5. Manner:
                Time and Place / 'undetermined';
                Character / 'unpredictable;
                Arrangement / 'parallel to microtubules'.